CW01072677

Hydrocolloid Applications

Hydrocolloid Applications

Gum technology in the food and other industries

A. NUSSINOVITCH
The Hebrew University of Jerusalem
Faculty of Agricultural, Food and Environmental Quality Sciences
Institute of Biochemistry, Food Science and Nutrition
Rehovot
Israel

BLACKIE ACADEMIC & PROFESSIONAL
An Imprint of Chapman & Hall
London · Weinheim · New York · Tokyo · Melbourne · Madras

Published by Blackie Academic & Professional, an imprint of Chapman & Hall, 2–6 Boundary Row, London SE1 8HN, UK

Chapman & Hall, 2–6 Boundary Row, London SE1 8HN, UK

Chapman & Hall GmbH, Pappelallee 3, 69469 Weinheim, Germany

Chapman & Hall USA, 115 Fifth Avenue, New York, NY 10003, USA

Chapman & Hall Japan, ITP-Japan, Kyowa Building, 3F, 2-2-1 Hirakawacho, Chiyoda-ku, Tokyo 102, Japan

DA Book (Aust.) Pty Ltd, 648 Whitehorse Road, Mitcham 3132, Victoria, Australia

Chapman & Hall India, R. Seshadri, 32 Second Main Road, CIT East, Madras 600 035, India

First edition 1997

Typeset in 10/12pt Times by Doyle Graphics, Tullamore, Ireland
Printed in Great Britain by T.J. Press, Padstow, Cornwall

ISBN 0 412 62120 7

A catalogue record for this book is available from the British Library

Library of Congress Catalog Card Number: 97–71428

Cover design: Ofer Tevel

∞ Printed on acid-free text paper, manufactured in accordance with ANSI/NISO Z39.48-1992 (Permanence of Paper).

In memory of my grandfathers and grandmothers Jacob and Sara Nusynowicz and Moses and Rosa Froman, who were murdered in concentration camps during the holocaust and with whom I never had the privilege of sharing my childhood.

Contents

Preface

Water-soluble gums are beneficial in many fields, including food, agriculture, adhesives, biotechnology, ceramics, cosmetics, explosives, paper, textiles and texturization, among many others. It is almost impossible to spend a day without directly or indirectly enjoying their qualities.

This book on hydrocolloid applications is divided into two major portions. The first is devoted to a few important gelling and non-gelling gums, their sources, the raw materials from which they are manufactured, their structures, functions and properties, followed by their food applications. The second part of the book details gums' industrial, non-food uses in a unique way: it assumes the reader's unfamiliarity with the many fields in which gums can be useful. It, therefore, provides a broad introduction to the development, technology and many aspects of gums' major non-food uses, as well as giving detailed explanations of where, when and how gums are incorporated into products in these industries. The text is also accompanied by a detailed index, designed to help the reader locate information easily.

I wish to thank the publishers for giving me the opportunity to write this book. Their patience is very much appreciated. I wish to thank my editor Camille Vainstein for working shoulder-to-shoulder with me when time was getting short and Dr Zippora Gershon for supporting me with references and good advice over the years. The patience, understanding and love of my wife Varda and my children Ya'ara, Eran and Yoav during the time the book was being written was essential to accomplishing the project. Last, but not least, I wish to thank the Hebrew University of Jerusalem for giving me the opportunity to work in my native country in a flourishing scientific and cultural atmosphere.

A. Nussinovitch

Rehovot, Israel, October 1996

1 Agar

1.1 Introduction and historical background

Agar was discovered in Japan in the mid-17th century (Yanagawa, 1942; Hayashi and Okazaki, 1970; Matsuhashi, 1978), although its name is Malayan (designating certain seaweeds and the jellies produced from them). The Japanese term for agar, *kanten*, means 'cold sky' and refers to the cold winter days or the cold weather in the mountains where such materials were manufactured. Following agar's exploitation in Japan, its use as a food was introduced to the natives by Chinese settlers (Tseng, 1946). The consumption of various kinds of agar-gel-like seaweed extracts probably dates back to prehistoric times in the coastal areas of Japan. Agar and its ability to produce fruit and vegetable jellies were introduced to Europe by Dutch people living in Indonesia. In 1882, Robert Koch introduced agar as a culture medium to the world. Its introduction to bacteriology followed, after its first application by Walter Hesse (who got the idea from his wife Frau Fanny Hesse) as a replacement for gelatin in culturing microbes. Agar has since become the single most important bacteriological medium. Today, Gelrite-gellan gum is used as a partial alternative to agar media and in related applications, particularly for culturing thermophilic microorganisms, since Gelrite gels are thermostable and can withstand prolonged incubations at high temperatures (Lin and Cassida, 1987).

Of the 15 000 varieties of seaweed, only 25 species have commercial value. Seaweeds are classified into four major categories, based on their pigments. These are red, brown, blue and blue-green; only red and brown seaweeds are important sources of hydrocolloids (Glicksman, 1983). Agar, carrageenan and furcellaran constitute a family of galactose polysaccharides extracted from red algae (Rhodophyceae) that has a common basic structure. They differ from one another in the proportions of D- and L-galactose, in the extent to which the galactose is modified to the 3,6-anhydroderivative, in the amount and position of sulfation and methylation of the individual sugar residues, and in the presence of other monosaccharides, such as xylose and uronic acid, and of substituents such as pyruvic acid and glycerol (Glicksman, 1983; Moirano, 1977). Sulfate content can be used to distinguish between red seaweeds. Agarose does not include any sulfate, whereas agaropectin has 3% to a maximal 10% and furcellaran has 8–19% sulfate (Percival, 1972).

1.2 Regulatory status and toxicity

Agar is described by the *Food Chemicals Codex III* (1981) as a 'dried hydrophilic colloidal polygalactoside' extracted from red algae of the class Rhodophyceae. Commercially produced, agar is sold in various nearly odorless forms, including strips, flakes and powders, in a range of colors from white to pale yellow. It is only soluble in boiling water. The maximum allowable limits for the presence of impurities are arsenic, not more than 3 ppm as arsenic; ash, not more than 6.5% on a dried basis; acid-insoluble ash, not more than 0.5% on a dried basis; residues of gelatin to pass a purity test; heavy metals such as lead, not more than 10 ppm; insoluble matter, not more than 1%; loss on drying, not more than 20%, and the limits of starch and water absorption to pass the test. Agar is considered GRAS (generally recognized as safe) by the FDA and is, therefore, permitted for use as a direct human additive in the USA (FDA, 1972, 1980). Agar is not degraded in the human digestive tract and can be found in the feces of animals and humans. It is harmless even when ingested in high doses. Experiments (Ariyama and Takahasi, 1931) leading to these conclusions were performed by feeding rats with 23% agar by weight for 70 days and comparing the results with rats fed a low-carbohydrate diet. The findings were confirmed in experiments with rabbits (Hove and Herndon, 1957). Food-grade agar was non-carcinogenic to mice and rats fed up to 50 000 ppm orally for 103 weeks, and no associated histopathological effects were found. Another study (Harmuth-Hoene, 1980) found no effect on calcium and copper uptakes in human subjects who received food-grade agar at a daily level of 22.5 g, which is much higher than the normal daily consumption of this hydrocolloid. In tests on humans, agar has been used as a laxative in daily doses of 4–15 g. The efficiency of agar is not great, with the laxative effect usually not being noticeable for nearly a week (Meer, 1980).

1.3 Collection of agar weed and its processing

1.3.1 General information

Japan is the major producer of agar, with Spain second. Together, they produce ~70% of the world supply. Recently, the cultivation of red algae has been initiated in Israel. *Gelidium* is one of the sources for the traditional manufacturing of agar in Japan, with as many as 24 local species (Segawa, 1965). Another source is the genus *Gracilaria* (Fig. 1.1), which became important after the invention of alkali pretreatment (Funaki, 1947) and is the most abundant agarophyte in the world (Matsuhashi, 1972; Armisen and Kain, 1995; Murano and Kain, 1995).

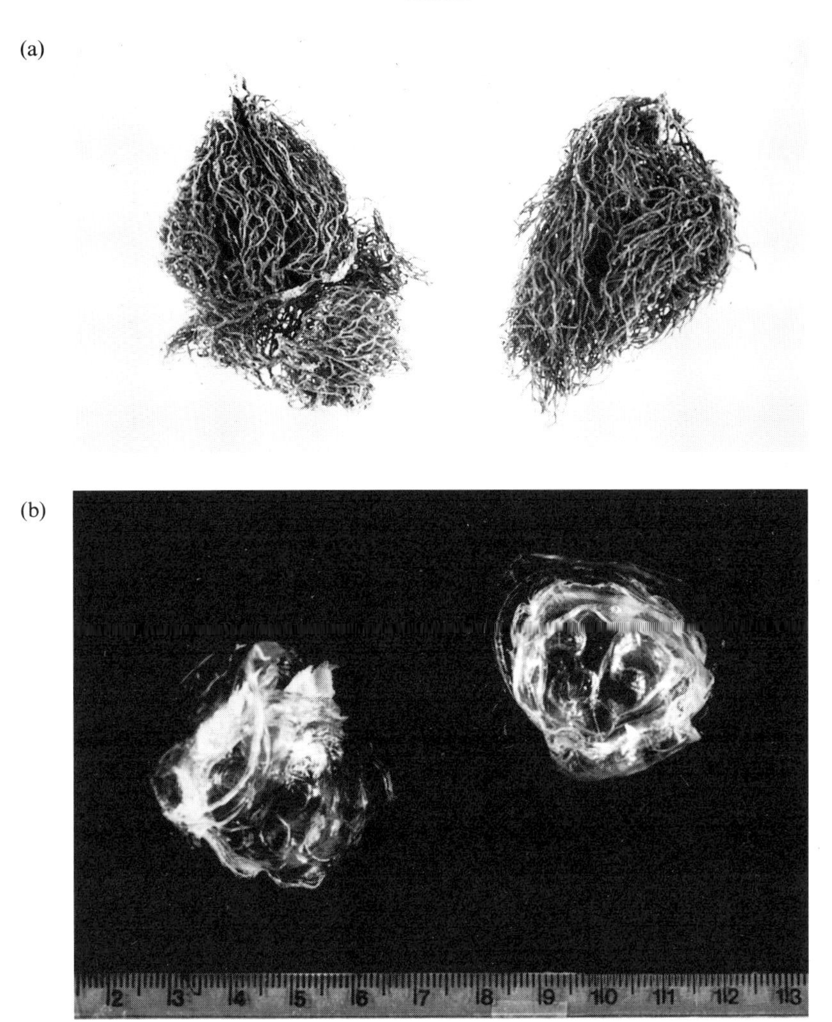

Figure 1.1 (a) A piece of dried *Gracilaria*, the botanical source of agar. (b) Laboratory freeze-dried agar 'pieces' extracted from *Gracilaria*.

1.3.2 Seaweed collection

In Japan, seaweed is collected by divers with no equipment other than goggles at water depths to 30 feet (9 m), or at below 60 feet by divers equipped with a diving apparatus (Newton, 1951). The seaweed is collected, stored in tubs or rafts and towed to shore, where it is dried and partially bleached, then taken for final processing. Deep-water seaweed is considered to yield the best gelling extracts, and the optimal harvesting period is April to September (Glicksman, 1969).

1.3.3 Traditional and modern manufacturing

For centuries, the extraction of agar from seaweed was based on simple boiling to obtain a jelly mass. An improved method of purification and preparation was, according to legend, discovered accidentally: an innkeeper who had prepared a dish of seaweed jelly for royal guests threw the leftovers outdoors where they froze during the night and thawed out the next day. This resulted in a dry, translucent substance that when reboiled in water yielded a clearer jelly of better quality. Thus, the purification of agar by commercial freezing and thawing became the preferred procedure, and it is still in use today, as will be discussed below.

Agar extraction from seaweed begins with mechanical cleaning of the red algae followed by washing with water. This is followed by a cooking process, in excess boiling water 15 to 20 times the volume of the seaweed (Yanagawa, 1942; Hayashi and Okazaki, 1970). The addition of 0.01–0.05% sulfuric acid or 0.05% acetic acid promotes better extraction. The algae is boiled for about 2 h and then simmered for 8–14 h. Calcium hypochlorite or sodium bisulfite can be introduced to bleach or decolorize the agar, yielding the highest possible quality product. The extract is filtered hot and the residual seaweed can be re-extracted. The filtrate is then cooled, gelled and cut. This is followed by dehydration by the freezing–drying method or by the pressing–dehydration method (generally for alkali-treated *Gracilaria* seaweed). The agar is marketed in dry, ground, packed form (powdered agar), in bars or in strings.

The finest quality agar is produced by the careful selection of raw materials. *Gelidium amansii* (a rigid seaweed) should be a major constituent in the seaweed extraction mixture, along with a minor amount of soft-type seaweed (e.g. *Ceramium* sp.). For rigid-type seaweed, extraction under pressure (gauge pressure of 1–2 kg cm^{-2} for 2–4 h) increases yields and reduces processing time. The exact conditions need to be determined to prevent destruction of the extracted agar. Alkali treatment at 85–90°C of agaroid-like mucilaginous substances was developed in Japan (Funaki, 1947). Proper processing of *Gracilaria* using unrefined seaweed coated with earth and sand results in high yields and increased gel strength.

Treatment of *Gracilaria* with alkali results in increased agarose content and reduced agaropectin and sulfate contents in the agarose fraction. The L-galactose 6-sulfate is assumed to be converted to 3,6-anhydro-L-galactose, as in the case of carrageenan (Rees, 1969; Guiseley *et al.*, 1980). Rigid-type agarophytes can also be extracted using polyphosphates or condensed phosphates. Laboratory experiments have shown that pretreating seaweed with enzymes, gamma-irradiation and extraction in ammonia media improves extraction, but none of these pretreatments has been applied in the industry. Evidence of a simplified agar extraction method by 'acid pretreatment' of agarophytes can be found in the literature (Matsuhashi, 1974). After extraction, filtration of the hot sol (through roughly meshed cloth bags or, in more advanced industries, by filter press) and gel dehydration are important steps.

Freezing the gel (Matsuhashi, 1981) before dehydration is common practice in the processing of agars: 48% of the water is eliminated from frozen agar gels by sublimation, 40% in the process of drip defrosting and 12% by vaporization. Freezing–dehydration can be either mechanical or natural and is applied, for example, to *Gelidium* extracts, for which pressing dehydration is not suitable.

There are several types of product. The bar-style agar Kaku-kanten appears as rectangular pieces of 7.5 g each on average. Its bulk density is $\sim$0.030–0.036 g cm^{-3}. It is sold for domestic use in bags containing one or two pieces. These bars are also produced in the Philippines by mechanical freezing and sun-drying but are less visually appealing (Matsuhashi, 1990). Agar is also sold in strings, i.e. Hoso-kanten. These strings are 28–36 cm in length, with a commercial unit having a net weight of 15–30 kg. Smaller quantities are sold to the public, and densely packed strings are used for overseas shipment. The biggest seller, however, is powdered (fine) agar (although agar flakes are also requested). Agar flakes are produced from *Gelidium* species by the freezing process, whereas the powder is produced from alkali-treated *Gracilaria* by the pressing–dehydration (non-freezing) method (Matsuhashi, 1990).

Irradiation of dried *Gelidiella*, *Gelidium*, *Gracilaria* and *Hypnea* species with 1000 Ci, produced by cobalt-60, in the range 0.9×10^4 to 6.4×10^4 rd g^{-1}, was reported to improve yield, gel strength and stability of the dry extracts (Smith and Montgomery, 1959). However, negative results have also been obtained with this treatment. Pretreatment of agarophytes with cellulolytic enzymes has, in some cases, accelerated extraction rates and increased gel strength and yields (Meer, 1980).

1.4 Structure

1.4.1 The agar sugar skeleton

The structure of agar that is extracted commercially from a number of red algae species has been studied for decades, primarily by Japanese scientists. A few of these have spent most of their professional lives in this field of research. Araki (1937) was the first to isolate agarose. Later work led to the conclusion that agar is composed of repeating units of D-galactose and 3,6-anhydro-L-galactose (Araki, 1958; Cottrell and Baird, 1980). Agar extract

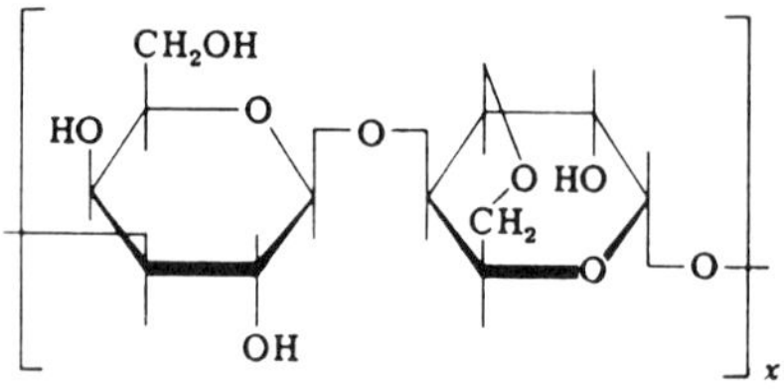

Figure 1.2 Structure of agarose (adapted from Araki, 1937).

comprises two groups of polysaccharides: agarose, the gelling component, is an essentially sulfate-free, neutral (non-ionic) polysaccharide (Fig. 1.2) and agaropectin is the non-gelling ionic (charged) polysaccharide. Agaropectin includes a small amount of sulfate ($\sim$2%) but has no commercial value and is largely discarded during the commercial production of agar. Agarose and agaropectin contents vary in different types of commercial agar (Araki, 1958, 1980). The percentage of agarose in agar-bearing seaweed can be 50–90% (Araki, 1937). The two polymer components can be fractionated by acetylation in chloroform to give soluble agarose acetate and insoluble agaropectin acetate (Araki, 1937), or by using quaternary ammonium salts to precipitate selectively agaropectin (Hjerten, 1962).

The basic sugar units of agarose (which consist of a linear structure with no branching) are D-galactose, L-galactose, 3,6-anhydro-L-galactose and D-xylose. Basic sugar units within agaropectin are D-galactose, L-galactose, 3,6-anhydro-L-galactose, D-xylose, galactose sulfate and pyruvic acid (Araki, 1958). Agarobiose is the common disaccharide structural unit of agar polysaccharides. The structure of agaropectin (which is a sulfated polysaccharide) is more complex than that of agarose and is less well understood. It is composed of agarose (its sugar skeleton is also composed of agarobiose) plus varying percentages of ester sulfates, D-glucuronic acid and small amounts of pyruvic acid. Hirase (1957) suggested that the pyruvic acid is attached in an acetal form to the D-galactose residues of the agarobiose. The sulfate content of the agar depends on the source of the raw material. According to Yaphe and Duckworth (1972), the term agar characterizes a family of polysaccharides with a backbone of alternate 1–3-linked and 1–4-linked D- and L-galactose residues. Many reports on the chemical composition of agars from newly reported agarophytes can be found elsewhere (Chirapart *et al.*, 1995).

1.4.2 The structure of agarose and agaropectin

Agarose has a linear structure with no branching. In the solid state it exists as a threefold, left-handed double helix with a pitch of 1.90 nm and a central cavity along the helical axis. This cavity (its interior is lined with hydroxyl groups that can participate in hydrogen bonding) accommodates water molecules without unfavorable steric clashes (Morris and Norton, 1983). Evidence for bound water within the helix has been found using ^{1}H-NMR relaxation studies (Ablett *et al.*, 1978). Agarose is essentially free of sulfate and consists of chains of alternating β1–3-linked D-galactose and α1–4-linked 3,6-anhydro-L-galactose (Fig. 1.2). 6-*O*-Methyl-D-galactose may also be present in variable amounts from about 1 to 20%, depending upon the algal species. Agar extracted from *G. amansii* was also found to contain 4-*O*-methyl-L-galatose (Araki, 1969). Agaropectin is a mixture of polysaccharides. It contains 3–10% sulfate (sulfated residues), glucoronic acid and sometimes pyruvic acid linked in acetyl linkages. Fractionation performed by Duckworth and Yaphe (1971a,b), showed that agar is not made up of natural (agarose) and charged (agaropectin) polysaccharides but rather comprises a series of related polysaccharides, ranging from a virtually neutral molecule to highly charged galactan.

1.4.3 Inorganic constituents of agar

The ash content of agar depends first and foremost on its origin (type of seaweed). The metallic and non-metallic constituents are important, especially in the bacteriological grade agars. Their concentrations (Seip, 1974) are: sulfur (inorganic), 1.0–1.5%; sodium, 0.6–1.2%; calcium, 0.15–0.25%; magnesium, 400–1200 ppm; potassium, 100–300 ppm; phosphorus, 10–80 ppm; iron, 5–20 ppm; manganese, 1–5 ppm; zinc, 5–20 ppm and strontium, 10–50 ppm.

1.5 Agar properties

The average molecular weights of six agar samples have been recently published (Tashiro *et al.*, 1996). They were determined by sedimentation equilibrium measurements in 0.1 M potassium chloride at 65°C. The wide molecular-weight distribution reported suggests the cleavage of agar molecules during alkali treatment. Although the gelation point depended on molecular weight in the case of samples from the same origin, this dependence did not exist in different algal species. In this study, the agar molecules were concluded to be in the form of random coils in 0.1 M potassium chloride at 65°C. The molecular shape of agar upon addition of a hydrogen-bonding inhibitor to the solution was also random coils, supporting the view that the sol–gel transition of agar results from the formation of hydrogen bonds (Tashiro *et al.*, 1996).

The physical properties of agar solutions and gels are important to the manufacturer and the scientist. Although agar is soluble in boiling water, when used in forms other than fine powder (i.e. bars, strings and flakes), an overnight soak in cold water helps achieve full dissolution. Even if the agar is of the soluble type, soaking for a short time (of the order of minutes) helps in reaching quick, good dissolution. The pH during soaking and boiling should be kept neutral.

Viscosity measurements on agar solutions should be performed at temperatures that are higher than the gelling temperatures (>40°C). A linear relationship was found between the logarithm of relative viscosity and agar concentration. The relative viscosity of agar sol versus its rotational velocity (determined from rotational viscometer measurements) exhibited an exponential decrease. The higher the temperature, the greater the decrease. For the same agar, a correlation was observed between the melting point of 1.5% agar and the relative viscosity of a 0.2% sol at 50°C. Moreover, evidence for a relationship between degree of viscosity and firmness of the agar gels was found. The presence of ions tends to reduce the viscosity of agar solutions (Matsuhashi, 1990).

Setting temperatures were found to increase with agar concentration, as did gelation temperatures of agarose. At a predetermined (constant) temperature, the higher the setting point of the sol, the higher the rigidity coefficient of the gel. The gelling temperature of agarose increases with increasing methoxyl content. The slower the cooling rate, the higher the temperature of agarose gelation.

The acoustic characteristics of agar are important because, although it is a solid material, its acoustic properties resemble those of water (Bouakkaz *et al.*, 1994). The uses of agar are, therefore, very diverse: within medicine (echography) it is used as a phantom. It is also used as an insulator to eliminate echo, which may be important with respect to informational echo. Therefore, a knowledge of agar's acoustic behavior is important and it is determined by measuring the attenuation coefficient, the speed of sound and the acoustic impedance (Bouakkaz *et al.*, 1994).

1.6 Gelation and melting of agar

Agar is unique among gelling agents because gelation occurs at temperatures far below the gel's melting temperature. Agar produces rigid gels at a concentration of ~1% (w/w). The sol sets to a gel at about 30–40°C. This gel is rigid and maintains its shape. Gels with self-supporting shapes can be formed with 0.1% agar (the rest being water). In the past the word 'brittle' best described the properties of an agar gel; today elasticity or rigidity can be achieved using different agars. The gel is melted by heating to ~85–95°C. There is a relationship between gel strength and melting point: both increase in the same direction, although some exceptions can be found. If the logarithmic agar concentration versus the inverse value of absolute temperature are graphed (Matsuhashi, 1972), the heat energy required to dissociate cross-linkage of the gel ($-\Delta H^\circ$ in kcal mol^{-1}) can be calculated as follows (R is the gas constant and $k = 2.303R$):

$$\log_{10} C = (\Delta H^\circ/2.303RT) + \text{const} \quad (1.1)$$

which can be converted to:

$$\Delta H^\circ = [k \log_{10}(C_1/C_2)/(1/T_1) - (1/T_2)] \quad (1.2)$$

This is significant because when agar is sensorily evaluated, firmness correlates with $-\Delta H^\circ$, in the sense that the larger the value, the firmer the sensory perception. Values found for $-\Delta H^\circ$ regularly range between tens and hundreds of kilocalories per mole, with an upper limit of 2000 kcal mol^{-1} (Matsuhashi, 1990).

Highly methylated agars were isolated from the red seaweed *Gracilaria eucheumoides*, harvested in Japan (Takano *et al.*, 1995). One of the extracted agars formed a thermo-reversible gel with a high melting point (up to 121°C) and was shown to consist of a regularly repeating structure. Melting points can be measured with an automatic device composed of a tube filled with the gel being tested (Kohyama *et al.*, 1989). A stainless-steel ball is placed on the upper surface of the gel, which is then heated at a constant rate, the gel temperature being measured with a thermocouple. Displacement of the steel ball as the gel melts is detected by an optical system: a digital videocamera employing a linear array of 256 photodiode cells (Kohyama *et al.*, 1989). A data processor transforms the signal from the camera into a signal that is proportional to the displacement of the ball. This technique and differential scanning calorimetry (DSC) were used in studies on agarose gels. The results showed that the measured melting point increases with increasing heating rate at rates above 0.5°C min^{-1}. The measured melting point depended on the weight of the steel ball, and melting point increased with increasing gel concentration. The authors also discussed the heat of reaction in the formation of network structures in agarose gels (Kohyama *et al.*, 1989).

The composition and thermal behavior of six sulfated polysaccharides were investigated using high-performance anion-exchange chromatography, IR spectroscopy and ^{13}C-NMR (nuclear magnetic resonance) (Lai *et al.*, 1994). The *Pterocladia capillacea* agar was a regular, alternating and nearly 'absolute' structure. The thermal behavior of the polysaccharides was examined by DSC. The thermogram of the *Pterocladia* polysaccharide gel showed three melting transition endotherms, at 40, 75 and 95°C, which might be attributed to a change in the double-helix coil conformation, the melting of the juncture zone and the disentanglement of the tightest cross-links, respectively. Similar viscometric and thermal behaviors among polysaccharides in the same family were observed, which could result from their acidic groups, structures and/or conformations; these are useful for algal taxonomy and applications (Lai *et al.*, 1994).

The definition of gel strength varies with corporation, country and individual scientist. This may be because of different methodologies, instruments, or testing conditions. Details on gel strength as defined by the Marine Colloids (Guiseley and Renn, 1977), the Meer Corporation (Meer, 1980) and Japanese companies can be found elsewhere. Furthermore, different gel testers have been invented to measure this gel property (Gifu, 1971, 1978). Apparent gel strengths are strengths at a non-defined concentration. If a linear correlation exists between apparent gel strength and gel concentration, then the value of the minus intercept divided by the slope corresponds to the minimum agar concentration that will form a gel (Matsuhashi, 1990).

The mechanical behavior of agar, carrageenan and alginate gels was evaluated for a wide range of gum and setting-agent concentrations, as well as for different gel-preparation methods (Nussinovitch *et al.*, 1990c). Similar modes of behavior were observed for agar gels (Fig. 1.3) and alginate set by diffusion (Fig. 2.7), i.e. linear relationship between the yield stress and deformability modulus versus the hydrocolloid concentration (Nussinovitch

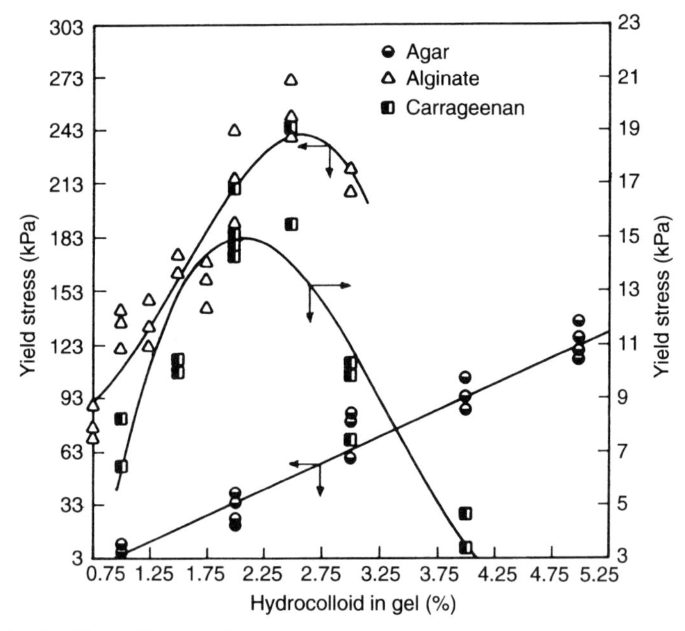

Figure 1.3 The effect of hydrocolloid concentration on the yield stress of agar, alginate and carrageenan gels. The yield stress values on the left apply to agar and alginate gels, those on the right to carrageenan. (From Nussinovitch *et al.*, 1990c; by permission of Oxford University Press.)

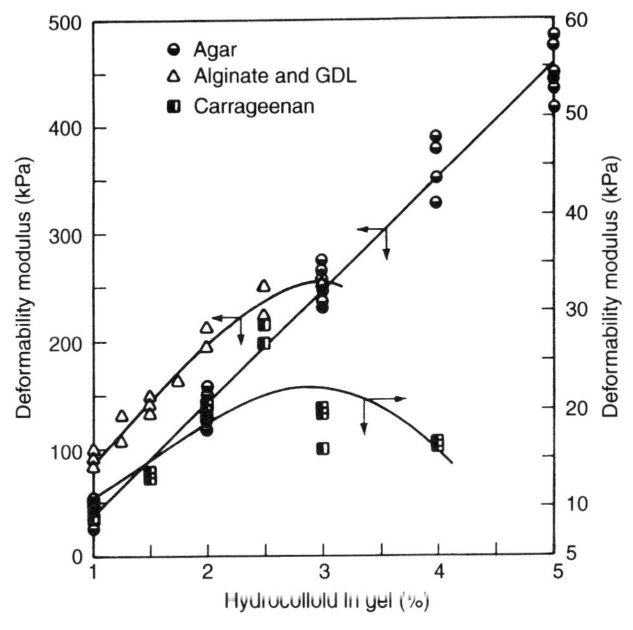

Figure 1.4 The effect of hydrocolloid concentration on the deformability modulus of agar, alginate and carrageenan gels. The deformability modulus values on the left apply to agar and alginate, those on the right to carrageenan. (From Nussinovitch *et al.*, 1990c; by permission of Oxford University Press.)

et al., 1990c). The higher the gum content, the stronger and stiffer the agar gel (Figs 1.3 and 1.4). A similar linear increase in agar gels, when the gum concentration is raised from 1 to 2%, was reported by Meer (1980). It should be noted that the preparation conditions might influence the measured mechanical properties, and the resultant perceived texture. Therefore, special attention needs to be paid to these conditions (Nussinovitch and Peleg, 1990; Nussinovitch *et al.*, 1990b). Stress relaxation of agar gels can be characterized by a modified Maxwell and non-exponential model (Nussinovitch *et al.*, 1989). Studies on agar-gel elasticity and the relationship between recoverable work and asymptotic relaxation modulus can be found elsewhere (Nussinovitch *et al.*, 1990a; Kaletunc *et al.*, 1991).

Other mechanical parameters used to describe the physical properties of agar gels are its rigidity coefficient and its breaking or cracking load. Information on very concentrated agarose gels has also been yielded by DSC, rheological, X-ray and NMR studies (Watase and Nishinari, 1986; Watase *et al.*, 1989). It is important to note that dissolution of agar in boiling acid solutions causes significant degradation. Gel stability is best achieved at pH values little over 7.0. The addition of $10\,\mathrm{mg\,l^{-1}}$ sodium carbonate to slightly acidic agar improves gel strength.

1.7 Agar gelation mechanism

The agarose component of the agar is responsible for gelation. At high temperature, agarose exists as a disordered 'random coil'; upon cooling, it forms a strong gel. Gelation has been reported to involve the adoption of an ordered double-helix state (Dea *et al.*, 1972; Hayashi *et al.*, 1978). The three equatorial hydrogen atoms on the 3,6-anhydro-α-L-galactose residues are responsible for the gelling of agar and agarose. These atoms, via steric effects, force the molecule into its double-helix shape. The gel's formed network hints at the presence of occasional 1–4-linked residues in native agarose, in which the anhydride bridge is absent. This allows the sugar ring to adopt the normal 'chair' shape. These 'kinking' residues terminate helix formation and enable the formation of a three-dimensional gel network. Each ordered junction zone involves between 7 and 11 double helices. The mechanism of gelation is believed to be similar to that of carrageenans and furcellaran (Glicksman, 1983; Rees, 1969). Morris (1986) reported X-ray diffraction analyses which suggested that linked bundles of associated right-handed double helices form the agarose gel, which sets as a result of a coil–double helix transition upon cooling. The interaction of the helices among themselves occurs at the 'junction zones' and these form a three-dimensional network capable of immobilizing water molecules in its interstices (Arnott *et al.*, 1974). The double-helix model for agarose gels was recently re-examined using optical rotation (Schafer and Stevens, 1995).

1.8 Gel syneresis

Agar gel shrinkage results in syneresis ('tearing'). The amount of syneresed water is inversely proportional to the square of the concentration for most practical concentrations. Agar gel syneresis is influenced by the concentration of the gel, its holding time, apparent gel strength, rigidity coefficient, pressurization and total sulfate content (Matsuhashi, 1990). Mechanically induced syneresis (while gel or gel-based food is being masticated) might have a bearing on the feeling of 'juiciness' in the mouth. The extent of this syneresis decreases with increasing gum concentration. Apparently, the higher osmotic pressure exerted by the increased gum concentration compensates for the mechanical pressure (Nussinovitch *et al.*, 1990c).

1.9 Gel clarity

Agarose sols are clear and colorless. Agar sols are hazy and sometimes yellowish. Agar gels are usually opaque and hazy, whereas agarose gels are generally much clearer. Upon freezing, agar gels (at around 0°C) collapse and do not recover their gel phase when thawed. However, the gel can be remelted and regelled to achieve nearly identical properties (Matsuhashi, 1990).

1.10 Effect of adding other materials on agar properties

The effect of inorganic salts on the time required to set agar sols has been thoroughly covered in the literature. Potassium sulfate has been shown to accelerate gelation the most, whereas iodine is the least effective. Ferric and ferrous salts (at 1 ppm) have been reported to reduce the transparency of a 1% agar gel. Sodium chloride (table salt) slightly increases gel strength (Takegawa, 1963). The addition of sugar (up to 60%) contributed to agar strength. The effects of sugars (sucrose, glucose, fructose, maltose, galactose, mannose and ribose) on the gel–sol transition of agarose and κ-carrageenan have been studied (Nishinari *et al.*, 1995). The effects of sugars were ascribed to hydrogen bonding between hydroxyl groups in polymers and sugars and/or structural changes in the solvent water.

A mathematical model was developed to predict the temperature profiles of cylindrical samples of sucrose-containing agar gels heated by microwave (Padua, 1993). Temperature was modeled in terms of the dielectric properties of the gels and the power absorbed by the sample. Experimental corroboration of the predicted temperatures was obtained by microwave-heating cylindrical gels of 2% agar containing 0, 40 and 60% sucrose. The levels of sucrose in the gels notably affected the temperature profiles in the cylinders. Samples with no sucrose showed a pronounced central heating effect, 40% sucrose samples showed an early uniform heating profile and 60% sucrose samples showed surface heating (Padua, 1993).

Other reports have shown a maximum strength at other sugar concentrations. Agar gel was strengthened by the addition of LBG (locust bean gum) (Selby and Wynne, 1973). If sodium alginate is added to an agar gel before setting, the alginate is later cross-linked by calcium chloride diffusing into the gel from the outside. Agarose–gelatin gels at high concentrations tend to interface with one another.

Other physiochemical gel properties include rheological properties of agarose gels with respect to molecular weight, thermal expansion of the gel, PMR (proton magnetic resonance) and NMR properties of the gel and film-forming ability.

1.11 Applications

The applications of agar depend upon its characteristics. By changing seaweed source and/or processing method, a full range of agar textures from the classical brittle to the very elastic can be achieved. Examples of a few applications include direct use of stringy agar, plain gels, fabricated gels and confectionery-type condensed jellies (Poppe, 1995; Carr *et al.*, 1995), its use as an agent to prevent stickiness and for the production of edible paper. Electrophoresis has been reported as a tool to identify red seaweeds used as food ingredients (Fleurence and Guyader, 1995).

1.11.1 Baking industry

Since agar gels can withstand high temperatures, they can be utilized in the baking industry. Agar is used as a stabilizer in pie fillings, icings, toppings, chiffon pies, meringues and other similar products. Icings are coatings or toppings of cakes and sweet goods that consist of shortening, milk solids, stabilizers, whipping agents, salt and flavoring. Four different types of icing exist: flat icings for sweet goods, fudge-type icings, cream icings and coating-type icings (Nash, 1960). Flat icings can suffer from problems of melting, disappearing or sticking to the cellophane wrapper in hot and humid weather; under other adverse conditions, the icing may crack and peel, diminishing its visual appeal (Glicksman, 1969). These problems of limited stability can be eliminated by decreasing the water content of the icings and by adding a gum to bind the free water. Agar or viscosity-formers such as LBG, alginate, carrageenan, pectin or karaya gum can be used for this purpose. The icing is made from the appropriate gum or gums, sugar, and water, and boiled. The hot syrup is mixed with icing sugar and applied at 130 ± 10°F (54 ± 6°C) to the baked product. The iced product is packaged after rapid evaporation from its surface. The resultant icing contains ~0.2% agar. Doughnut glazes tend to crack as a result of sugar crystallization: the addition of agar as a stabilizer at levels of 0.5–1.0%, depending upon the sugar content, increases the viscosity of the glaze and, in combination with glucose and invert sugar, changes its crystallization properties. The glaze adheres better to the doughnut, is more flexible, resists melting and prevents adherence of the baked products (Glicksman, 1969). An improved icing stabilizer is obtained by blending the agar with surface-active agents such as sorbitan esters of high fatty acids (Steiner and Rothe, 1949).

1.11.2 Agar in confections

Agar can be used in confections at concentrations of ~0.3–1.8%, and as a filler in candy bars (Fig. 1.5). In jelly candies, starch and pectin are used in larger proportions relative to the agar, which is widely used in its powdered form because of its solubility and the strength it confers upon the product. Good results are achieved if the agar is soaked for a few hours before being cooked with sugar, corn syrup, invert sugar and monosodium citrate as an

Figure 1.5 Ready-to-eat multilayered sweetened agar gel. (Courtesy of O. Ben-Zion.)

acidulant, with added color and flavoring. After mixing, the jelly is deposited into starch or paper-lined slabs. Other sweet confections based on agar and MC (methylcellulose) or invert sugar and sugar-heated solutions have been proposed. Once they are combined, the two solutions are boiled and cooled.

1.11.3 Meat and fish products

Agar at levels of 0.5–2.0% is used in the preparation of soft meat and fish products to gel them, thereby eliminating transit damage to brittle tissues and preventing possible textural loss. Agar is preferred to gelatin or carrageenan because it has a higher melting temperature and gel strength. In the 1960s, the Japanese exported large chunks of canned tuna preserved in agar jelly to western Europe. Gelatin and some other natural thickening agents are used in canned corned beef (Mueller and Steibing, 1993; Shehata *et al.*, 1994).

1.11.4 Other uses

Agar, as well as carrageenan, is used to design films, including water-soluble antibiotics, to coat and extend the shelf life of poultry. Agar is also used in the production of dehydrated powdered fish extract to be used in soups and flavoring preparations. Agar levels of 0.05–0.85% are used to improve the stability and sensory evaluation of cheeses and cream cheese, but this use is not limited to agar. Agar is also used to gel cream or milk in order to form a solid material that can be dissolved in hot coffee or tea. Agar at levels of ~0.05–0.15% is used in the fining of wines. Many uses of agar in vegetarian or health products can be found, e.g. as a bulking agent in cereals, in the preparation of starchless desserts, in aspic salads and in puddings, fruit, butter, jams and preserves.

References

Ablett, S., Lillford, P.J., Baghdadi, S.M.A. and Derbyshire, W. (1978) *J. Colloid Interface Sci.*, **67**, 355–9.

Araki, C. (1937) Agar–agar. III. Acetylation of the agar-like substance of *Gelidium amansii L. J. Chem. Soc. Japan*, **58**, 1338–50.

Araki, C. (1958) Seaweed polysaccharide, in *Proc. 4th Int. Cong. Biochem.*, Vienna, Pergamon Press, London, pp. 15–30.

Araki, C. (1969) Studies on agar skeleton of agaropectin. *Mem. Shijonawate-gakuen Women's Coll.*, **3**, 1.

Araki, C. (1980) Carbohydrates of agar, in *Jikken Kagaku Koza*, vol. 22. Chemical Society of Japan, Tokyo, 468–87.

Ariyama, H. and Takahasi, K. (1931) The relative nutritional value of various carbohydrates and related compounds. *Bull. Agric. Chem. Soc. Japan*, **6**, 1–5.

Armisen, R. and Kain, J.M. (1995) World-wide use and importance of *Gracilaria. J. Appl. Phycol.*, **7**(3), 231–43.

Arnott, S., Fulner, A., Scott, W.E. *et al.* (1974) The agarose double helix and its function in agarose gel structure. *J. Mol. Biol.*, **90**(2), 269–84.

Bouakkaz, A., Cachard, C. and Gimenez, G. (1994) Evaluation of agar, solid material with acoustic properties equivalent to water. *J. Phys.* **4**(5), 1221–4.

Carr, J.M., Sufferling, K. and Poppe, J. (1995) Hydrocolloids and their use in the confectionery industry. *Food Technol.* **49**(7), 41–2, 44.

Chirapart, A., Ohno, M., Ukeda, H. *et al.* (1995) Chemical compositions of agars from a newly reported Japanese agarophyte, *Gracilariopsis lemaneiformis. J. Appl. Phycol.*, **7**(4), 359–65.

Cottrell, I.W. and Baird, J.K. (1980) Gums, in *Kirk-Othmer Encyclopedia of Chemical Technology*, vol. 12, 3rd edn, Wiley Interscience, New York, pp. 45–66.

Dea, I.C.A., McKinnon, A.A. and Rees, D.A. (1972) Tertiary and quaternary structure in aqueous polysaccharide systems which model cell wall cohesion: reversible changes in conformation and association of agarose, carrageenan and galactomannans. *J. Mol. Biol.* **68**, 153–72.

Duckworth, M. and Yaphe, W. (1971a) The structure of agar. Part I. Fractionation of a complex mixture of polysaccharides. *Carbohydr. Res.*, **16**, 189–97.

Duckworth, M. and Yaphe, W. (1971b) The structure of agar. Part II. The use of a bacterial agarose to elucidate structural features of the charged polysaccharides in agar. *Carbohydr. Res.* **16**, 435–45.

FDA (1972) *GRAS Food Ingredients: Agar–Agar*, PB 221–225, NTIS, US Department of Commerce, Washington, DC.

FDA (1980) *Agar–Agar*, 21 CFR 184.1115, No. 57,912,15, US Department of Commerce, Washington, DC.

Fleurence, J. and Guyader, O. (1995) Contribution of electrophoresis to the identification of red seaweeds (*Gracilaria* sp.) used as food ingredients. *Sc. Aliments*, **15**(1), 43–8.

Food Chemicals Codex III (1981) *Agar*, National Academy Press, Washington, DC, pp. 11–12.

Funaki, K. (1947) *Manufacturing Method of Agar from Seaweeds.* Japanese Patent No. 175290.

Gifu Pref. Agar Research Laboratory (1971) *A Study on Processing of Agar from Agarophytes other than Gelidium Species.* Special report of GPARL for 1968–70 fiscal year.

Gifu Pref. Agar Research Laboratory (1978) *Studies on Preservation of Agar and Agar Seaweeds.* Special report of GPARL for the 1968–70 fiscal year.

Glicksman, M. (1969) *Gum Technology in the Food Industry*, ch. 8, Academic Press, New York, pp. 199–266.
Glicksman, M. (1983) *Food Hydrocolloids*, vol 2, *Seaweed Extracts*, CRC Press, Boca Raton, FL, pp. 63–73.
Guiseley, K.B. and Renn, D.W. (1977). Agarose: purification, properties and biochemical applications, in *Marine Colloids*, FMC Bio-Products, Rockland, ME.
Guiseley, K.B., Stanley, N.F. and Whitehouse, P.A. (1980) Carrageenan, in *Handbook of Water-Soluble Gums and Resins* (ed. R.L. Davidson), McGraw-Hill, New York, pp. 5–11.
Harmuth-Hoene, A.E. (1980) Effects of dietary guar flour and agar on N balance, mineral and trace element uptake and digestive energy in humans. *Berichte der Bundeforchungsanstalt für Ernahrung*, No. 5, Karlsruhe, Germany.
Hayashi, A., Kinoshita, K., Kuwano, M. *et al.* (1978) Studies of the agarose gelling system by the fluorescence polarization method. *Polym. J.*, **10**(5), 485–94.
Hayashi, K. and Okazaki, A. (1970) *'Kanten' Handbook*, Korin-shoin, Tokyo, pp. 1–534.
Hirase, S. (1957) Chemical constitution of agar–agar. XIX. Pyruvic acid as a constituent of agar–agar. Identification and estimation of pyruvic acid in the hydrolysis of agar. *Bull. Chem. Soc. Japan*, **30**, 68–70. (Chem. Abstr. **52**, 9479i).
Hjerten, S. (1962) A new method for preparation of agarose for gel electrophoresis. *Biochim. Biophys. Acta*, **62**, 445–9.
Hove, E.L. and Herndon, J.F. (1957) Growth of rabbits on purified diets. *J. Nutr.*, **63**(2), 193–9.
Kaletunc, G., Normand, M.D., Nussinovitch, A. *et al.* (1991) Determination of gels elasticity by successive compression decompression cycles. *Food Hydrocolloids*, **5**, 237–47.
Kohyama, K., Ishikawa, Y., Nishinari, K. *et al.* (1989) An automatic measurement of gel melting point. *Sci. Aliments*, **9**(2), 227–37.
Lai, M.F., Li, C.F. and Li, C.Y. (1994) Characterization and thermal behaviour of 6 sulfated polysaccharides from seaweeds. *Food Hydrocolloids*, **8**(3–4), 215–32.
Lin, C.C. and Cassida, L.E., Jr (1987) Gelrite as a gelling agent in media for the growth of thermophilic microorganisms. *Appl. Environ. Microbiol.*, **47**, 427–30.
Matsuhashi, T. (1972) Firmness of agar gel, with respect to heat energy required to dissociate cross linkage of gel, in *Proc. 7th Int. Seaweed Symp.* (ed. T. Nishizawa), University of Tokyo Press, Japan, p. 460.
Matsuhashi, T. (1974) *Processing Method of Tokoroten and Agar from Seaweeds.* Japanese Patent No. 739750.
Matsuhashi, T. (1978) Fundamental studies on the manufacture of agar. PhD thesis, Tokyo University of Agriculture, Tokyo.
Matsuhashi, T. (1981) Kanten, in *New Edition Handbook of Refrigeration and Air-conditioning*, vol. *Applications*, Japanese Association of Refrigeration, Tokyo, p. 946.
Matsuhashi, T. (1990) Agar, in *Food Gels* (ed. P. Harris), Elsevier Applied Science, London, pp. 1–53.
Meer, W. (1980) Agar, in *Handbook of Water-soluble Gums and Resins* (ed. R.L. Davidson), McGraw-Hill, New York, pp. 7.2–7.14.
Moirano, A.L. (1977) Sulfated seaweed polysaccharides, in *Food Colloids*, ch. 8 (ed. H.D. Graham), Avi Publishing, Westport, CT.
Morris, E.R. and Norton, I.T. (1983) Polysaccharide aggregation in solutions and gels, in *Aggregation Processes in Solution* (eds W. Jones and J. Gormally), Elsevier, Amsterdam.
Morris, V.J. (1986) Gelation of polysaccharides, in *Functional Properties of Food Macromolecules* (eds J.R. Mitchell and D.A. Ledward), Elsevier, Amsterdam, pp. 121–70.
Mueller, W.D. and Steibing, A. (1993) Suitability of plant and animal gelling agents for manufacture of canned corned beef. *Fleischwirtschaft*, **73**(11), 1307–11.
Murano, E. and Kain, J.M. (1995) *Gracilaria* and its cultivation. *J. Appl. Phycol*, **7**(3), 245–54.
Nash, N.H. (1960) Functional aspects of hydrocolloids in controlling crystal structure in foods, in *Physical Functions of Hydrocolloids*, American Chemical Society, Washington, DC, pp. 45–58.
Newton, L. (1951) *Seaweed Utilization*, Sampson Low, London, pp. 107–8.
Nishinari, K., Watase, M., Miyoshi, E. *et al.* (1995) Effects of sugar on the gel–sol transition of agarose and κ-carrageenan. *Food Technol. Chicago*, **49**, 10, 90, 92–6.
Nussinovitch, A., Kaletunc, G., Normand, M.D. *et al.* (1990a) Recoverable work vs. asymptotic relaxation modulus in agar, carrageenan and gellan gels. *J. Texture Studies*, **21**, 427–38.

Nussinovitch, A., Kopelman, I.J. and Mizrahi, S. (1990b) Evaluation of force deformation data as indices to hydrocolloid gel strength and perceived texture. *Int. J. Food Sci. Technol.* **25**, 692–8.

Nussinovitch, A., Kopelman, I.J. and Mizrahi, S. (1990c) Effect of hydrocolloid and minerals content on the mechanical properties of gels. *Food Hydrocolloids*, **4**(4), 257–65.

Nussinovitch, A. and Peleg, M. (1990) Strength–time relationship of agar and alginate gels. *J. Texture Studies*, **21**, 51–60.

Nussinovitch, A., Peleg, M. and Normand, M.D. (1989) A modified Maxwell and a non-exponential model for characterization of the stress relaxation of agar and alginate gels. *J. Food Sci.*, **54**, 1013–16.

Padua, G.W. (1993) Microwave heating of agar gels containing sucrose. *J. Food Sci.* **58**(6), 1426–8.

Percival, E. (1972) Chemistry of agaroids, carrageenans and furcellaran. *J. Sci. Food Agric.*, **23**, 933–40.

Poppe, J. (1995) New approaches to gelling agents in confectionery. *Manufacturing-Confectioner*, **75**(5), 119–26.

Rees, D.A. (1969) Structure, conformation and mechanism in formation of polysaccharide gels and networks, in *Advances in Carbohydrate Chemistry and Biochemistry* (eds M.L. Wolform and R.S. Tipson), Academic Press, New York, pp. 267–232.

Schafer, S.E. and Stevens, E.S. (1995) A reexamination of the double-helix model for agarose gels using optical rotation. *Biopolymers*, **36**(1), 103–8.

Segawa, S. (1965) *Genshoku Nippon Kaiso Zukan* (Natural-Color Picture Book of Marine Seaweeds), Hoikusha, Tokyo.

Seip, W.F. (1974) Specifications and experience in the use of bacteriological grade agar–agar by a leading manufacturer of dehydrated media, in *Proc. 8th Int. Seaweed Symp.*, Bangor, Wales, August 17–24.

Selby, H.H. and Wynne, W.H. (1973) Agar, in *Industrial Gums* (ed. R.L. Whistler), Academic Press, New York, pp. 19–48.

Shehata, H.A., Shalaby, M.T. and Hassan, A.M. (1994) Gelatin and some other natural thickening agents for use in canned corned beef. *J. Food Sci. Technol. India*, **31**(4), 298–301.

Smith, F. and Montgomery, R. (1959) *The Chemistry of Plant Gums and Mucilage*, Reinhold, New York, p. 426.

Steiner, A. and Rothe, L.B. (1949) *Stabilizer for icings*. US Patent 2,823,129.

Takano, R., Hayashi, K. and Hara, S. (1995) Highly methylated agars with a high gel-melting point from the red seaweed, *Gracilaria eucheumoide. Phytochemistry*, **40**(2), 487–90.

Takegawa, O. (1963) *Strength of Agar Gels*, Nippon Kaiso-kogio Res. Lab. Report, No. 2.

Tashiro, Y., Mochizuki, Y., Ogawa, H. *et al.* (1996) Molecular weight determination of agar by sedimentation equilibrium measurements. *Fisheries Sci.* **62**(1), 80–3.

Tseng, C.K. (1946) Phycolloids: useful seaweed polysaccharides, in *Colloid Chemistry*, vol. VI, (ed. J. Alexander), Reinhold, New York, p. 630.

Watase, M. and Nishinari, K. (1986) Rheology, DSC and volume or weight change induced by immersion in solvents for agarose and kappa-carrageenan gels. *Polym. J.*, **18**, 1017–25.

Watase, M., Nishinari, K., Clark, A.H. *et al.* (1989) Differential scanning calorimetry, rheological, X-ray and NMR of very concentrated agarose gels. *Macromolecules*, **22**(3), 1196–201.

Yanagawa, T. (1942) *Kogyo-tosho*, Tokyo, pp. 1–352.

Yaphe, W. and Duckworth, M. (1972) The relationship between structures and biological properties of agars, in *Proc. 7th Int. Seaweed Symp.* (ed. T. Nishizawa), University of Tokyo Press, Japan, pp. 15–22.

2 Alginates

2.1 Introduction

Alginate was first discovered by E.C.C. Stanford in 1881, while searching for useful products from kelp. He developed the process of alkali extraction of a viscous material, 'algin', from the algae and later precipitated it using mineral acid (Stanford, 1883, 1884). Algin was isolated 15 years later by Krefting (Krefting, 1896). In 1929, the commercial production of algin was initiated by the Kelco Co. in California. The extracted material was first used as a boiler compound and for can-sealing purposes. In 1934, the use of alginate for foods (as an ice-cream stabilizer) became important. In 1944, propylene glycol alginate (PGA) was developed and produced commercially. Later, alginate-production plants were established in the USA, Europe and Japan (McNeely, 1959).

2.2 Sources

Alginates are a group of naturally occurring polysaccharides, extracted from brown seaweed (they differ from agar and carrageenan, which are extracted from red seaweed). Of the many different species of brown seaweed, the most widely used are *Laminaria hyperborea*, *Macrocystis pyrifera* and *Ascophyllum nodosum* (Fig. 2.1). Most of the alginates in the USA are extracted from *M. pyrifera*, which is a giant kelp found in sea beds of 50 feet (15 m) to 1 mile (1.6 km) wide, several miles long and 25–80 feet (7.5–25.0 m) deep. The kelp attaches itself to the rocky bottom with a rootlike structure called a holdfast, because of the strong ocean currents. The kelp is harvested 3 feet (1 m) below the surface by mechanical means. This permits sunlight penetration and promotes algal growth. The cut kelp is processed shortly thereafter. In North America, several other *Laminaria* species can be found and are used for domestic production including *Laminaria digitata*, *Laminaria cloustoni* and *Laminaria saccharina*. In Europe, *Laminaria* species and *A. nodosum* are used for production, whereas the Japanese harvest *Ecklonia cava* and the South Africans *Ecklonia maxima*. In the cell wall (natural state) and intercellular spaces, alginate is found as a mixed calcium/sodium/potassium salt of alginic acid. The alginate molecules provide the strength and flexibility necessary for algal growth in the sea. The

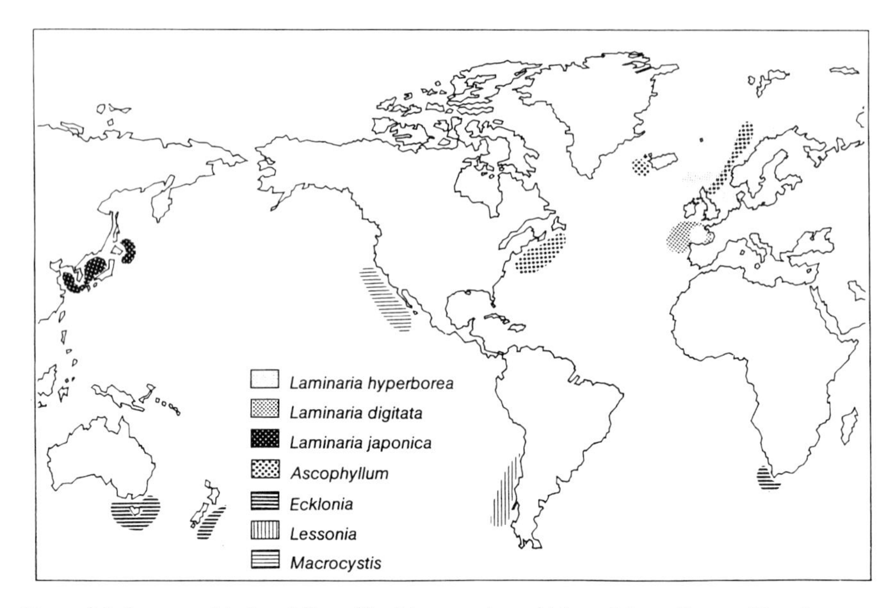

Figure 2.1 Sources of industrially utilized brown algae. (Adapted from Proton Biopolymers A/S, 1990.)

gum, generally sold as sodium alginate, is water soluble and used as a thickener as well as a gel producer in the presence of calcium and/or other polyvalent metal ions.

Experiments to extract alginic acid from residual algae by the hot-water method were reported (Nishide *et al.*, 1992). As a result of this comprehensive study it was concluded that residual alga is unsuitable for extracting alginic acid because results indicated that this alginic acid is strongly affected by the sulfated-polysaccharide extraction procedure (Nishide *et al.*, 1992). Alginate-like polysaccharides can also be synthesized by *Pseudomonas fluorescens* and *Pseudomonas putida* grown in batch cultures on glucose and fructose as carbon sources (Conti *et al.*, 1994). Promising results of alginate batch production by immobilized *Azotobacter vinelandii* have also been reported. The amount of alginate produced represented ~60% of that recovered from a cell-free culture (Lebrun *et al.*, 1994).

2.3 Structure

In a first attempt to define alginate's structure, Stanford suggested it to be a nitrogenous material, with the formula $C_{76}H_{76}O_{22}(NH_2)_2$. However, more modern isolation methods subsequently showed the pure product to be nitrogen free. Up until the 1950s, alginic acid was thought to be a polymer of anhydro-1–4-β-D-mannuronic acid (Steiner and McNeely, 1954),

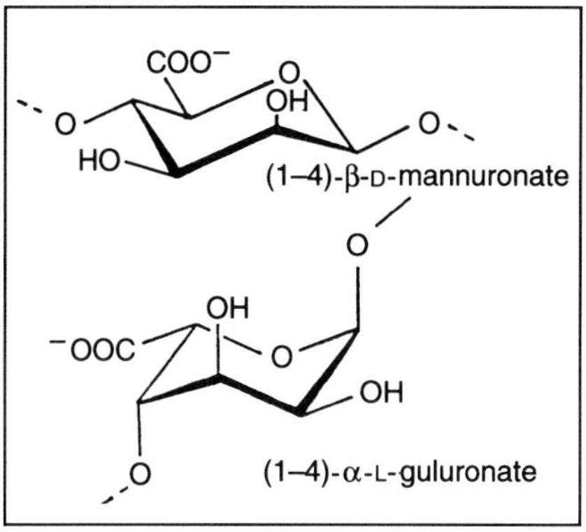

Figure 2.2 The alginate monomer units. (Adapted from Proton Biopolymers A/S, 1990.)

and only in the 1960s was L-guluronic acid also shown to be present (Fig. 2.2). Alginic acid is, therefore, a linear copolymer composed of D-mannuronic acid (M) and L-guluronic acid (G) (Whistler and Kirby, 1959; Hirst and Rees, 1965). Regions can consist of one unit or the other, or both monomers in alternating sequence, i.e. M blocks, G blocks or heteropolymeric MG blocks, respectively (Fig. 2.3). The monomers tend to settle into their most energetically favorable structure in the polymer chain. For

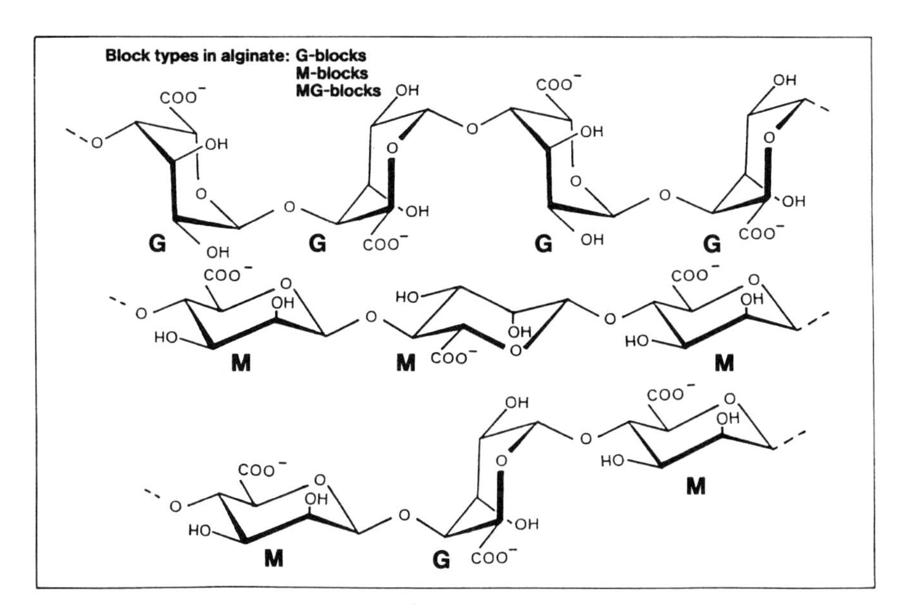

Figure 2.3 Block types in alginate. (Adapted from Proton Biopolymers A/S, 1990.)

G–G this is the $^1C_4$ chair-form linked by an $\alpha(1–4)$ glycosidic bond. For M–M it is the $^4C_1$ chair-form linked by a $\beta(1–4)$ glycosidic bond (Onsøyen, 1992). The carboxylic group is responsible for an equatorial/equatorial glycosidic bond in M–M, an axial/axial glycosidic bond in G–G and an equatorial/axial glycosidic bond in M–G. The geometries of the M, G and alternating blocks depend on the particular shapes of the monomers and their mode of linkage in the polymer. M blocks have an extended ribbon shape (flexible), G blocks are buckled (stiff) and the MG-block regions are of intermediate stiffness. Information on the chemical composition of alginate has also been obtained by NMR spectroscopy (Grasdalen, 1983). The molecular weight, the proportion and the arrangement of M and G units affect the behavior of a particular alginate. The composition of commercial brown seaweed has been studied (Haug, 1964), and the percentages of M range from 61% in *M. pyrifera* to 31% in *L. hyperborea*. The content of the alternating segments ranges from 26.8% in *Laminaria* to 41.7% in *Macrocystis*. Different monomer ratios have been found in 'high M' (e.g. *Macrocystis* and *Ascophyllum*) and 'high G' (*Laminaria*) alginates. Molecular weights of commercial alginates are in the range 32 000–200 000, corresponding to a degree of polymerization of 180–930 (Glicksman, 1969).

2.4 Alginate sources and production

2.4.1 *Raw materials*

Only a few of the many species of brown seaweed are suitable for commercial production. The three previously mentioned seaweeds from North Atlantic coastal regions (*L. hyperborea*), from the west coast of America (*M. pyrifera*) and from northern Europe and Canada (*A. nodosum*) are harvested by beach collection, mechanical harvesting or cutting by hand. The seaweed is then towed to the processing plant where it can be processed either wet or after drying.

2.4.2 *Processing and production of PGA*

Current methods of production are based on modifications of processes invented by Green (1936), and by Le Gloahec and Herter (1938). Green's process is carried out at 10°C. The fresh kelp is leached in dilute hydrochloric acid for a few hours, drained and shredded. The weed is then digested with 2–2.5% soda ash at pH 10. The produced mass is redigested with additional soda ash and pulverized in a hammer mill. Water is added at a volumetric proportion of 6:1 and pH is maintained in the range of 9.6 to 11. The product is then clarified and filtered, precipitated with 10% calcium chloride to get calcium alginate, and treated with 5% hydrochloric acid.

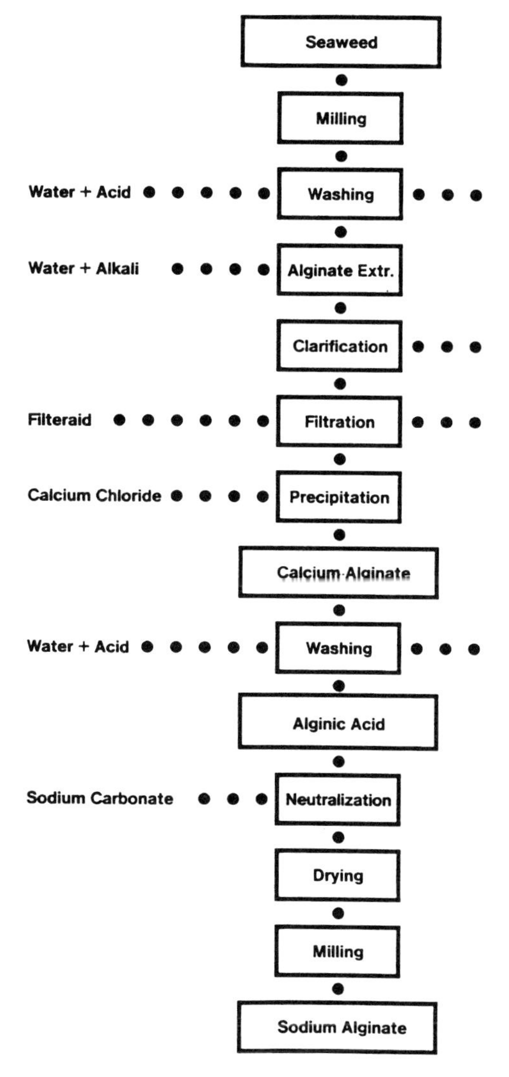

Figure 2.4 Flow sheet of alginate production. (Adapted from Proton Biopolymers A/S, 1990.)

After washing and a purification treatment with carbonate, oxide or hydroxide to obtain the desired salt, the product is dried. In the Le Gloahec and Herter process, fresh or dried alga is leached with dilute calcium chloride, washed to remove soluble material and then washed with dilute hydrochloric acid to remove any residual alkaline salts. The weed is digested at 40°C for 2 h with two volumes of a 4% soda-ash solution and mixed to a paste. The paste is then diluted with water and aerated. The produced emulsion separates during setting and the cellulose in the top layer is

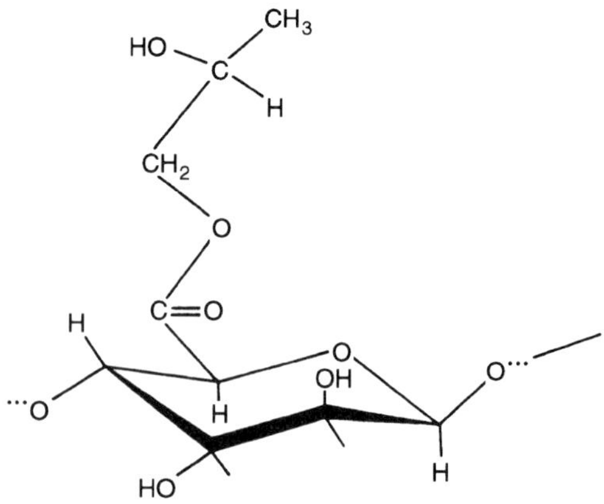

Figure 2.5 Structure of PGA.

removed. The remaining solution, containing the crude sodium alginate, is decolorized, clarified and finally treated with hydrochloric acid to precipitate the alginic acid, followed by treatment with metallic carbonate, oxide or hydroxide to achieve the desired alginate salt. Although in the many existing extraction plants modifications and improvements in processing can be noted, Stanford remains responsible for the basic method.

The alginate extraction steps are depicted schematically in Fig. 2.4. The process is based on ion-exchange. The first step is the milling of the seaweed, followed by washing and treating it with strong alkali and heating to extract the alginate, which is precipitated with calcium chloride and then treated with acid to form alginic acid. The alginic acid can be treated with sodium carbonate (or other bases) to produce the alginate salts, or reacted with propylene oxide to produce PGA (Fig. 2.5). After the reaction, up to 90% of the carboxyl groups are esterified, and the remaining groups either remain free or are neutralized with sodium or calcium. It is possible to purchase different PGAs designed for specific applications.

2.4.3 Alginate-like polymers

In addition to the well-known traditional gum extraction from brown seaweed, alginate-like polymers are produced by some bacteria as an exocellular secretion. Alginate from *Pseudomonas aeruginosa* – a partially acetylated product from a secondary pathogen of patients with cystic fibrosis (Doggett and Harrison, 1969) – contains variable amounts of M and G units. A similar polymer is produced by the soil bacteria *Azotobacter vinelandii*. Other non-pathogenic species can be utilized as potential producers of alginate. Enzymatic modification with mannuronan-C-5-epimerase (an enzyme) from *Azotobacter* yields alginates with high G content. This technique is not yet used commercially.

2.5 Commercial aspects

Alginates are sold for specific applications. The contents of M and G residues and their ratio are not measured by the manufacturer. Instead, they are estimated from data on the seaweed source and the blended quantities used to achieve the desired gel strength and solution viscosity. Sodium alginate is a commercially available product sold in bulk quantities; calcium, potassium and ammonium alginates, as well as alginic acid, are also available. It is important to note that alginic acid, the free-acid form of alginate, has limited stability: to stabilize it and make it water soluble, the acid is transformed into a commercial alginate by the incorporation of salts such as sodium carbonate, potassium carbonate, ammonium hydroxide, magnesium hydroxide, calcium chloride or propylene oxide.

2.5.1 Alginate solution preparation procedures

Alginic acid and calcium alginate are insoluble in water. They can swell and form a paste-like material but do not form the smooth solution produced by sodium alginate in water with a very low level of calcium. Alginate must be dispersed before dissolution can take place. High-shear stirring and dry mixing with other formulation ingredients such as starch or sugar, or dispersion in vegetable oil or glycerol promote the separation of alginate particles and subsequently their dissolution, provided these steps are taken before water is added. If neither of these solutions is practical, several specially prepared, easily dispersible more expensive alginate types are sold. If dispersion is not performed properly, clumps that are swollen on the outside prevent contact between the water molecules and their center. In this case, a very high-shear mixing must be performed to resolve the problem. Care must be taken to avoid overheating during mixing, which might cause degradation of the alginate, resulting in decreased viscosity or gel strength. Soft water is recommended to avoid any undesired cross-linking of the alginate by calcium.

After dissolving alginate in water, a smooth solution with long flow properties is produced. Solution viscosity is checked by rotational viscometers designed for the study of non-Newtonian liquids. In practice, it is difficult to distinguish between thickening and gelling, since a very weak gel may appear as a thick solution, and thickening may often be the result of limited calcium–alginate cross-linking. Chemical and physical variables affecting the flow characteristics of alginate solutions include the presence of salts, sequestering agents and polyvalent cations, polymer size, temperature, shear rate, concentration of the gum in solution and the presence of other miscible solvents. Solution viscosity should be related to molecular weight, but it is also influenced by the level of residual calcium from the manufacturing process. It depends on temperature (the higher the temperature the lower the viscosity), as well as on thermal depolymerization, which, when it occurs, is dependent on time, temperature and pH. (Depolymerization can be deliberately achieved via enzyme systems (Takeuchi *et al.*, 1994) for industrial, medicinal or other purposes.) Cooling of the gum solution results in increased viscosity but does not result in gel formation. Frozen alginate solution maintains its viscosity after thawing. The addition of alcohols and glycols (water-miscible solvents) results in increased viscosity.

The pH level influences alginate solutions differently, depending on the type of alginate used. Sodium alginate solutions are not stable above pH 10, PGA is more stable at acidic pH and sodium alginate precipitates at pH < 3.5. Sequestering agents prevent the calcium that is inherent in the alginate from reacting with it or prevent alginate's reaction with polyvalent ions in the solution. The addition of sequestering agents reduces viscosity compared with that seen without these agents. This is true for both M- and G-type alginates. Monovalent salts also reduce the viscosity of dilute sodium alginate. The alginate polymer contracts as the ionic strength of the solution increases. Maximal viscosity is achieved at a salt level of 0.1 N. This effect can be reduced by increasing the gum concentration, keeping the calcium level low. The effect of salt is pronounced after a long storage period and is dependent upon the alginate source, the degree of polymerization, gum concentration in the solution and the character of the monovalent salt used.

2.6 Alginate mechanism of gelation

Alginates form gels with a number of divalent cations (McDowell, 1960). For food purposes, calcium is particularly suitable because of its non-toxicity. Borax can be used to produce gels for non-food applications. Until comparatively recently, gels were generally thought to be produced by ionic bridging with calcium ions of two carboxyl groups on adjacent polymer chains. However, although these bonds are important for gelation, they are not considered sufficiently energetically favorable to account for the gelation of the gum (Rees, 1969). In poly-L-guluronate segments (with chain lengths of over 20 residues), enhanced binding of calcium ions occurs and a cooperative mechanism is involved in the gelation. Cross-linking takes place via carboxyl groups by primary valences and via hydroxyl groups by secondary valences. Coordinate bonds extend to two nearby hydroxyl groups of a third unit that may be in the same molecular chain to retain the macromolecule's coiled shape (Glicksman, 1969) or the unit may be in another chain, which results in the formation of a huge molecule with a three-dimensional netlike structure (Glicksman, 1969). No such effects are observed for chains made of M blocks or alternating blocks. Gel formation and the resulting gel strength obtained from an alginate is, therefore, very closely connected to the amount of G blocks and the average G-block

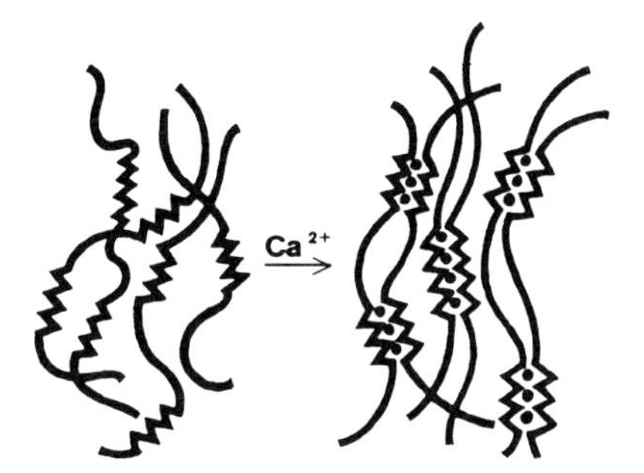

Figure 2.6 The 'egg-box' model for alginate gel formation. (Adapted from Proton Biopolymers A/S, 1990.)

length. High G content and long G blocks give alginates high calcium reactivity and the strongest gel-forming potential. The degree of polymerization should be >200 to achieve optimal gel strength. A mechanism whereby a cavity that acts as a binding site for calcium ions between two diaxially linked G residues is formed produces a three-dimensional structure, called an 'egg-box' (Grant *et al.*, 1973), in which the calcium interacts with the carboxyls and with the electronegative oxygen atoms of the hydroxyl groups (Fig. 2.6). Gelation involves chain dimerization and later aggregation of the dimers. The alginate gel is considered a semi-solid material, and the junction zones at which the alginate polymers are bound represent the solid state. After gelation, the water molecules are physically entrapped by the alginate matrix or network but retain their ability to migrate. This has very important consequences for various applications. Reacting soluble sodium alginate with calcium results in the production of gels with various consistencies. To achieve a stiff texture, a small amount of calcium can be sufficient. In fact, the stoichiometric amount is 0.75 mg calcium per gram of algin (depending on the viscosity grade of the selected sodium alginate). Optimal complexing of the alginate and calcium is obtained by using approximately 40% of the stoichiometric amount, which varies with calcium source, the presence of other soluble solids and pH, among other things (Messina and Pape, 1966). A mathematical model was developed to describe the kinetics of formation of calcium alginate gel in thermally set carrier gels (Chavez *et al.*, 1994). The Sharp interface model was used to describe the cross-linking kinetics. The model assumes diffusion of calcium ions through a preformed gel of sodium alginate and a selected carrier, and an instantaneous reaction between the calcium and sodium alginate. Such a model can be used to predict processing effects on food gels (Chavez *et al.*, 1994). Small-angle X-ray scattering was used to investigate the structural changes in an alginate solution during the sol–gel transition induced by two divalent cations, calcium and copper. Purified alginates exhibited polyelectrolytic behavior (Zheng *et al.*, 1995). Copper ions showed

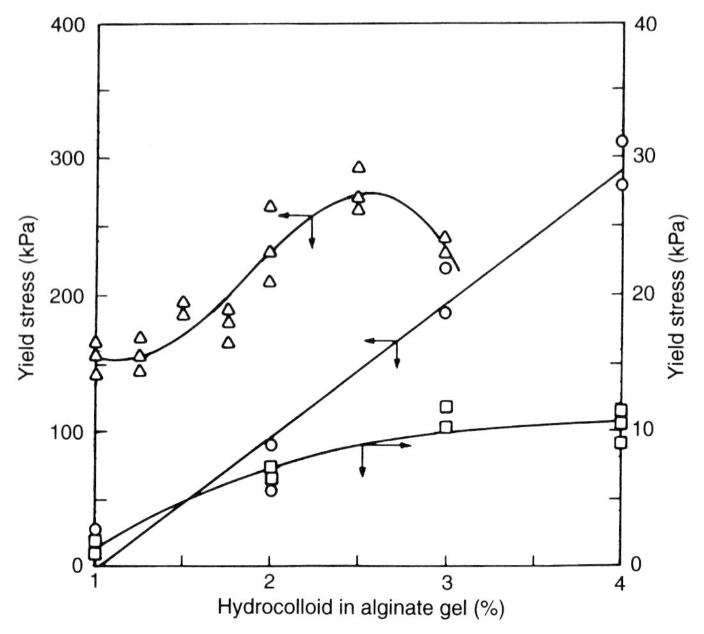

Figure 2.7 The effect of hydrocolloid concentration on the yield stress of alginate gels: alginate set by diffusion (○); alginate plus GDL (△); alginate plus SHMP (□). The yield stress values at the left apply to alginate by dialysis and alginate plus GDL, those on the right to alginate plus SHMP. (From Nussinovitch *et al.*, 1990; by permission of Oxford University Press.)

a higher affinity for alginate than calcium ions. In the calcium alginate system, the derived molecular parameters were correlated with the macroscopic properties of gelation. This correlation was not observed in the copper alginate system, suggesting that different gelation mechanisms occur with calcium alginate and copper alginate (Zheng *et al.*, 1995).

The effect of gum concentration on the gel yield stress and on the deformability modulus of glucono-δ-lactone (GDL) alginate and carrageenan gels was characterized by a curve having a maximum yield stress for a particular concentration of hydrocolloid (Fig 2.7). Increasing the gum concentration beyond that point decreases the yield stress and the deformability modulus (Fig. 28.9) (Nussinovitch *et al.*, 1990). Alginate gel setting by sodium hexametaphosphate (SHMP) exhibited a similar mechanical behavior, in contrast to an alginate gel set by calcium diffusion, which is characterized by a linear relationship between the yield stress and the gum concentration (Nussinovitch *et al.*, 1990). The stress relaxation properties of alginate gels have also been studied and will affect their potential contribution to future foods (Nussinovitch *et al.*, 1989).

A strength–time study of alginate gels showed that the compressive strength increases rapidly for ~15 h, then tends to stabilize asymptotically (Nussinovitch and Peleg, 1990). An empirical, two-parameter mathematical

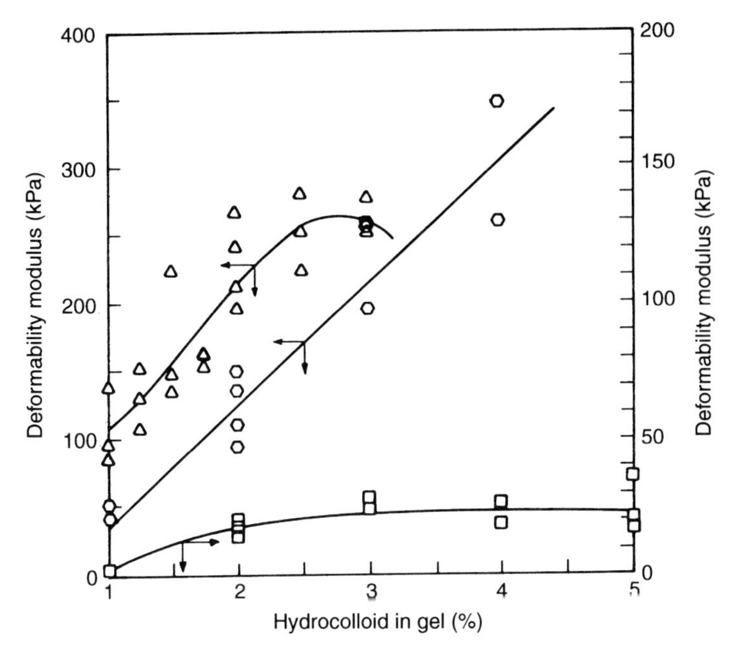

Figure 2.8 The effect of hydrocolloid concentration on the deformability modulus of alginate gels: alginate plus GDL (△); alginate plus SHMP (□); alginate set by diffusion (⬡). The deformability modulus values on the left apply to alginate plus GDL and alginate set by diffusion, those on the right to alginate plus SHMP. (From Nussinovitch *et al.*, 1990; by permission of Oxford University Press.)

model was used to estimate the asymptotic value on the basis of data obtained during the first 20 h. The common procedure of testing gels after 24 h was found to be satisfactory, and a similar two-parameter model enabled a prediction of the gel's final weight with reasonable accuracy (Nussinovitch and Peleg, 1990).

2.7 Gel preparation

2.7.1 Degree of conversion, thixotropy and alginate–pectin gels

The properties of alginate gels depend on the type of alginate used, the degree of conversion to calcium alginate, the source of the calcium ions and the method of preparation. Strong brittle gels are formed with high-G alginates. Weaker, more elastic gels that are less prone to syneresis are formed with high-M alginates. The degree of calcium conversion to calcium alginate (replacing sodium with calcium) is responsible for the tested effect. A low level of conversion results in increased viscosity. Higher conversion results in the development of gel structures. Still further conversion results in increased gel strength and thixotropic behavior, whereby the produced gel converts into a fluid upon applying shear and reforms when the shearing stops. With additional calcium, irreversible gels are produced. Following shearing, those gels do not reform. Combinations of alginate with high-methoxy pectin (exhibiting synergism) can produce a gel at low solid contents and a wide range of pHs, in contrast to high-methoxy pectins that produce gels with high solid sugar contents. Since fruits are rich in pectin, the addition of alginate to cooked fruits (e.g. apples) results in a rigid gel when alginates with high G content are used (Toft, 1982; Toft *et al.*, 1986).

2.7.2 Diffusion setting

Alginate gels can be produced by diffusion setting. Beads can be achieved by simply dropping alginate solution into a calcium chloride bath. With objects of small diameters or thicknesses (such as beads or thin films), the required time for calcium to diffuse into the objects is not a limiting factor, especially if its concentration is deliberately increased. Concentrations of the order of 15% can be used for short times (of the order of seconds). For food purposes, less bitter calcium salts such as calcium lactate or calcium acetate can be used. Acidic baths can be used when a calcium salt that is soluble only under acidic conditions (e.g. calcium hydrogen orthophosphate) is solubilized within the alginate solution before extrusion. Diffusion occurs both in and out of a gel piece, a fact that needs to be taken into account when preparing a product containing ingredients with low molecular weights, such as sucrose, acids, etc. Internal setting occurs when calcium is released under controlled conditions from within the system. The rate of calcium release is controlled by pH and the solubility of the calcium salt. To eliminate too rapid a reaction between the alginate and the calcium (or other cross-linking agent), a sequestering agent is added to the system and, depending upon pH, the system's ability to entrap or free the ions is changed. Thus, it is possible to make a preparation at a neutral pH and to change the pH abruptly to an acidic one, inducing the appropriate sequestering agent to release its entrapped ingredient, which then reacts with the alginate. Calcium release should be monitored using a special ion-selective calcium ion electrode. Full knowledge of the calcium concentration at any given stage of the process enables the development of better structured products.

2.7.3 Establishing the differences between alginate gels

Gel properties such as gel strength can be tested using a universal testing machine (Instron or J. J. Lloyd), a marine colloids' gel tester, the FIRA jelly tester, or other such instruments. Gel strength is defined as the maximal force in newtons needed to break a gel, or the stress that breaks a gel (stress at failure). Gel strength is a measure of gel 'hardness'. The main factors influencing gel properties are temperature, pH and the presence of proteins and neutral polymers. Alginate gels are thermostable. In other words, once setting occurs, they do not melt upon reheating. Once the alginate is internally set and all of its components have been mixed in, the system contains everything necessary for gelation. However, if the solution temperature is increased, the alginate chains gain extra thermal energy to permit alignment, and gelation is postponed. Associations then occur when the solution is cooled. In such systems, because of the calcium's availability to all alginate molecules, a lesser degree of syneresis is observed. This is in contrast to systems in which gelation is induced by diffusion: there the molecules closest to the calcium source react first, resulting in instability, which induces greater gel shrinkage and syneresis. The higher the alginate and calcium concentrations, the higher the temperature needed to prevent gelation. At 0.6% sodium alginate, and enough calcium to give 60% conversion (a molar ratio of 0.5 is theoretically sufficient for 100% replacement of the sodium), a temperature of 80°C is required to prevent gelation. Production of a gel made of alginic acid (without calcium ions) is possible by maintaining the pH at below 3.5 (setting acid bath). However, the resultant gels are grainy and suffer from syneresis. In calcium setting, reducing pH lowers the amount of calcium required to obtain a gel. Under acidic conditions, protein is also capable of interacting with sodium alginate (Imeson, 1984) by electrostatic interaction. A reduction in pH increases this interaction's strength. At acidic pH, proteins are destabilized and denatured in the presence of alginate, and high-molecular-weight complexes are also produced.

2.8 Applications

Several applications of alginate in foods have been reviewed (Cottrell and Kovacs, 1980; Nussinovitch, 1993). They include fruit texturization (Chapter 20), preparation of diffusion-set gels, uses in protein extrusion, extension of the storage life of potatoes, immobilization of banana enzymes, use in minced fish patties, use in meat products and many others. Cooking yield, pH, binding and textural properties of alginate-structured beef rolls were studied (Shand *et al.*, 1993). Cooking yields of algin–calcium products were improved with gellan gum. Other gums (xanthan, guar, pectin, carboxymethylcellulose (CMC)) improved water holding but had detrimental effects on product texture. Another study detailed the optimal use of ingredients (algin, calcium carbonate, encapsulated lactic acid, dry lactic acid, calcium lactate, palm oil and beef) that affect the functionality and economics of meat products (Ensor *et al.*, 1990). The uses of alginate are expected to increase in the future.

2.8.1 Fruit-like products

Fruit products are discussed in Chapter 5 and other texturized products in Chapter 20. A few additional points on this subject are mentioned here. The first to develop a process for the production of artificial cherries was Peschardt (1946). He dropped a flavored, colored, alginate–sugar solution into a bath of soluble calcium salt. Following the instantaneous formation of a calcium alginate skin, gelation of the interior of the 'cherries' was induced by slow diffusion of calcium into the spherical particle and cross-linking with the alginate inside. The thermostability of the artificial cherries enabled their inclusion in baked goods. In the 1960s in England, this product was used in fruitcake. The product was successfully marketed in Europe and included natural cherry purée, corn syrup, alginate, flavoring and color. These imitation fruits were prepared in five different sizes and since they included cherry purée, they were called cherry balls rather than imitation cherries (Kneeland, 1961).

Fruit analogs can be successfully prepared by internal setting, using a two-mix process. In the first, alginate, anhydrous dicalcium phosphate, disodium hydrogen orthophosphate, glucose, sucrose and water are mixed, and in the second, fruit purée, sucrose, glucose, citric acid and sodium citrate dihydrate are blended. The anhydrous dicalcium phosphate is insoluble at neutral pH. The two mixtures are pumped through a high-shear mixer then passed through under no-shear conditions. The calcium ions released from the anhydrous dicalcium phosphate when the pH is lowered are responsible for the gelling reaction. If a continuous process is preferred to the batch method, the mix can be extruded onto a conveyor belt. The velocity of the conveyor belt is adjusted so that setting will occur at the end of the belt, where the structured slab is diced into pieces of the desired shape and size. Unilever patent BP 1,484,562 describes a co-extrusion system for preparing re-formed blackcurrants. The basic idea behind this process is the maintenance of a continuous flow of alginate mix (with high M content, sodium alginate and water) in an outer tube. Fruit purée mix is pumped intermittently through an inner coaxial tube. The fruit stream is coated with the alginate solution, which is later cross-linked and gelled in a calcium setting bath. The fruit purée mix contains blackcurrant purée, citric acid, calcium lactate, sugar, cross-linked potato starch, CMC and water, and the setting bath includes calcium lactate and sugar dissolved in water.

2.8.2 Water dessert gels

Attractive, edible gels or jellies can be produced with alginates. Although calcium is preferred as the metal ion for cross-linking, other divalent or trivalent metal ions can be used (Glicksmam, 1962a). Reaction rates depend on the selection of calcium ions, on concentration and pH. Gel formation that is too rapid produces a grainy, discontinuous gel, whereas a very slow process results in very soft gels. The controlled release of calcium ions can be achieved by using a soluble calcium salt with the desired degree of ionization sparingly, by properly selecting the salt to control its solubility and by using sequestering or retarding agents for the controlled release of calcium ions. These systems have been the subject of many patents (Steiner, 1948) and have been used for the production of fruit jams and jellies, jellied salads and broths, dessert gels and candied jellies. Their production includes several ingredients: water-soluble alginate, a calcium salt such as tricalcium phosphate (which produces calcium ions with the addition of acid), weak acid or an acid-releasing agent such as GDL, and a gel-retarding agent such as SHMP to entrap the calcium ions until the acid is added and then to liberate the cations slowly, thereby yielding a smooth gel. Several patents are related to these ideas, including those of McDowell (1960) and Angermeier (1951) for edible alginate-jelly preparations, and the spray-drying of a blend consisting of sodium alginate, SHMP and sodium citrate, recommended by Poarch and Twieg (1957) to improve algin's dispersibility and solubility in a cold-water gelling system of this type. Dry-mix systems have also been developed by Morton and McDowell (1960). Since alginate gels are rigid and consequently do not melt in the mouth, Kohler and Dierichs (1959) claimed that heat-reversible alginate gels could be produced with partial amides of alginic acid (70% carboxyl groups converted to carbonamide groups, and 30% in the form of ammonium salts). High-viscosity carboxymethyl alginate was proposed by Rocks (1960) as a cold-water dessert gel that is relatively unaffected by the varying mineral content of tap water. Such a system was also used by Glicksman (1962b) to produce a frozen dessert gel that is less texturally influenced by freeze–thaw cycles.

2.8.3 Milk puddings, ice-cream stabilizers and other dairy products

The first milk puddings were of inferior quality, because of alginate's imperfect solubility in milk, the development of granular structures and a lack of gel strength and firmness. These products have since been improved (Anon., 1958). The preparation of a good-quality milk pudding based on a specially treated blend of a water-soluble alkali metal alginate, a mild alkali and a small quantity of calcium salt was proposed by Gibsen (1957). A simpler composition was used to form a good gel in water that could be admixed with skim-milk powder to form a milk pudding when dissolved in cold water (Ohling, 1959). Other improvements were suggested by McDowell and Boyle (1960a,b) and Hunter and Rocks (1960). Alginate can be used for the retardation and regulation of ice-crystal growth in ice creams to achieve a smooth texture. The same addition can also contribute to retardation of the product's meltdown and assist in controlling overrun. Small amounts of sodium alginate (0.1 to 0.5%) are most often used as ice-cream stabilizers. Since the stabilizers have water-holding properties and dispersive ability, they contribute to good body properties and texture protection. Sodium alginate may react with the calcium in the ice-cream mix, thereby reducing the concentration of the calcium ions in the water. This is beneficial in eliminating the clumping of fatty globules in the product (Boyle, 1959). The use of alginates in ice-cream stabilization has been the subject of many patents (Steiner, 1949; Shiotani and Hara, 1955; Kohler and Dierichs, 1958; Miller, 1960). The water and oil in soft cheese spreads can separate: a small inclusion of sodium alginate can be used to prevent this, in addition to minimizing surface hardening and improving the texture of the processed cheese. The addition of $\sim 0.15\%$ sodium alginate is sufficient to thicken whipped cream and to act as a stabilizer upon whipping (Krigsman, 1957). The production of synthetic creams can be improved by adding alginate (to give a quicker whip, greater tolerance to overwhipping, stabilized overrun and freedom from syneresis) and methylcellulose (as a foam-forming colloid) (Boyle, 1959).

2.8.4 Fish and meat preservation and sausage casings

Oxidative rancidity of fatty fish such as mackerel and herring occurs, even at low temperatures after quick freezing. This can be prevented by block freezing the fish in alginate jelly. The film formed around each fish piece excludes air, making rancidity almost impossible (Helgerud and Olsen, 1955, 1958; Olsen, 1955). The film also retains moisture, in addition to preserving the fish and its appearance. Frozen fish can be easily separated when thawed, since the jelly thaws first. Regulation of the jelly's salt content and viscosity can reduce the freezing time by 25%. Storage with the jelly coating also helps reduce the off-flavors and unpleasant smells associated with fish. This process was extended to other fishes and shellfish (Anon., 1957). Alginates have been used to dust herring (Henning, 1957), as an aid in its canning. Curing preservation of salted and dried mackerel and extension of the shelf life of these products by different film-forming gums, namely sodium alginate and tamarind kernel powder, guar gum and agar–agar, were reported (Shetty *et al.*, 1996).

Calcium alginate films have been used to coat whole beef cuts prior to freezing, by dipping in a 10–15% sodium alginate solution, then 3.5–5% calcium chloride, followed by 10–20% glycerol (as a plasticizer). The meat is frozen in salt brine at $-20°C$ and, upon defrosting, it reabsorbs the released juices (Berlin, 1957; Slepchencnko *et al.*, 1956). Reports can be found on the use of calcium alginate films to coat poultry parts (Mountney and Winter, 1961), as a carrier for proteolytic enzymes to tenderize meat (Pressman, 1957), to prevent salt rust of sausage and prolong its shelf life (Childs, 1957) and as an undercoating on sausages beneath an overcoating of ethyl cellulose (Weingand, 1959). An edible meat coating of sodium alginate and a cornstarch slurry was used to coat beef steaks, pork chops and skinned chicken drumsticks to improve texture and juiciness, color, appearance and odor. Details on using alginates for restructured meat products can be found in the literature (Means and Schmidt, 1986a; Schmidt and Means, 1986; Means *et al.*, 1987; Ernst *et al.*, 1989; Trout, 1989; Trout *et al.*, 1990). Alginates have been used as a food-packaging coating material and more extensively as a synthetic sausage casing (Wolff & Co., 1951). In 1955, the Visking Corporation introduced alginate casings, under the name of Tasti-Jax. The casings were not successful in the USA but were widely used in Germany. A typical casing is produced from 6% sodium alginate extruded into an acidic solution of 10% calcium chloride; it is then plasticized by washing with 15% glycerol and 8% calcium lactate (Weingand, 1957). The adhesive qualities of the inside surface can be improved by washing with a dilute solution of multivalent cations such as aluminum or zirconium (Weingand, 1957). On the one hand the casings are more elastic, hygienic and easier to handle than natural casings. On the other, if an interaction with sodium salts occurs, the casings swell, become shiny, weaker and less attractive, and fail to shrink during cooking. A few attempts at eliminating these disadvantages have been proposed by Langmaack (1961) and Hartwig (1960).

2.8.5 *Bakery toppings, fillings, beverages and salad dressings*

Alginates have been used to prepare icings for sweet yeast-dough products, and to stabilize other types of icing (Glabau, 1943). The addition of alginate to icing formulations produces non-sticky icings that do not crack (Gibsen and Rothe, 1955). Alginates have been found effective at maintaining foam height and aerated structures (Glabau, 1943). The texture of whipped sugar toppings is improved by the additon of alginates (General Foods Corp., 1954). Alginates can reduce or prevent syneresis in baking jellies. Also advantageous is their stability at high baking temperatures (Dorner and Tessmer, 1953). Sour-cherry pie fillers are stabilized with alginates (Strachan *et al.*, 1960). Improved clarity in canned and frozen peach and cherry pie fillings can be achieved by combining PGA with a modified waxy maize starch (Kunz and Robinson, 1962).

Pulp sedimentation in fruit drinks can be prevented by adding sodium alginate or PGA (Moncrieff, 1953). In fermented fruit–milk beverages, PGA was more efficient than other gums when used as a stabilizer. It was found that the higher the degree of esterification, the better the stabilization achieved. In chocolate-milk drinks, alginate blended with phosphate is effective as a stabilizer for ~0.15% cocoa powder (Krigsman, 1957). Acidic drinks based on an aqueous mixture of a vegetable extract, juice or infusion and milk products can be stabilized by a combination of pectin, polyol alginate and a fatty acid ester (Lelli and Ferrero, 1995). Beer foam can be stabilized by adding 40–50 ppm PGA. Sodium alginate has also been used for the clarification of wine and the removal of tannins, coloring material and nitrogenous substances.

Sodium alginate and PGA are common gums used in salad dressings. Tragacanth is also used because of its resistance to acid degradation. However, because of the latter's variable quality and availability, price fluctuations and the frequent fumigation needed to reduce its natural microbial contamination, PGA is favored. Dressings are basically a vegetable oil emulsion that also contains salt, sugar, spices and flavorings. The addition of PGA slows separation of the oil and water phases, making the dressings or sauces stable at high room temperatures or in the refrigerator and conferring a longer shelf life. Adding PGA to salad dressing stabilized with cornstarch improves its stabilization, since retrogradation of starch may be prevented, and the final product is a soft, smooth-textured gel that does not crack or allow oil separation upon standing. Salt as a dissolved electrolyte was found to stabilize oil-in-water emulsions in the presence of PGA, xanthan gum and/or polysorbate (Yilmazer *et al.*, 1991; Yilmazer and Kokini, 1992). Salt affected the stability of the ternary system, depending on the emulsifier : stabilizer ratio. Creep measurements indicated that low salt concentrations are particularly effective when PGA–xanthan combinations are used in the presence of polysorbate-60 (Yilmazer and Kokini, 1992).

2.8.6 Other applications

Alginate gel technology can be used in the extrusion of proteins (Oates *et al.*, 1987), to extend the storage life of potatoes (An-qi, 1987), to immobilize banana enzymes, as an addition to minced fish to improve its texture and in meat products such as structured meats, because alginate thermostability could be advantageous for microwave treatment (Means and Schmidt, 1986b). Alginates are preferred in the production of Turkish delight jellies, and PGA has been found advantageous in the production of dietetic salad dressings. Levonor (1-phenyl-2-aminopropane alginate) lowered the feeding reflex in animals, with negligible undesirable side effects. Alginates enabled the production of novel fried foods free from the taste of the cohesive foodstuff, and also kept the oil clean, a characteristic suitable for mass production (Harada and Ikeda, 1995). Alginates have also been used in synthetic potato chips, dough products and meat-like products, macaroni and spaghetti products to give an increased rate of extrusion, and as a preservative of vitamin A in margarine. Yeast cells have been immobilized in alginate beads for beer and ethanol production and in the second fermentation of champagne (Onsøyen, 1990, 1992). Chapter 15 describes other biotechnological aspects of alginate and other hydrocolloids.

References

Angermeier, H.F. (1951) *Edible Alginate Jelly*. US Patent 2,536,708.

Anon. (1957) Quick frozen scampi. *Food Trade Rev.*, **27**(8), 15–17.

Anon. (1958) Puddings and pie fillings. *SeaKem Extracts*, **5**(1), 2 (published by Marine Colloids, Inc., Springfield, NJ).

An-qi, Xu (1987) Paper given at the *7th World Congr. Food Sci. Technol.* Singapore.

Berlin, A. (1957) Calcium alginate films and their application for meats used for freezing. *Myasn Industr. SSSR*, **28**, 44. (Chem. Abstr. **51**, 17007b).

Boyle, J.L. (1959) The stabilization of ice cream and ice lollies. *Food Technol. Australia*, **11**, 543.

Chavez, M.S., Luna, J.A. and Garrote, R.L. (1994) Cross linking kinetics of thermally preset alginate gels. *J. Food Sci.*, **59**(5), 1108–10.

Childs, W.H. (1957) *Coated Sausage*. US Patent No. 2,811,453.

Conti, E., Flaibani, A., Regan, O. *et al.* (1994) Alginate from *Pseudomonas fluorescens* and *P. putida:* production and properties. *Microbiology-Reading*, **140**(5), 1125–32.

Cottrell, I.W. and Kovacs, P. (1980) Alginates, in *Handbook of Water-soluble Gums and Resins*, Ch. 2 (ed. R.L. Davidson), McGraw Hill, New York, pp. 1–43.

Doggett, R.C. and Harrison, G.M. (1969) In *Proc. 5th Int. Cystic Fibrosis Conf.* (ed. D. Lawson), Cambridge University Press, Cambridge, pp. 175-8.

Dorner, H. and Tessmer, E. (1953) Thickening agents. *Brot Gebaeck*, **7**, 163.

Ensor, S.A. Sofos, J.N. and Schmidt, G.R. (1990) Optimization of algin/calcium binder in restructured beef. *J. Muscle Foods*, **1**(3), 197–206.

Ernst, E.A., Ensor, S.A., Sofos, J.N. *et al.* (1989) Shelf life of algin/calcium restructured turkey products held under aerobic and anaerobic conditions. *J. Food Sci.*, **54**(5), 1147–54.

General Foods Corp. (1954) *Confectionery Product*. British Patent No. 2,485,043.

Gibsen, K.F. (1957) *Alginate Composition for Making Milk Puddings.* U.S. Patent No. 2,808, 337.

Gibsen, K.F. and Rothe, L.B. (1955) Algin, versatile food improver. *Food Eng.*, **27**(10), 87–9.

Glabau, C.A. (1943) How to overcome sticky icings. *Baker's Weekly*, Dec. 20, p. 49.

Glicksman, M. (1962a) Utilization of natural polysaccharide gums in the food industry. *Adv. Food Res.*, **11**, 109–200.

Glicksman, M. (1962b) *Freezable Gels*. US Patent No. 3,060,032.

Glicksman, M. (1969) Seaweed extracts, in *Gum Technology in the Food Industry*, ch. 8, Academic Press, New York, pp. 199–273.

Grant, G.T., Morris, E.R., Rees, D.A. *et al.* (1973) Biological interactions between polysaccharides and divalent cations: the egg-box model. *FEBS Lett.*, **32**, 195.

Grasdalen, H. (1983) Highfield ^{1}H-NMR spectroscopy of alginate. Sequential structure and linkage conformations. *Carbohydr. Res.*, **118**, 255–60.

Green, H.C. (1936) *Fibrous Alginic Acid*. US Patent No. 2,036,934.

Harada, S. and Ikeda, M. (1995) *Fried Food and Process for Producing Same*. European Patent Application, EP 0 603,879 A2.

Hartwig, M. (1960) *Method of Reducing the Swelling Capacity of Synthetic Alginate Skins.* US Patent No. 2,965,498.

Haug, A. (1964) *Composition and Properties of Alginates, Rep. No. 30*, Norwegian Institute Seaweed Research, Trondheim, Norway.

Helgerud, O. and Olsen, A. (1955) *A Method for the Preservation of Food*. British Patent No. 728,168.

Helgerud, O. and Olsen, A. (1958) *Block Freezing of Foods (Fish)*. US Patent No. 2,763,557.

Henning, W. (1957) *Canned Herring*. Germam Patent No. 1,004,470.

Hirst, E.L. and Rees, D.A. (1965) The structure of alginic acid. Part V. Isolation and unambiguous characterization of some hydrolysis products of the methylated polysaccharide. *J. Chem. Soc.*, 1182–7.

Hunter, A.R. and Rocks, J.K. (1960) *Cold Milk Puddings and Method of Producing the Same.* US Patent No. 2,949,366.

Imeson, A.P. (1984) In *Gums and Stabilizers for the Food Industry 2*, (eds G.O. Phillips, D.J. Wedlock and P.A. Williams), Pergamon Press, Oxford, p. 189.

Kneeland, R.F., Jr (1961) Letter to Ohio Products Co., August 22 in: Glicksman, M. (1969) Seaweed extracts, in *Gum Technology in the Food Industry*, Academic Press, New York, p. 269.
Kohler, R. and Dierichs, W. (1958) *Stabilizers for Frozen Desserts*. US Patent No. 2,854,340.
Kohler, R. and Dierichs, W. (1959) *Product and Process for the Production of Aqueous Gels*. US Patent No. 2,919,198.
Krefting, A. (1896) *An Improved Method of Treating Seaweed to obtain Valuable Products Therefrom*. British Patent No. 11,538.
Krigsman, J.G. (1957) Alginic acid and the alginates applied to the food industry. *Food Technol. Australia*, **9**, 183–5.
Kunz, C.E. and Robinson, W.B. (1962) Hydrophilic colloids in fruit pie fillings. *Food Technol.*, **16**(7), 100–2.
Langmaack, L. (1961) *Method of Producing Synthetic Sausage Casings*. US Patent No. 2,973,274.
Lebrun, L., Junter, G.A., Jouenne, T. *et al.* (1994) Exopolysaccharide production by free and immobilized microbial cultures. *Enz. Microb. Technol.*, **16**(12), 1048–54.
Le Gloahec, V.C.E. and Herter, J.R. (1938) *Treating Seaweed*. US Patent No. 2,138,551.
Lelli, A. and Ferrero, P. (1995) *Stabilizer for Acidic Milk Drinks*. EP 0639,335 A1.
McDowell, R.H. (1960) Applications of alginates. *Rev. Pure Appl. Chem.*, **10**(1), 1–15.
McDowell, R.H. and Boyle, J.L. (1960a) *Powdered Alginate Jelly Composition and Method of Preparing Same*. US Patent 2,935,409.
McDowell, R.H. and Boyle, J.L. (1960b) *Gelling of Milk with Alginate*. British Patent No. 839,767.
McNeely, W.H. (1959). Algin, in *Industrial Gums* (ed. R.L. Whistler), Academic Press, New York, pp. 55–82.
Means, W.J., Clarke, A.D., Sofos, J.N. *et al.* (1987) Binding, sensory and storage properties of algin, calcium structured beef steaks. *J. Food Sci.*, **52**(2), 252–62.
Means, W.J. and Schmidt, G.R. (1986a) Algin/calcium gel as a raw and cooked binder in structured beef steaks. *J. Food Sci.*, **51**(1), 60–5.
Means, W.J. and Schmidt, G.R. (1986b) *Process for Preparing Algin/calcium Gel Structured Meat Products*. US Patent No. 4,603,054.
Merton, R.R. and McDowell, R.H. (1960) *Powdered Alginate Jellies*. US Patent No. 2,930,701; British Patent No. 828,350.
Messina, B.T. and Pape, D. (1966) Ingredient cuts heat-process time. *Food Eng.*, **8**(4), 48–51.
Miller, A. (1960) *Composition and Method for Improving Frozen Confections*. US Patent No. 2,935,406.
Moncrieff, R.W. (1953) Stabilizing fruit drinks. *Food*, **22**, 498–9.
Mountney, G.J. and Winter, A.R. (1961) The use of a calcium alginate film for coating cut-up poultry. *Poultry Sci.*, **40**, 28.
Nishide, E., Mishima, A., Anzai, H. *et al.* (1992) Properties of alginic acid from sulfated polysaccharides extracted from residual algae by the hot water method. *Bull. College Agric. Vet. Med.*, Nihon University, No. 49, 140–2.
Nussinovitch, A. (1993) Gum-based texturized products, in *Yearbook of Science and Technology*, McGraw-Hill, New York, pp. 138–140.
Nussinovitch, A., Kopelman, I.J. and Mizrahi, S. (1990) Effect of hydrocolloids and minerals content on the mechanical properties of gels. *Food Hydrocolloids*, **4**(4), 257–65.
Nussinovitch, A. and Peleg, M. (1990) Strength–time relationships of agar and alginate gels. *J. Texture Studies*, **21**(1), 51–60.
Nussinovitch, A., Peleg, M. and Normand, M.D. (1989) A modified Maxwell and a non-exponential model for characterization of the stress relaxation of agar and alginate gels. *J. Food Sci.*, **54**, 1013–16.
Oates, C.G., Ledward, D.A. and Mitchell, J.R. (1987) Paper given at the *7th World Congr. Food Sci. Technol.*, Singapore.
Ohling, R.A.G. (1959) *Food Preparation by Cold Gelation*. Canadian Patent No. 574,261.
Olsen, A. (1955) Freezing fish in alginate jelly. *Food Manuf.*, **30**(7), 267–70, 285.
Onsøyen, E. (1990) Marine hydrocolloids in biotechnological applications, in *Advances in Fisheries Technology and Biotechnology for Increased Profitability*, papers from the 34th Atlantic Fisheries Technol. Conf. Seafood Biotechnol. Workshop (eds M.N. Voight and J.R. Botta), Technomic, Lancaster, PA, pp. 265–86.

Onsøyen, E. (1992) Alginates, in *Thickening and Gelling Agents for Food*, ch. 1 (ed. A. Imeson), Chapman & Hall, Glasgow, pp. 1–25.
Peschardt, W.J.S. (1946) *Manufacturing Artificial Edible Cherries, Soft Sheets, and the like*. US Patent No. 2,403,547.
Poarch, A.E. and Tweig, G.W. (1957) *Gel-forming Compositions*. US Patent No. 2,809,893.
Pressman, R. (1957) *Coatings for Meat-wrapping Sheets*. US Patent No. 2,811,454.
Proton Biopolymers A/S (1990) Technical information – Alginates.
Rees, D.A. (1969) Structure, conformation and mechanism in formation of polysaccharide gels and networks, in *Advances in Carbohydrate Chemistry and Biochemistry*, vol. 24 (eds M.L. Wolform and R.S. Tipson), Academic Press, New York, pp. 267–332.
Rocks, J.K. (1960) *Method and Composition for Preparing Cold Water Desserts*. US Patent No. 2,925,343.
Schmidt, G.R. and Means, W.J. (1986) *Process for Preparing Algin/calcium Gel Structured Meat Products*. US Patent No. 4,603,054.
Shand, P.J., Sofos, J.N. and Schmidt, G.R. (1993) Properties of algin calcium and salt phosphate structured beef rolls with added gums. *J. Food Sci.*, **58**(6), 1224–30.
Shetty, C.S., Bhaskar, N., Bhandary, M.H. *et al.* (1996) Effect of film forming gums in the preservation of salted and dried mackerel. *J. Sci. Food Agric.*, **70**(4), 453–60.
Shiotani, H. and Hara, M. (1955) *Ice Cream Stabilizer*. Japanese Patent No. 3031 (*Chem. Abstr.*, **51**, 13263e).
Slepchencnko, I.R., Knizhnik, E.B. and Piraeva, L.A. (1956) Production of calcium alginate films and their utilization in the freezing of meat. *Sbornik Stud. Rabot. Moskov. Teknol. Inst. Myasnoi i Molch. Prom.*, No. 4, 39 (*Chem. Abstr.*, **53**, 13440f).
Stanford, E.C.C. (1883) On algin: a new substance obtained from some of the commoner species of marine algae. *Chem. News*, **47**, 254.
Stanford, E.C.C. (1884) On the economic applications of seaweeds. *J. Soc. Arts*, **32**, 717.
Steiner, A.B. (1948) *Algin Gel-forming Compositions*. US Patents No. 2,441,729.
Steiner, A.B. (1949) *Alginate Ice Cream Stabilizing Composition*. US Patent No. 2,485,934.
Steiner, A.B. and McNeely, W.H. (1954) Algin in review. *Adv. Chem.*, **11**, 68–82.
Strachan, C.C., Moyls, A.W., Atkinson, F.E. *et al.* (1960) Commercial canning of fruit pie fillings. *Can. Dept. Agr. Publ.* 1062.
Takeuchi, T., Murata, K. and Kusakabe, I. (1994) A method for depolymerization of alginate using the enzyme system of *Flavobacterium multivolum*. *J. Jap. Soc. Food Sci. Technol.*, **41**(7), 505–11.
Toft, K. (1982) Interactions between pectins and alginates. *Prog. Food Nutr. Sci.*, **6**, 89–96.
Toft, K., Grasdalen, H. and Smidsrod, O. (1986) Synergistic gelation of alginates and pectins, in: *ACS Symp. Ser. No. 310 Chem. Funct. Pectins* (eds M.L. Fishman and J.J. Jen), American Chemical Society, Washington, DC.
Trout, G.R. (1989) Color and bind strength of restructured pork chops: effect of calcium carbonate and sodium alginate concentration. *J. Food Sci.*, **54**(6), 1466–70.
Trout, G.R., Chen, C.M. and Dale, S. (1990) Effect of calcium carbonate and sodium alginate on the textural characteristics, color and color stability of restructured pork chops. *J. Food Sci.*, **55**(1), 38–42.
Weingand, R. (1957) *Synthetic Sausage Casings*. US Patent No. 2,802,744.
Weingand, R. (1959) *Process for Producing Synthetic Sausage Casing from Alginates*. US Patent No. 2,897,547.
Whistler, R.L. and Kirby, K.W. (1959) Composition of alginic acid of *Macrocystis pyrifera*. *Hoppe-Seyler's Z. Physiol. Chem.*, **314**, 46.
Wolff & Co. (1951) *Sausage Casings*. British Patent No. 711,437.
Yilmazer, G. and Kokini, J.L. (1992) Effect of salt on the stability of propylene glycol alginate/xanthan gum/polysorbate-60 stabilized oil in water emulsions. *J. Text. Studies*, **23**(2), 195–213.
Yilmazer, G., Carrillo, A.R. and Kokini, J.L. (1991) Effect of propylene glycol alginate and xanthan gum on stability of O/W emulsions. *J. Food Sci.*, **56**(2), 513–17.
Zheng, Y.W., White, J.W., Konno, M. *et al.* (1995) A small angle X-ray scattering study of alginate solution and its sol–gel transition by adding of divalent cations. *Biopolymers*, **35**(2), 227–38.

3 Carrageenans

3.1 Introduction and historical background

Carrageenan is present as an intercellular matrix component in numerous species of red seaweed. Inside the alga, this gum material fills the voids in the cellulosic plant structure. The gum creates flexible structures within the alga that help it to cope with wave motion and vibrant currents. In industrial applications, carrageenans are used in small proportions to induce major changes as thickening or gelling agents. Chinese records exist from 600 BC on the uses of red seaweed. One of the uses for raw sources of carrageenan (the small red, cold-water alga *Chondrus crispus*) consisted of boiling in milk to yield a thickened product: this practice has spread over the centuries from Ireland, the first place mentioned to host such activities (Towle, 1973). Cooked seaweed as a component of a dish called 'St Patrick's soup' was an outcome of the mid-19th century Irish potato famine. The dried, bleached plants were imported to the USA from Ireland until, in 1835, the occurrence of these same plants in Scituate, Massachusetts was verified (Glicksman, 1969). In 1844, the first successful extraction of carrageenan from *Chondrus crispus* was attributed to Schmidt, followed by a patentable process of extraction by alcohol precipitation (Bourgade, 1871). The resultant accelerated production brought into existence many seaweed and gum suppliers. Seaweeds, which have traditionally been used by the western food industry for their polysaccharide extract – alginate, carrageenan and agar – also contain compounds with nutritional benefits. Therefore, seaweeds have recently been approved in France for human consumption (as vegetables and condiments), thus creating new opportunities for the food industry (Mabeau and Fleurence, 1993).

3.2 Structure

An understanding of the structure of carrageenan was gained gradually starting with a determination of the presence of D-galactose and continuing to the verification of the presence of ester-sulfate groups and the beginning of speculations that carrageenan comprises two fractions (Haas, 1921). Later, major steps were taken by Smith and Cook (1953) to isolate and separate these two fractions (κ and λ): only κ-carrageenan was sensitive to

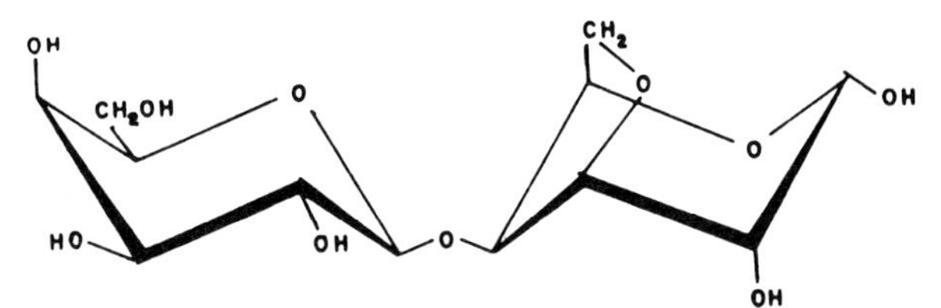

Figure 3.1 Basic repeating unit of carrageenans.

potassium salt precipitation. Further study and characterization of these fractions followed (O'Neill, 1955). In 1967, the structure of a third carrageenan (ι) was elucidated (Mueller and Rees, 1968).

Red seaweeds contain a family of gums, including agar, furcellaran and the three types of carrageenan (i.e. κ, ι and λ). Carrageenans are linear polysaccharides composed of alternating $\beta(1-3)$- and $\alpha(1-4)$-linked galactose residues. The basic repeating unit of carrageenans is carrabiose (a disaccharide). The 1–4-linked residues are commonly, but not invariably, present as 3,6 anhydride. The repeating unit is presented in Fig. 3.1. Variations of this basic structure can result from substitutions (either anionic or non-ionic) on the hydroxyl groups of the sugar residues, and from the absence of a 3,6-ether linkage (Stanley, 1990).

Carrageenans are highly sulfated. The 1–3-linked D-galactose residues occur as 2- and 4-sulfate, or are occasionally non-sulfated, whereas the 1–4-linked residues occur as 2-sulfate, 6-sulfate, 2,6-disulfate, 3,6-anhydride and 3,6-anhydride 2-sulfate (Stanley, 1990). Carrageenans extracted from *Gigartina* species sometimes include pyruvate residues (Hirase and Watanabe, 1972; DiNinno *et al.*, 1979; McCandless and Gretz, 1984). Sulfated galactans from the Grateloupiaceae family may contain methoxyl groups, although their alternating carrageenan structure is unclear (Parolis, 1981).

Fractionation of the carrageenan from *Chondrus crispus* with potassium chloride (Smith and Cook, 1953) resulted in the isolation of κ- and λ-carrageenan: κ-carrageenan was precipitated by potassium chloride, while the λ fraction remained in solution. In these fractions, half of the sugar units in κ-carrageenan were 3,6-anhydro-D-galactose, whereas λ-carrageenan contained little or none of this sugar (Smith *et al.*, 1955). Carrageenans include a small number of ideal or limited polysaccharides (Rees, 1963; Dolan and Rees, 1965; Anderson *et al.*, 1968, 1973; Lawson *et al.*, 1973; Penman and Rees, 1973a, b). The names μ, κ, ν, ι, λ, θ and ξ are presently applied to these limited carrageenans (Fig. 3.2). The μ- and ν-carrageenans are presumed to be precursors in the biosynthesis of κ and ι derivatives, via the enzyme 'dekinkase' (Hirase *et al.*, 1967; Stancioff and Stanley, 1969; Lawson and Rees, 1970), or when industrial processing involves base-catalysed S_N2

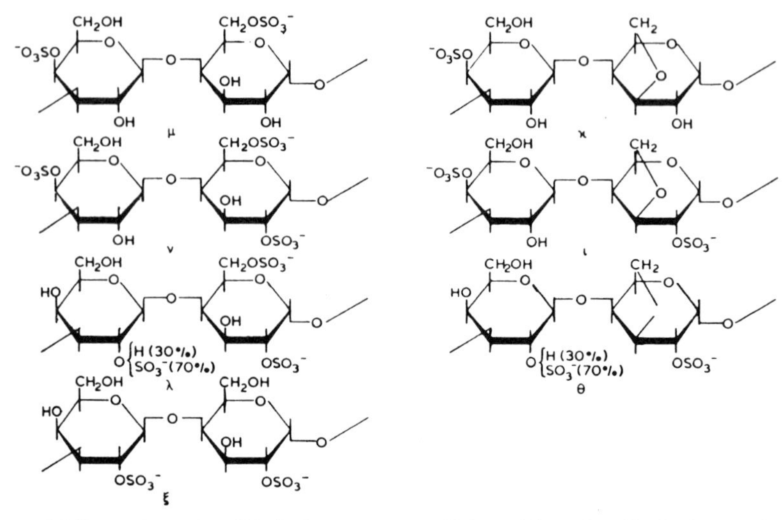

Figure 3.2 Repeating units of limited carrageenans. (Adapted from *Handbook of Water-Soluble Gums and Resins*, McGraw-Hill, NY, 1980, R.L. Davidson, ed.)

elimination of 6-sulfate (Stanley, 1963). Carrageenans produced from *Eucheuma gelatinae* are of the β-type and analogous to κ-carrageenan except for their lack of sulfate on the 1–3-linked residues. The position of β-carrageenan in the carrageenan system is similar to the relationship between agarose and the agar polysaccharides system (Greer and Yaphe, 1984). Furcellaran, the 'Danish agar' used in milk-pudding powders (Bjerre-Petersen *et al.*, 1973; Stanley, 1987), has the carrabiose backbone and is like κ-carrageenan except for its amount of half-ester sulfate. Information on its composition can be found elsewhere, but the distribution of sulfate along the molecular chain is not completely known (Painter, 1966).

3.3 Sources and production

Carrageenan-yielding species occur in at least seven families: Solieriaceae, Gigartinaceae, Furcellariaceae, Phyllophoraceae, Hypneaceae, Rhabdoniaceae and Rhodophyllidaceae (Stanley, 1990; Deslandes *et al.*, 1985). Carrageenans can be extracted from both diploid and haploid plants. Primary red seaweed sources of carrageenan are *Chondrus crispus*, a small cold-water seaweed producing κ and λ types; the warm-water *Eucheuma* species producing κ and ι types (Fig. 3.3); and the large cold-water *Gigartina* species from which κ and λ types are produced (Thomas, 1992). The cold-water seaweeds are harvested once a year, whereas the warm-water seaweeds grow on a 3-month cycle. Carrageenan-yielding algae are grown in the

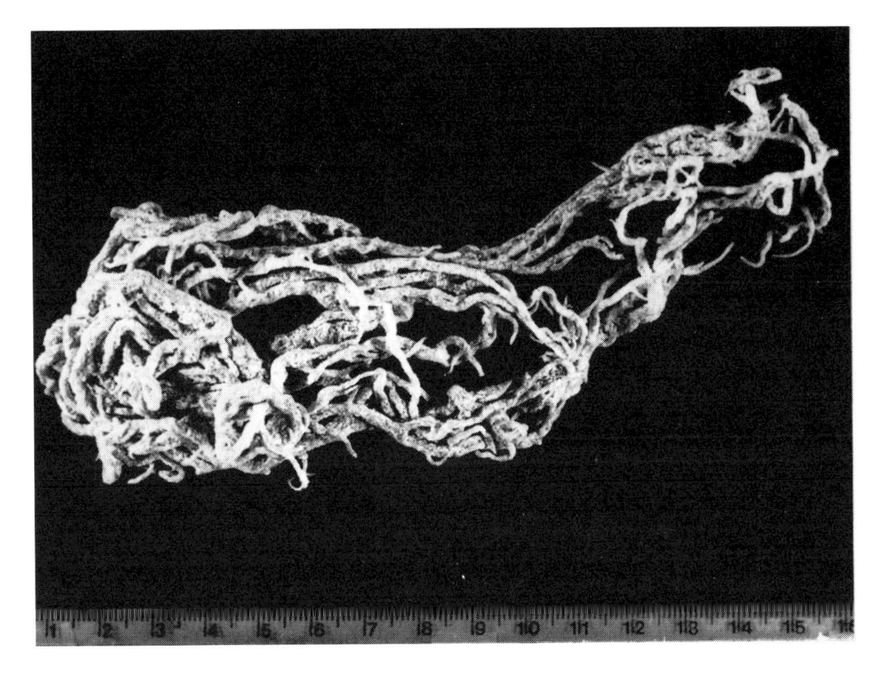

Figure 3.3 A dried piece of *E. cottonii* from which carrageenan is extracted.

Philippines, Indonesia, Canada, the USA, France, Korea, Spain, Portugal, Morocco, Mexico, Chile, Denmark and Brazil.

The time and conditions under which the seaweeds are harvested and their rapid drying to the appropriate moisture content optimize the chances of preserving the quality of the carrageenan. After shipping, the dried seaweed is stored in warehouses until extraction. For manufacturing, seaweeds are tested and lots are selected to obtain the desired preplanned product. Blending is necessary to achieve consistent quality. A single manufacturer may market ~ 200–300 blends (Thomas, 1992). To produce the hydrocolloid, the carrageenan, which exists in a gel-like state in the alga at room temperature, needs to be heated in water to a temperature above its gel's melting point (Fig. 3.4). The seaweed should be cleaned to remove any sand before extraction proceeds. Alkaline conditions are necessary to reduce or eliminate acid-catalysed depolymerization processes or modification of the galactan structure, resulting in better gelling ability. Modifications as well as differences in extraction techniques exist. Three processes are most commonly used: alcohol precipitation, freeze–thawing and gel press. In the alcohol process, the seaweed is washed to remove dirt and impurities. Extraction is then performed with dilute alkali for several hours. Residues are separated by centrifugation followed by filtration through porous silica

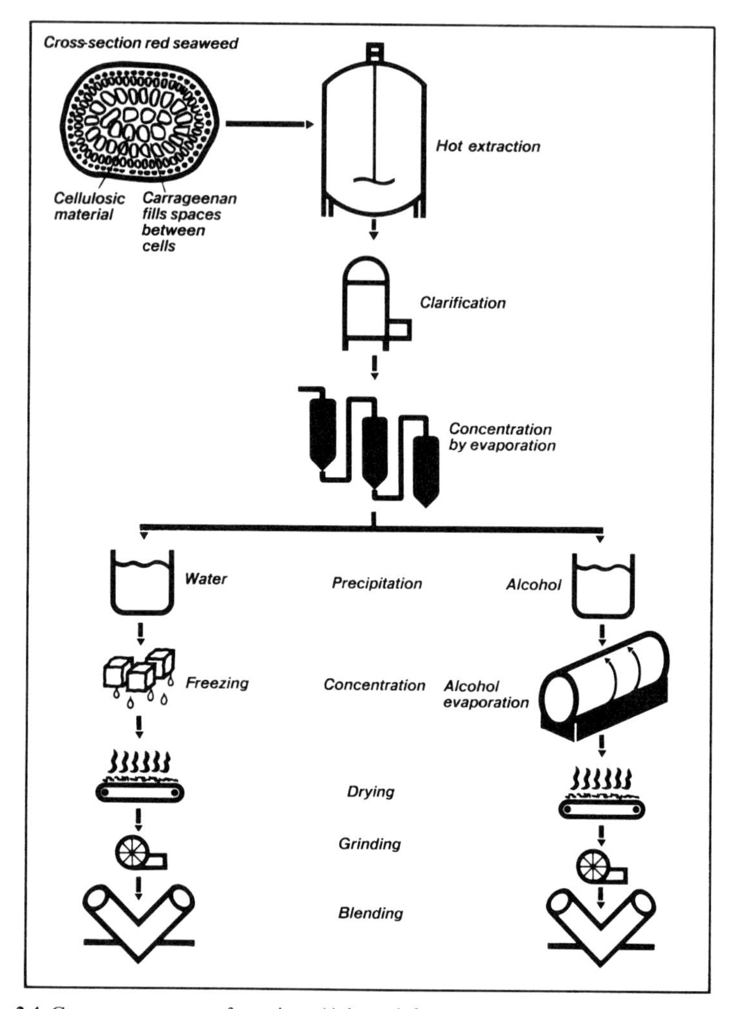

Figure 3.4 Carrageenan manufacturing. (Adapted from Marine Colloids, The Carrageenan People, Introduction Bulletin A-1.)

or activated charcoal to yield a 1–2% carrageenan solution. Evaporators are used to concentrate the carrageenan extracts 1.5- to 2-fold and isopropyl alcohol is used to obtain a coagulant that is pressed dry by removing the highest possible amount of excess liquid. The precipitated carrageenan is dried in a steam-heated drier, followed by particle-size reduction to the required state. The other two processes (freeze–thawing and gel press) are similar to the first, except for the way in which the carrageenan is recovered from solution.

In the freeze–thaw process, after concentration by evaporation, the material is extruded through spinnerets into a cold solution of potassium

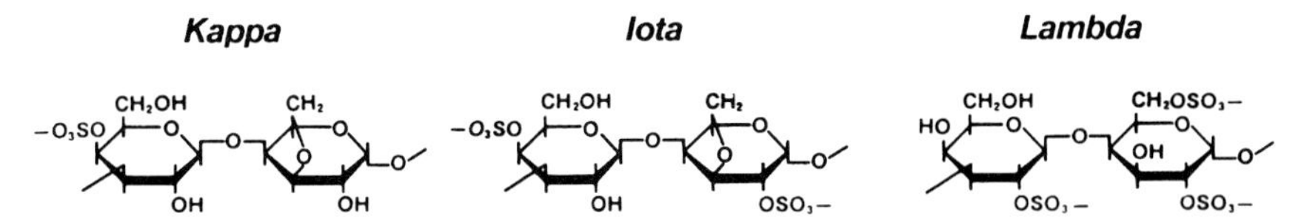

Figure 3.5 Three types of carrageenan. (Adapted from Marine Colloids, The Carrageenan People, Introduction Bulletin A-1.)

chloride. The gelled threads are further dewatered by subsequent potassium chloride washes followed by pressing. The gel passes through freezing, thawing, chopping and rewashing in a potassium chloride solution followed by air drying. Because potassium chloride is used, the process is limited to furcellaran and κ-carrageenan, both of which gel with potassium and exhibit marked syneresis. The gel press method omits the freeze–thaw cycle and makes use of pressure to dewater the gel.

3.4 Available types of carrageenan

Carrageenans are anionic polymers (they contain half-ester sulfate groups). The free acid is unstable, and carrageenans are commonly sold as a mixture of sodium, potassium and calcium salts. Commercial carrageenans are usually sold in κ, ι or λ forms. In fact, each form contains varying amounts of the other two types. For the purpose of viscosity production, λ-carrageenan is used, whereas for gelling applications, mixed calcium and potassium salts of the κ and ι types are used (Fig. 3.5).

3.5 Regulatory aspects

In the USA, carrageenan has GRAS (21 CFR 182.7255) status and is approved as a food additive (21 CFR 172.620) by the Food and Drug Administration. In Europe, carrageenan is recognized for its safety and usefulness (having the designation E407). Since carrageenan is a 'natural product', in Japan it is not subjected to food-additive regulations other than those of good manufacturing practice (GMP). In 1984, technologists from around the world confirmed that carrageenans could be used in foods and that it is unnecessary to specify acceptable daily intake (ADI). However, the safety of carrageenan in the presence of acid has been questioned, because of its depolymerization. Degraded carrageenan, also known as poligeenan, has been used to treat peptic ulcers in the past and is used as a dispersing aid and deflocculent for barium sulfate suspensions in diagnostic X-ray testing (Manning, 1990).

3.6 Molecular weight and consistency

Carrageenan is generally considered a high-molecular-weight linear polysaccharide. It comprises repeating galactose units and 3,6-anhydrogalactose (3,6-AG), both sulfated and non-sulfated, joined by alternating $\alpha(1-3)$- and $\beta(1-4)$-glycosidic linkages (Thomas, 1992). Carrageenans are polydisperse, in other words they do not have sharply defined molecular weights. To present average molecular weights, a few definitions are used: M_n is the number average molecular weight and M_w is the weight average molecular weight. Commercial food-grade carrageenans typically have M_n values of $\sim 200 \times 10^3$ to 400×10^3. At masses below 10^5 Da, the functionality of the carrageenans is largely lost.

To sell a more or less uniform carrageenan, the raw materials need to be blended or standardized and adjusted via the addition of sugars to give constant values of gel strength in water and milk, and constant viscosities and stabilizing efficiency in milk. After blending, carrageenans are packaged and sold in 90 kg polyethylene-lined fiber drums. Precautions are taken to minimize caking of the powder caused by moisture absorbance.

3.7 Carrageenan solutions

3.7.1 Properties

If κ-carrageenan is dispersed in cold water, it swells to a great extent (except the sodium salt) but dissolves only slightly. The degree of swelling depends on the method of preparation. The dissolution temperature is not constant and depends on gum concentration and on the cations carried with it, but the average dissolution temperature is $\sim 120°F$ (49°C) (Glicksman, 1969). Cold dispersions of ι-carrageenan swell and take on thixotropic properties. Heat is necessary for dissolution. Lambda-carrageenan is fully soluble in cold water, regardless of the cations associated with it. Organic solvents such as methanol, ethanol, acetone and glycerin slow the solubility of the carrageenans. Sufficient quantities of salts also prevent the hydration and dissolution of κ- and λ-carrageenans (Glicksman, 1969). Iota-carrageenan is unique in its ability to dissolve in hot salt solutions but is less soluble in sugar solutions than are κ- and λ-carrageenans (Glicksman, 1969).

Following carrageenan dissolution, a highly viscous solution results. The viscosity is a result of its linear macromolecular structure and polyelectrolytic nature. Mutual repulsion among sulfate groups along the polymer chain causes the molecule to extend, and its hydrophilicity causes it to be surrounded by a sheath of immobilized water molecules. Both factors contribute to resistance to flow. Viscosity depends on concentration, temperature, the presence of other solutes, the molecular weight of the carrageenan and its type.

3.7.2 *Preparation*

During the preparation of carrageenan solutions, precautions are taken to eliminate clumping. Cold water and milk should be used for dispersion purposes before heating. If equipment such as a high-shear mixer is used, the powder particles are added slowly and gradually to the vortex. The carrageenan powder should be blended with the sugar before its dispersion in water. Sometimes liquids such as alcohols are used to wet the carrageenan powder before its dispersal in water. Carrageenans are soluble in water at temperatures above 75°C. Conventional mixing equipment can be used to deal with solutions of ~10% soluble carrageenan. Sodium salts of κ- and ι-carrageenan are soluble in cold water, whereas salts with calcium and potassium exhibit a varying degree of swelling and do not dissolve completely. As already noted, λ-carrageenan is fully soluble in cold water.

3.7.3 *Viscosity*

Commercial carrageenan solutions are available in viscosities of ~5–800 mPa s, as measured at 75°C and a concentration of 1.5% (w/w). If the measured viscosities are less than 100 mPa s they have flow properties very close to Newtonian. Carrageenan solutions generally exhibit pseudoplastic (shear-thinning) properties. In other words, with solutions it is important to specify at what shear rate the viscosity measurements have been performed, and this rate should be suited to the specific application. The pseudoplastic behavior of carrageenan can be described by a power-law equation:

$$\log \eta_a = \log a + (b - 1) \log \mathring{\gamma} \tag{3.1}$$

where η_a is the apparent viscosity, $\mathring{\gamma}$ is the shear rate, a is the consistency index, and b is the flow behavior index. The value of b is either unity, representing Newtonian flow, or less than unity (shear thinning). To gain a better fit to a wider range of shear rates, a quadratic term can be added:

$$\log \eta_a = \log a + (b - 1) \log \mathring{\gamma} + c \log^2 \mathring{\gamma} \tag{3.2}$$

In this equation, the value of c is either zero or negative (Stanley, 1990). In general, viscosity increases nearly exponentially with concentration. This behavior is typical of carrageenan and other linear charged polymers as a consequence of interactions between such polymer chains. The presence of salts reduces the viscosity of carrageenan (ionic macromolecule) solutions through a reduction in electrostatic repulsion. Carrageenans such as types κ and ι may gel in the presence of potassium and calcium. As a general rule the higher the temperature, the lower the viscosity of the solution. Precautions to avoid thermal degradation of polymers need to be taken. At temperatures above the gelling point, both gelling and non-gelling solutions behave similarly. However, for gelling carrageenans, if ions are used as gel promoters, then at cooling there is an increase in the measured apparent viscosity.

3.7.4 Effect of molecular weight

The relationship between apparent viscosity $[\eta]$ and the average molecular weight M are in accordance with the Mark–Houwink equation:

$$[\eta] = KM^{\alpha}$$

where K and α are constants. Depolymerization can occur by acid-catalysed hydrolysis, and this is related to 3,6-anhydride content. Cleavage occurs preferentially at the 1–3-glycoside linkage and is promoted by the strained ring system of the anhydride. Sulfation at C-2 of the 1–4-linked units appears to mitigate attack by acids, degradation rates for ι-carrageenan being roughly one-half those for κ-carrageenan under the same conditions. Generally, gelled hydrocolloids are more stable to acid than those in the sol state. Depolymerization follows pseudo-zero-order kinetics in either state, but the rate for the gel state is about one-sixth that for the sol state. The secondary and tertiary structures developed upon gelation may shield the glycosidic bonds from attack, enabling the use of carrageenans in low pH systems, especially if enough potassium chloride or other potassium salt is present to maintain the gel state (Stanley, 1990). Carrageenans are mostly stable at around pH 9. Depolymerization depends on hydrogen ion activity and temperatures and rates can be estimated by simple mathematical relationships (Stanley, 1990). The degradation of κ-carrageenan by hydrolysis in lithium chloride/hydrochloric acid pH 2 buffer was studied at 35, 45 and 55°C (Singh and Jacobsson, 1994). The reduction in the value of M_W was measured over time by size-exclusion chromatography with multi-angle laser light scattering and refractive index detection. The hydrolysis process was also followed by measurement of gel-breaking strength and low-shear viscosity. The results may be of future use in process design and optimization, as well as for control purposes (Singh and Jacobsson, 1994).

3.8 Gel preparation and mechanical properties

Gel formation may be likened to crystallization or precipitation from solution. Both κ- and ι-carrageenans need heat for dissolution. After cooling, and in the presence of positively charged ions (e.g. potassium or calcium), gelation occurs. Gel preparation involves dispersing the carrageenan plus the salt (potassium chloride) in distilled water and heating to ~ 80°C, then adding the water evaporated during the heating process and slowly cooling to room temperature to induce gelation. Gel strength can be measured with a Marine Colloids Gel Tester or other similar instruments such as a Universal Testing Machine. Kappa-carrageenan produces strong, rigid gels exhibiting some syneresis. It forms helices with potassium ions, and calcium ions cause the helices to aggregate and the gel to contract and

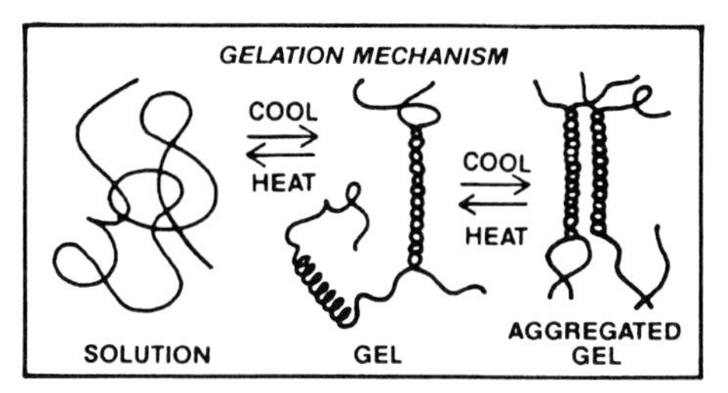

Figure 3.6 Gelation mechanism of carrageenan. (Adapted from Marine Colloids, The Carrageenan People, Introduction Bulletin A-1.)

become brittle. The produced gel is slightly opaque but becomes clear with sugar. It contains $\sim$25% ester sulfate and 34% 3,6-anhydrogalactose. Iota-carrageenan produces elastic gels. It forms helices with calcium ions, and limited aggregation contributes to elasticity, with no syneresis. The gel is clear and stable to freeze–thawing. Iota-carrageenan contains $\sim$32% ester sulfate and 30% 3,6-anhydrogalactose (Stanley, 1990). The interaction of manganese ions with carrageenans in aqueous gels was studied by electron spin resonance (ESR) (Tsutsumi *et al.*, 1993). ESR signals of the manganese ions were diminished in the presence of gellan gum and ι-carrageenan. No change was observed for κ-carrageenan. It is suggested that manganese ions interact strongly with gellan gum and ι-carrageenan. Analysis of the ESR signal intensity of varying manganese concentrations gave the dissociation constant and the number of binding sites for the manganese–polymer complex. Results indicate that one manganese ion binds two residues of gellan gum, whereas nearly equimolar binding occurs in ι-carrageenan gels (Tsutsumi *et al.*, 1993).

The association of molecular chains into double helices was proposed by Rees (1972) as a mechanism of gel formation (Fig. 3.6). Evidence of double helices was found in X-ray diffraction studies (Arnott *et al.*, 1974) and from the dimerization that occurred during the cooling of solutions containing carrageenan segments that were too short to form an extended network (Bryce *et al.*, 1974, 1982). At temperatures higher than the gelling point of the sol, the polymer exists in solution as random coils. The domain model is a recently advanced description of gelation: gelation is proposed to occur when domains aggregate to form a three-dimensional network. Both ι- and κ-carrageenans need potassium or other gel-promoting cations to create a three-dimensional network. The presence of potassium (which has a favorable radius in the hydrated state compared with bulkier cations) leads to decreased repulsion between the chains. Another model by Bayley (based on a study using X-ray diffraction) proposed an 'egg-box' structure in which cations of favored sizes lock adjacent chains into an aggregate (Bayley, 1955). However, this assumed model of carrageenan primary structure is now known to be incorrect. The domain model proposes that cations are locked together in helical regions of adjacent domains to build up a network. This is regarded as a revival of Bayley's (1955) concept, but on a tertiary, rather than secondary level (Painter, 1966). The nature of this cation specificity is still under study. An alternative model of carrageenan gelation postulates cation-induced aggregation of single helices (Paoletti *et al.*, 1984). The presence of 6-sulfated 1–4-linked residues in the polymer chain of κ- or ι-carrageenan or of furcellaran has been shown to detract from the strength of their gels. This is a result of kinks in the chain that inhibit the formation of double helices. These kinks can be removed by conversion of 6-sulfated residues into 3,6-anhydride by alkaline treatment, the melting temperatures of carrageenan gels being higher than their gelling temperatures. This is a function of charge density and is called hysteresis. It has been studied by optical rotation and is explained by thermodynamics involving terms of a free energy surface with two minima (Rees *et al.*, 1969; Stanley, 1990).

The effect of gum concentration on gel yield stress and deformability modulus was studied for carrageenan, agar and alginate gels (Nussinovitch *et al.*, 1990c). Carrageenan is characterized by a maximum yield stress for a certain particular hydrocolloid concentration. Increasing the gum concentration beyond that point will decrease the yield stress. Maximum yield stress is obtained for carrageenan gels at ~1.5% potassium chloride.

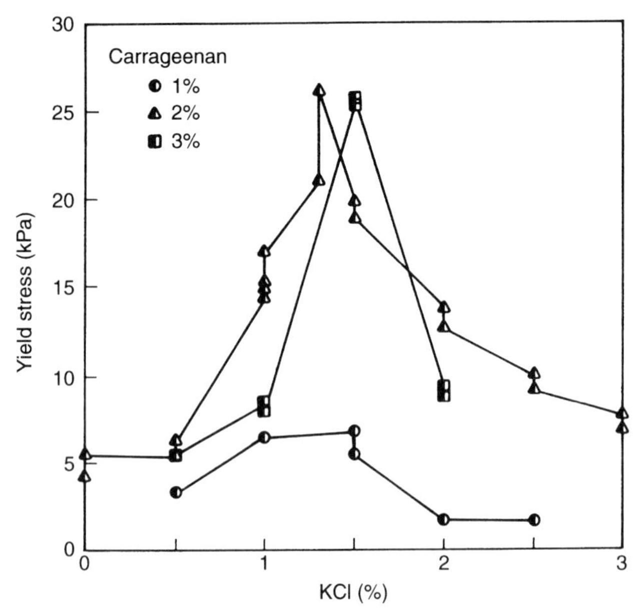

Figure 3.7 The effect of potassium chloride concentration on the yield stress of carrageenan gels. (From Nussinovitch *et al.*, 1990c; by permission of Oxford University Press.)

Maximum strength was obtained at more or less the same potassium chloride concentration (Fig. 3.7) for 1, 2, and 3% carrageenan gels (Nussinovitch *et al.*, 1990c). The force–deformation data can serve as an index for carrageenan gel strength and perceived texture (Nussinovitch *et al.*, 1990b). The proposed method can even be used for carrageenan gels that do not fracture under normal compression tests, for example the unyielding gel produced from 2% κ-carrageenan, 2% LBG and 0.2% potassium chloride (Rotbart, 1984; Rotbart *et al.*, 1988). Studies on the elasticity of carrageenan gels, and the relationship between their recoverable work and asymptotic relaxation modulus, can be found elsewhere (Nussinovitch *et al.*, 1990a).

3.8.1 Transition temperatures

Transition temperatures, determined by optical rotation and light scattering, can be derived using the function $\theta_g = a + bc^{0.5}$, where θ_g is the gelling temperature, c is the concentration of the gelling cation and a and b are constants. The gelling temperature depends on the square root of the ion concentration and indicates that Coulombic rather than salting-out effects control the sol-to-gel transition. Melting temperatures are similarly related to cation concentration. The influence of temperature, in addition to ionic content, is an important factor when considering the functional properties of carrageenans. Both κ- and ι-carrageenans set to various gel textures from 40 to 70°C, depending on cation content. These gels are stable at room temperature but can be remelted by heating to temperatures 5 to 10°C higher than their gelling temperatures. Repeated cycles of heating and cooling result in gel reformation (Stanley, 1990). The effects of sugars on the gel–sol transition and gel properties of agarose and κ-carrageenan have been discussed by Nishinari *et al.* (1995). The effects of sugars are ascribed to hydrogen bonding between hydroxyl groups in polymers and sugars and/or structural changes in the solvent water (Nishinari *et al.*, 1995).

Syneresis generally occurs after gelation. This common effect can be avoided by incorporating LBG or konjac glucomannan into the gel. The smooth regions of these galactomannans are presumed to bind to the double helices of the κ-carrageenan or furcellaran to reduce the tendency towards producing tightly packed aggregates. Guar does not produce such formations because of a higher density of galactose side-groups.

The ι-carrageenan gels show very little tendency to undergo syneresis because 2-sulfate groups on the 3,6-anhydride residues may act as wedges preventing tightly packed aggregation, which could be the reason for the κ-gels' rigidity (Stanley, 1990). The effects of LBG and konjac glucomannan on the rheological properties of κ-carrageenan and agarose gels were investigated. At a high LBG concentration (1% (w/w)), some gel-like characteristics were induced in κ-carrageenan systems (Goycoolea *et al.*, 1995). A reduction in viscosity was observed in dilute LBG solutions

(0.085% (w/w)), at predetermined conditions. The addition of LBG or konjac mannan to agarose appeared to augment the polysaccharide network. It was concluded that LBG or konjac mannan chains promote conformational ordering of carrageenan or agarose by binding to the double helix as it forms (Goycoolea *et al.*, 1995). In κ-carrageenan–LBG systems studied by using dynamic and viscosimetric measurements, profound differences were found between κ-carrageenan gels and 4:1 κ-carrageenan–LBG mixed gels in terms of gelation and melting temperatures and other rheological parameters (Fernandes *et al.*, 1991). Information on the rheological behavior of κ-carrageenan–LBG mixtures at a very low level of the gum and a description of the minimum gelling conditions of such systems can be found elsewhere (Fernandes *et al.*, 1994a,b).

3.8.2 Reactivity with proteins

In the intercellular matrix, carrageenans are covalently bound to protein moieties and found as proteoglycans. Such proteins can be removed via alkaline extraction of the polysaccharide from the algae. A similar situation can be found with agars. Analyses of the gum powder evidenced the presence of up to 1.0% nitrogen, which was proven to be of protein origin; therefore, as much as 6% protein is bound to the carrageenans. The extraction and partial removal of bound protein results in a more regular polymer, enabling S_N2 chain-straightening mechanisms to occur. Electrostatic interactions between negatively charged carrageenans and positively charged ions on proteins (as in milk) are possible. Therefore, carrageenans can be considered stabilizers for proteinaceous systems other than milk, protein precipitants, enzyme inhibitors, blood anticoagulants and lipemia-clearing agents, and stabilizers in immunologically active substances. The interactions involved can be specific or non-specific (Stanley, 1990). Information on the interaction of carrageenan and starch in cream desserts can be found elsewhere (Nadison, 1995).

One of the aspects of carrageenans' reactivity with milk is gelation. This occurs when κ-casein and carrageenan interact to form a complex that aggregates into a three-dimensional gel network. Specific interactions with κ-casein occur at pH levels above the isoelectric point of the protein (Lin, 1977; Hansen, 1982; Snoeren, 1976; Stainsby, 1980). (Non-specific interactions can occur below this point (Fig. 3.8).) Carrageenans may interact with α_{s1}- and β-caseins, but binding is much weaker than with κ-casein and does not result in gel formation. The electrostatic interaction (attraction) between negatively charged sulfate groups and the positively charged region in the peptide chain of κ-casein has been described, demonstrated and discussed in many studies (Payens, 1972; Snoeren *et al.*, 1975). Kappa-casein isolated from a milk-protein system occurs as independent spherical aggregates (submicelles) that are about 20 nm in diameter. When carrageenan is present

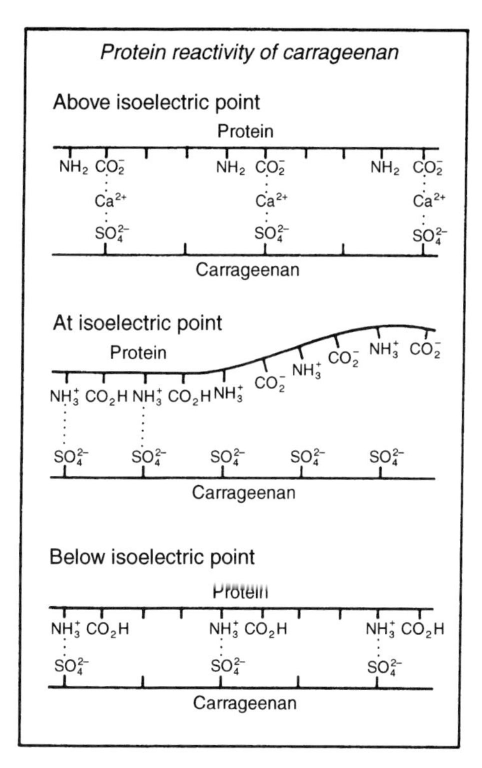

Figure 3.8 Protein reactivity of carrageenan. (From Marine Colloids, The Carrageenan People, Introduction Bulletin A-1.)

they become aligned into thread-like structures ~20 nm in width. In comparison, the carrageenan chains are ~2 nm wide (Snoeren, 1976; Snoeren *et al.*, 1976). The κ-casein particles are attached to the hydrocolloid chain by electrostatic attraction, primarily with κ- and ι-carrageenans. Those carrageenans (and not λ-carrageenan) are effective for suspending cocoa particles in chocolate milk, perhaps because the formation of a gel network is necessary in parallel with the fact that suspension ability holds only for high-molecular-weight carrageenans. Kappa-casein makes up 8–15% of the casein micelle. Part of the outer surface of the micelle is exposed and can interact with carrageenan (Schmidt and Payens, 1976). The ultrastructure of carrageenan–milk sols and gels has been reported (Hood and Allen, 1977). In the micelle, κ-casein shields the calcium-sensitive caseins from contact with calcium ions, thus probably protecting the α_{s1}- and β-caseins. This protective role is shared with the carrageenans. Electron microscopy studies revealed a κ-carrageenan/α_s-casein system in which α_s-casein in aggregate sizes of 100 to 500 nm was entrapped by interconnecting strands of κ-carrageenan while almost no protein was attached to it. More specific information on the interaction of sulfated polysaccharides with proteins can be obtained elsewhere (Lin, 1977). The possibility that no complex is formed between α_s-casein and carrageenan in the absence of calcium ions was suggested following electrophoresis and sedimentation studies and verified by physicochemical techniques (Hansen, 1968; Skura and Nakai, 1980). In fact, results suggested that calcium ions serve as mediators in ionic interactions. However, Lin (1977) suggested that, since the stability of the complex is independent of ionic strength, the cations neutralize charged ester sulfates of carrageenan and promote aggregation of the gum chains into a double-helix formation. Lambda-carrageenan is not effective in the stabilization of α_s-casein systems, perhaps because of its structure (Lin, 1977); however, alkaline modification of λ-carrageenan to form non-gelling θ-carrageenan changes this situation, although not to the extent achieved by κ-carrageenan (Stanley, 1990). Others proteins such as β-casein, paracasein and those from soy, peanut, cottonseed and coconut can be stabilized against precipitation with calcium ions by carrageenans. Studies of other sulfated polysaccharides as stabilizers of α_s-casein can be found in Lin (1977). Non-specific interactions, such as the precipitation reaction of carrageenan with gelatin (MacMullan and Eirich, 1963) and the interaction between carrageenan and β-lactoglobulin have also been reported (Hidalgo and Hansen, 1969). The effects of hydrocolloid stabilizers on calcium equilibrium in skim milk were studied, using a resin contact time method and equilibrium dialysis (Sui-Chen-Lee and Hansen, 1995). The presence of stabilizers caused an apparent increase in exchangeable calcium for short contact times. Addition of ι-carrageenan (0.05%) caused an $\sim$20% reduction in the dialysable form of calcium, whereas other stabilizers (xanthan, CMC, LBG, alginate) had no measurable effect (Sui-Chen-Lee and Hansen, 1995).

3.8.3 Other synergistic effects

A gelatin-like texture can be produced by combining LBG with κ-carrageenan, thereby decreasing the cost of producing a firm gel. Strong elastic gels (at least four times stronger than κ-carrageenan alone) can be produced by exploiting the synergism between konjac flour and κ-carrageenan. This interaction can also be used to form heat-stable gels at temperatures above the boiling point. The synergistic interactions between carrageenan and milk proteins have already been discussed. In fact, carrageenan was first used in milk gels, as a stabilizer in evaporated milk and ice-cream mixes. In such cases, κ-carrageenan forms a weak gel in the water portion of the milk system and also interacts with the surface area of the casein micelles.

3.9 Applications

Carrageenans are used as binders and gelling, thickening and stabilizing agents. Typical dairy applications of carrageenans are in milk gels, puddings and pie fillings, whipped products, cold-prepared milks, acidified milks, frozen desserts, pasteurized and sterilized milk products and infant formulations. These gums are available as part of ready-made preparations, or as part of a powder to be added to milk. Brittle gels or creamier products can be achieved, depending upon the combination of κ- and ι-carrageenans and their combinations with polyphosphates. In chocolate milks, 0.025–0.035% carrageenan is added prior to pasteurization and cooling, when the interactions produce a delicate structure that keeps the cocoa particles suspended; the drink is richer and has a better oral quality. In sherbets and ice creams, carrageenans are used as stabilizers to prevent the separation of fat or other solids within the products (Stanley, 1990; Thomas, 1992). Changes in consumer demand within the confectionary industry have promoted a review of the ingredients needed to create low-fat and reduced-calorie products (Izzo *et al.*, 1995). Ingredients are required to provide viscosity and stability as gel-forming fat substitutes in fat- and calorie-reduced systems. The use of cellulose gum and carrageenan, either alone or in combination, is essential for lowering fat and calories in confections. Cellulose gel in chocolate coatings, carrageenan in gummy confections and combined ingredients in fat free and reduced-fat caramels and heat-stable marshmallow products can be helpful in meeting consumer demand (Izzo *et al.*, 1995).

3.9.1 Milk applications

Carrageenans are used as a dry ingredient in powders to be added to milk and milk products. In milkshakes and instant-breakfast powders, carrageenan is used to suspend the ingredients, making the drink richer with more body. A fine-mesh λ-carrageenan is used for rapid cold-solubility (Guiseley *et al.*, 1980). Creamy flans and custards are achieved by including κ-carrageenan within the product. Combinations of κ- and ι-carrageenan together with tetrasodium pyrophosphate (TSPP) help not only to induce gelation but also to increase firmness and creaminess of the products. Lambda-carrageenan is preferred for cold-prepared mixes and in combination with ι-carrageenan and TSPP to avoid or decrease syneresis. In Dutch desserts, a combination of starch and carrageenan is used to achieve a pleasing texture after long periods of storage (Anon., 1988a). The use of κ-carrageenan instead of the starch-carrageenan blend results in a more uniform set (Guiseley *et al.*, 1980). In chocolate-milk preparations composed of about 1% cocoa and 6% sugar, vanillin is added as a flavor modifier and carrageenan reacts with κ-casein after pasteurization and during cooling to produce a stable suspension, creating a tastier, richer drink. The addition of carrageenans at 0.025–0.040% is enough to suspend calcium phosphate in chocolate milks. The hydrocolloid addition inhibits whey-off without imparting any palpable gel structure or flavor to the milk (Anon., 1988a). In ice creams and sherbets, which can be defined as partly frozen foams, LBG, guar gum and CMC are used as primary stabilizers, i.e. to prevent the separation or uneven distribution of fat and other solids and to prevent or decrease the formation of ice and lactose crystals. Stabilizer composed of 10–30% (w/w) guar gum, 5–20% (w/w) carrageenan and/or xanthan gum and 60–80% (w/w) of an emulsifier (selected from fatty acid mono- or diglycerides, lactic or citric acid esters of fatty acid mono- or diglycerides, sorbitan mono- or tristearates and lecithins) has been claimed to enable the production of a pourable, aerated dairy dessert having a shelf life of 3 weeks without evidence of sagging or separating-out (Tilly, 1995).

As a secondary stabilizer, κ-carrageenan can prevent whey-off during storage before the mix has been frozen, and later when it passes through a freeze–thaw cycle (Grindrod and Nickerson, 1968; Arbuckle, 1972; Keeney and Kroger, 1974; Nielsen, 1976). Whole milk has a better appearance and oral texture if low levels of carrageenan are added as emulsified fat stabilizers. Carrageenans have also been used in sterilized milks to prevent fat and protein separation (Glicksman, 1983). Evaporated milks contain as little as 0.005% κ-carrageenan to prevent fat and protein separation. In products to be diluted with water before consumption, such as infant formulae, κ-carrageenan, partially modified with alkali, is used to prevent fat and protein separation. In cottage-cheese production, carrageenan and LBG are used to prevent separation and to provide the curd with 'cling' (Glicksman, 1983). In the production of hard cheeses, carrageenan improves yield by altering the precipitation of whey proteins. In imitation cheeses, ~2.5% κ-carrageenan is used as a binder (gelling agent) of water, hydrogenated fat and sodium caseinate. It also serves to improve sliceability and shreddability (Anon., 1988a). For traditional cheese, a dry mixture of carrageenans and microcrystalline cellulose (MCC) is dispersed in a portion of the milk before pasteurization. Use of calcium chloride and an increased amount of rennet is recommended, as is doubling the quantity of the starter culture. Annatto and lipolytic enzymes can be added to improve color and flavor (Bullens *et al.*, 1995). Incorporation of carrageenan or gellan gum at 250 and 500 mg kg^{-1} increased yields (fresh weight) of Cheddar cheese by 4 to 17% compared with conventional methods using no gum. Gellan gum gave higher yields than carrageenan. Cheese treated with carrageenan contained higher levels of nitrogen than that with or without added gellan. Carrageenan gum caused reductions of 10 and 22–26% in total solids and nitrogen, respectively, relative to controls (Kanombirira and Kailasapathy, 1995). A new frozen yoghurt has been reported to combine the technological characteristics of ice cream and the qualities of yoghurt, including freshness, viable lactic acid bacteria and a distinctive flavor. This product was prepared from fresh, semi-skim, pasteurized and homogenized milk, to which selected starter cultures, sugar, guar gum, thickening agents, carrageenan and mono- and diglycerides of fatty acids had been added (Spano, 1994). An alternative gelling/stabilizing system for better mousse texture was developed, based on pectin, starch and carrageenan (Jensen, 1995).

3.9.2 *Water applications*

Typical water applications are as a water-gelation agent, fat stabilizer, thickener and for suspension, bodying and pulping effects. Iota-carrageenan can be used to mimic gelatin gels, except for their high melting temperature. This is an advantage where refrigeration is not an option to create the gel, as is the case in preparing desserts based on gelatins (Glicksman, 1966). To improve the texture of carrageenan-based gels, agar and LBG can be added (Baker, 1949, 1954). To maintain the transparent appearance of carrageenan gels, methods to clarify LBG have been developed. To eliminate the problem of insolubility in cold water, sodium salts of κ- and ι-carrageenans, which have the capacity for solubilization in the cold, are used (Glicksman *et al.*, 1970). Instead of the more conventional fruit jellies based on sugar and pectin combinations, mixtures of κ- and ι-carrageenan and non-caloric sweetener are used to build a low-calorie product. To eliminate acid hydrolysis, the acid is added as late in the process as possible (Anon., 1988b). Sorbet with a smooth texture can be achieved using a blend of carrageenan, LBG and guar or pectin. The carrageenan:sorbet ratio should be optimized to eliminate the possibilities of flavor masking (too high a ratio) or the production of a grainy product (too low an amount of additives) (Anon., 1987a). In combination with guar gum, κ-carrageenan is used to manufacture pet foods. Seaweed flour is used to replace the extracted carrageenans in such preparations (Anon., 1987b; Lewis *et al.*, 1988). In sauces and relishes, κ- or ι-carrageenans provide texture, sheen and improved adhesiveness (Anon., 1988c). In reduced-calorie dressings, carrageenans are used to impart the oral sensation of a high-oil system (Anon., 1988d), for high-protein dressings additional protective colloids are required (Kampmann, 1995). In combination with xanthan, carrageenan is used to stabilize oil-in-water emulsions, or reduced-calorie spreads (Anon., 1988d). Other water applications include pie fillings (Anon., 1988b). With added sodium chloride, sweetness decreased in κ-carrageenan systems, possibly because the content of endogenous cations (calcium, potassium and sodium) influenced sodium ion mobility. The sweetest systems, containing lactose and/or xanthan, underwent the greatest flavor enhancement by sodium chloride (Barisas *et al.*, 1995).

3.9.3 *Meat and fish*

In gefilte fish, carrageenans (κ- and ι-) and their combination with LBG help achieve the desired flavor and texture (Guiseley *et al.*, 1980; Glicksman, 1983). A combination of κ-carrageenan, LBG and potassium chloride is used to coat fish in order to eliminate freeze burn and mechanical disintegration during processing. Kappa-carrageenan is used in frankfurters to reduce their fat content without affecting their taste (Anon., 1988e). In the production of ham and during pumping and tumbling, κ-carrageenan interacts with protein and binds free water to maintain moisture and soluble solids content. The pumping solution is composed of brine, in which gums are dispersed after salt dissolution. Gum dissolution occurs during cooking (Anon., 1987c). Poultry processing involves a few problems, such as protein denaturation, textural changes, moisture loss, fat oxidation, breakdown of meat texture, etc. These problems can be reduced by incorporating phosphates, salts, starches, proteins and carrageenan into the brine solution, which is then introduced into the meat by injection or tumbling (Anon, 1987c). Low-fat ground-beef patties gained good taste, texture and high acceptability scores over control patties (ground beef and carrageenan alone), when a blend of ι- and κ-carrageenans having various viscosity and gelling characteristics together with other water binders (taken from xanthan/LBG, pea flour, algin or modified starch) were used. Thus, non-meat ingredients helped produce improved low-fat beef products (Bullock *et al.*, 1995). The use of carrageenan, starch and milk-soy proteins to serve in low-fat, high-added-water bologna and to improve its overall acceptability, has also been reported (Dexter *et al.*, 1993).

References

Anderson, N.S., Dolan, T.C.S. and Rees, D.A. (1968) Carrageenans. Part III. Oxidative hydrolysis of methylated κ-carrageenan and evidence for a masked repeating structure. *J. Chem. Soc.* (C), 596–601.

Anderson, N.S., Dolan, T.C.S. and Rees, D.A. (1973) Carrageenans. Part VII. Polysaccharides from *Eucheuma spinosum and Eucheuma cottonii.* The covalent structure of ι-carrageenan. *J. Chem. Soc., Perkin Trans. 1*, 2173–6.

Anon. (1987a) *Sorbet.* Application Bulletin B-49, FMC Corp., Marine Colloids Div.

Anon. (1987b) *Processed Poultry Products.* Application Bulletin B-51, FMC Corp., Marine Colloids Div.

Anon. (1987c) *Meat Application* – ham pumping/*tumbling.* Application Bulletin B-41, FMC Corp., Marine Colloids Div.

Anon. (1988a) *Calcium Fortified Fluid Milk.* Application Bulletin E-53, FMC Corp., Marine Colloids Div.

Anon. (1988b) *Pumpkin Pie Fillings.* Application Bulletin D-17, FMC Corp., Marine Colloids Div.

Anon. (1988c) *Imitation Mayonnaise.* Application Bulletin C-61, FMC Corp., Marine Colloids Div.

Anon. (1988d) *No Oil/low oil Pourable Italian Type Dressing.* Application Bulletin C-5, FMC Corp., Marine Colloids Div.

Anon. (1988e) *Low-fat Frankfurters.* Application Bulletin B-60, FMC Corp., Marine Colloids Div.

Arbuckle, W.S. (ed.) (1972) *Ice Cream*, 2nd edn, Avi Publishing, Westport, CT.

Arnott, S., Scott, W.E., Rees, D.A. *et al.* (1974) ι-carrageenan molecular structure and packing of polysaccharide double helices in oriented fibres of divalent cation salts. *J. Mol. Biol.*, **90**, 253–67.

Baker, G.L. (1949) *Edible Gelling Compositions containing Irish Moss Extract, Locust Bean Gum and an Edible Salt.* US Patent No. 2,466,146.

Baker, G.L. (1954) *Gelling Compositions.* US Patent No. 2,669,519.

Barisas, L., Rosett, T.R., Gao, Y. *et al.* (1995) Enhanced sweetness in sweetener-NaCl-gum systems. *J. Food Sci.*, **60**(3), 523–7.

Bayley, S.T. (1955) X-ray and infrared studies of κ-carrageenan. *Biochim. Biophys. Acta*, **17**, 194–205.
Bjerre-Petersen, E., Christensen, J. and Hemmingsen, P. (1973) Furcellaran, in *Industrial Gums*, 2nd edn (eds R.L. Whistler and J.N. BeMiller), Academic Press, New York, pp. 123–36.
Bourgade, G. (1871) *Improvement in Treating Marine Plants to Obtain Gelatin.* US Patent No. 112,535.
Bryce, T.A., Clark, A.H., Rees, D.A. *et al.* (1982) Concentration dependence of the order–disorder transition of carrageenans. Further confirmatory evidence for the double helix in solution. *Eur. J. Biochem.*, **122**, 63–9.
Bryce, T.A., McKinnon, A., Morris, E.R. *et al.* (1974) Chain conformations in the sol–gel transitions, and their characterisation by spectroscopic methods. *J. Chem. Soc., Faraday Disc.*, **57**, 221–9.
Bullens, C., Krawczyk, G. and Geithman, L. (1995) Cheese products with reduced fat content involving use of carrageenans and microcrystalline cellulose. *Latte*, **20**(2), 177–80.
Bullock, K.B., Huffman, D.L., Egbert, W.R. *et al.* (1995) Nonmeat ingredients for low-fat ground beef patties. *J. Muscle Foods*, **6**(1), 37–46.
Deslandes, E., Floch, J.Y., Bodeau-Bellion, C. *et al.* (1985) Evidence for λ-carrageenan in *Soliera chordalis* (Solieriaceae) and *Callibepharis jubata, Callibepharis ciliata, Cystoclonium purpureum* (Rhodophyllidaceae). *Bot. Mar.*, **28**, 317–18.
Dexter, D.R., Sofos, J.N. and Schmidt, G.R. (1993) Quality characteristics of turkey bologna formulated with carrageenan, starch, milk and soy protein. *J. Muscle Foods*, **4**(3), 207–23.
DiNinno, V.L., McCandles, E.L. and Bell, R.A. (1979) Pyruvic acid derivative of a carrageenan from a marine red alga. *Carbohydr. Res.*, **71**, C1–C4.
Dolan, T.C.S. and Rees, D.A. (1965) The carrageenans. Part II. The positions of the glycosidic linkages and sulphate esters in λ-carrageenan. *J. Chem. Soc.* 3534–9.
Fernandes, P.B., Goncalves, M.P. and Doublier, J.L. (1991) Phase diagrams in κ-carrageenan/locust bean gum systems. *Food Hydrocolloids*, **5**(1/2), 71–3.
Fernandes, P.B., Goncalves, M.P. and Doublier, J.L. (1994a) Rheological behavior of κ-carrageenan/galactomannan mixtures at a very low level of κ-carrageenan. *J. Texture Studies*, **25**(3), 267–83.
Fernandes, P.B., Goncalves, M.P. and Doublier, J.L. (1994b) Rheological description at the minimum gelling conditions of κ-carrageenan/LBG systems. *Food Hydrocolloids*, **8**(3/4), 345–9.
Glicksman, M. (1966) *Frozen gels of* Eucheuma. US Patent No. 3,250,62.
Glicksman, M. (1969) *Gum Technology in the Food Industry*. Academic Press, New York.
Glicksman, M. (1983) Red seaweed extracts (agar, carrageenans, furcellaran), in *Food Hydrocolloids*, vol. 2 (ed. M. Glicksman), CRC Press, Boca Raton, FL, pp. 73–113.
Glicksman, M., Farkas, E. and Klose, R.E. (1970) *Cold-water-soluble* Eucheuma *gel mixtures*. US Patent No. 3,502,483.
Goycoolea, F.M., Richardson, R.K., Morris, E.R. *et al.* (1995) Effect of locust bean gum and konjac glucomannan on the conformation and rheology of agarose and κ-carrageenan. *Biopolymers*, **36**(5), 643–58.
Greer, C.W. and Yaphe, W. (1984) Characterization of hybrid (beta-kappa-gamma) carrageenan from *Eucheuma gelatinae* J. Agardh (Rhodophyta, Solieriaceae) using carrageenases, infrared and ^{13}C-nuclear magnetic resonance spectroscopy. *Bot. Mar.*, **27**, 473–8.
Grindrod, J. and Nickerson, T.A. (1968) Effect of various gums on skim milk and purified milk proteins. *J. Dairy Sci.*, **51**, 834–41.
Guiseley, K.B., Stanley, N.F. and Whitehous, P.A. (1980) Carrageenan, in *Handbook of Water-soluble Gums and Resins* (ed. R.L. Davidson), McGraw-Hill, New York, pp. 5.1–5.30.
Haas, P. (1921) The nature and composition of Irish moss mucilage. *Pharm. J.*, **106**, 485.
Hansen, P.M.T. (1968) Stabilization of α_s-casein by carrageenan. *J. Dairy Sci.*, **51**, 192–5.
Hansen, P.M.T. (1972) Stabilization of α_s-casein by carrageenan. *J. Dairy Sci.*, **51**(2), 192–5.
Hansen, P.M.T. (1982) Hydrocolloid–protein interactions: relationship to stabilization of fluid milk products. A review, in *Gums and Stabilizers for the Food Industry* (eds G.O. Phillips, D.J. Wedlock and P.A. Williams), Pergamon Press, Oxford, pp. 127–38.
Hidalgo, J. and Hansen, P.M.T. (1969) Interactions between food stabilizers and β-lactoglobulin. *Agric. Food Chem.*, **17**, 1089–92.

Hirase, S. and Watanabe, K. (1972) The presence of pyruvate residues in ι-carrageenan and similar polysaccharides. *Bull. Inst. Chem. Res.*, **71**, C1–C4.

Hirase, S., Araki, C. and Watanabe, K. (1967) Component sugars of the polysaccharide of the red seaweed *Grateloupia elliptica*. *Bull. Chem. Soc. Japan*, **40**, 1445–8.

Hood, L.F. and Allen, J.E. (1977) Ultrastructure of carrageenan–milk sols and gels. *J. Food Sci.*, **42**, 1062–5.

Izzo, M., Stahl, C. and Tuazon, M. (1995) Using cellulose gel and carrageenan to lower fat and calories in confections. *Food Technol.*, **49**(7), 45–6, 48–9.

Jensen, T.W. (1995) Alternative gelling system for mousse. *Scand. Dairy Inform.*, **9**(4), 22–3.

Kampmann, R. (1995) Carrageenans in dressings. *Int. Z. Lebensmittel-Technik, Marketing, Verpackung Analytik*, **46**(10), 44, 46–7.

Kanombirira, S. and Kailasapathy, K. (1995) Effects of interactions of carrageenan and gellan gum on yields, textural and sensory attributes of Cheddar cheese. *Milchwissenschaft*, **50**(8), 452–8.

Keeney, P.G. and Kroger, M. (1974) Frozen dairy products, in *Fundamentals of Dairy Chemistry*, 2nd edn (eds B.H. Webb, A.H. Johnson and J.A. Alford). Avi Publishing, Westport, CT, pp. 873–913.

Lawson, C.J. and Rees, D.A. (1970) An enzyme for the metabolic control of polysaccharide conformation and function. *Nature*, **227**, 390–3.

Lawson, C.J., Rees, D.A., Stancioff, D.J. and Stanley, N.F. (1973) Carrageenans. Part VIII. Repeating structures of galactan sulphates from *Furcellaria fastigiata, Gigartina cancliculata, Gigartina chamissoi, Gigartina atropurpurea, Ahnfeltia durvillaei, Gymnogongrus furcellatus, Eucheuma cottonii, Eucheuma spinosum, Eucheuma isiforme, Eucheuma uncinatum, Agardhiella tenera, Pachymenia hymantophora* and *Gloiopeltis cervicornis*. *J. Chem. Soc., Perkin Trans* 1, 2177–82.

Lewis, J.G., Stanley, N.F. and Guist, G.G. (1988) Commercial production and applications of algal hydrocolloids, in *Algae and Human Affairs* (eds C.A. Lembi and J.R. Waaland), Cambridge University Press, Cambridge, pp. 205–36.

Lin, C.F. (1977) Interaction of sulfated polysaccharides with proteins, in *Food Colloids* (ed. H. D. Graham), Avi. Publishing, Westport, CT, pp. 320–46.

Mabeau, S. and Fleurence, J. (1993) Seaweed in food products: biochemical and nutritional aspects. *Trends Food Sci. Technol.*, **4**(4), 103–7.

MacMullan, E.A. and Eirich, F.K. (1963) The precipitation reaction of carrageenan with gelatin. *J. Colloid Sci.*, **18**, 526–37.

Manning, D.W. (1990) Carrageenans regulatory aspects (section 7e in *Carrageenans* by N.F. Stanley), in *Food Gels* (ed. P. Harris), Elsevier Applied Science, London, pp. 79–119.

McCandless, E.L. and Gretz, M.R. (1984) Biochemical and immunochemical analysis of carrageenans of the Gigartinaceae and Phyllophoraceae. *Hydrobiologia*, **116**/**117**, 175–8.

Mueller, G.P. and Rees, R.A. (1968) Current structural views of red seaweed polysaccharides, in *Proc. Drugs from the sea* (ed. H.D. Freudenthal), Marine Technology Society, Washington, DC, pp. 241–55.

Nadison, J. (1995) The interaction of carrageenan and starch in cream desserts. *Scand. Dairy Inform.*, **9**(2), 24–5.

Nielsen, B.J. (1976) Function and evaluation of emulsifiers in ice cream and whipped emulsions. *Gordian*, **76**, 200–25.

Nishinari, K., Watase, M., Miyoshi, E. *et al.* (1995) Effects of sugar on the gel–sol transition of agarose and κ-carrageenan. *Food Technol. Chicago*, **49**(10), 90, 92–6.

Nussinovitch, A., Kaletunc, G., Normand, M.D. *et al.* (1990a) Recoverable work vs asymptotic relaxation modulus in agar, carrageenan and gellan gels. *J. Texture Studies*, **21**, 427–38.

Nussinovitch, A., Kopelman, I.J. and Mizrahi, S. (1990b) Evaluation of force deformation data as indices to hydrocolloid gel strength and perceived texture. *Int. J. Food Sci. Technol.*, **25**, 692–8.

Nussinovitch, A., Kopelman, I.J. and Mizrahi, S (1990c) Effect of hydrocolloid and minerals content on the mechanical properties of gels. *Food Hydrocolloids*, **4**(4), 257–65.

O'Neill, A.N. (1955) Derivatives of 4-*O*-β-D-galactopyranosyl-3,6-anhydro-D-galactose from κ-carrageenan. *J. Am. Chem. Soc.*, **77**, 6324–6.

Painter, T.J. (1966) The location of the sulphate half-ester groups in furcellaran and κ-carrageenan, in *Proc. 5th Int. Seaweed Symp.* (eds E.G. Young and J.L. McLachlan), Pergamon Press, London, pp. 305–13.

Paoletti, S., Smidsrod, O. and Grasdalen, H. (1984) Thermodynamic stability of the ordered conformation of carrageenan polyelectrolytes. *Biopolymers*, **23**, 1771–94.

Parolis, H. (1981) The polysaccharides of *Phyllymenia hieroglyphica* and *Pachymenia hymantophra. Carbohydr. Res.*, **93**, 261–7.

Payens, T.A.J. (1972) Light scattering of protein reactivity of polysaccharides especially of carrageenans. *J. Dairy Sci.*, **55**, 141–50.

Penman, A. and Rees, D.A. (1973a) Carrageenans. Part X. Synthesis of 3,6-di-*O*-methyl-D-galactose, a new sugar from the methylation analysis of polysaccharides related to ζ-carrageenan. *J. Chem. Soc., Perkin Trans. 1*, 2188–91.

Penman, A. and Rees, D.A. (1973b) Carrageenans. Part XI. Mild oxidative hydrolysis of κ- and λ-carrageenans and the characterization of oligosaccharide sulphates. *J. Chem. Soc., Perkin Trans. 1*, 2191–6.

Rees, D.A. (1963) The carrageenan system of polysaccharides. Part I. The relation between the κ- and λ-components. *J. Chem. Soc.*, 1821–32.

Rees, D.A. (1972) Mechanism of gelation in polysaccharides systems, in *Gelation and Gelling Agents, British Food Manufacturing Industries Research Association, Symp. Proc.* No. 13, London, pp. 7–12.

Rees, D.A., Steele, I.W. and Williamson, F.B. (1969) Conformational analysis of polysaccharides. III. The relation between stereochemistry and properties of some natural polysaccharides. *J. Polym. Sci., Part C*, **28**, 261–76.

Rotbart, M. (1984) Carrageenan isolation, characterization and upgrading of functional properties. MSc thesis, Technion-Israel Institute of Technology, Israel.

Rotbart, M., Neeman, I., Nussinovitch, A. *et al.* (1988) The extraction of carrageenan and its effects on the gel texture. *Int. J. Food Sci. Technol.*, **23**, 591–9.

Schmidt, C. (1844) Uber phlanzeschleim und bassorin. *Annal. Chem. Pharm.*, **51**, 29–62.

Schmidt, D.G. and Payens, T.A.J. (1976) Micellar aspects of casein. *Surf. Colloid Sci.*, **9**, 162–229.

Singh, S.K. and Jacobsson, S.P. (1994) Kinetics of acid hydrolysis of κ-carrageenan as determined by molecular weight, gel breaking strength and viscosity measurements. *Carbohydr. Polym.*, **23**(2), 89–103.

Skura, B.J. and Nakai, S. (1980) Physicochemical verification of the nonexistence of α_{s1}-casein–κ-carrageenan interaction in calcium-free systems. *J. Food Sci.*, **45**, 582–91.

Smith, D.B. and Cook, W.H. (1953) Fractionation of carrageenan. *Arch. Biochem. Biophys.*, **45**, 232–3.

Smith, D.B., O'Neill, A.N. and Perlin, A.S. (1955) Studies on the heterogeneity of carrageenan. *Can. J. Chem.*, **32**, 1352–60.

Snoeren, Th.H.M. (1976) κ-Carrageenan. A study on its physicochemical properties, sol–gel transition and interaction with milk proteins, Thesis, Nederlands Instituut voor Zuivelonderzoek, Ede, the Nederlands.

Snoeren, Th.H.M., Both, P. and Schmidt, D.G. (1976) An electron-microscope study of carrageenan and its interaction with κ-casein. *Neth. Milk Dairy J.*, **30**, 132–41.

Snoeren, Th.H.M., Payens, T.A.J. *et al.* (1975) Electrostatic interaction between κ-carrageenan and κ-casein. *Milchwissenschaft*, **30**, 393–6.

Spano, A. (1994) A new product for the ice cream manufacturer. *Latte*, **19**(4), 396–402.

Stainsby, G. (1980) Proteinaceous gelling systems and their complexes with polysaccharides. *Food Chem.*, **6**, 3–14.

Stancioff, D.J. and Stanley, N.F. (1969) Infrared and chemical studies on algal polysaccharides, in *Proc. XIth Int. Seaweed Symp.* (ed. R. Mrgalef), Subsecretaria de la Marina Mercante, Madrid, pp. 595–609.

Stanley, N.F. (1963) *Process for Treating a Polysaccharide of Seaweeds of the Gigartinaceae and Solieriaceae Families.* US Patent 3,094,517.

Stanley, N.F. (1987) Production, properties and use of carrageenans, in *Production and Utilization of Products from Commercial Seaweeds* (ed. D. J. McHugh), FAOUN, Rome, pp. 97–147.

Stanley, N.F. (1990) Carrageenans, in *Food Gels* (ed. P. Harris), Elsevier Applied Science, London, pp. 79–119.
Sui-Chen-Lee and Hansen, P.M.T. (1995) Effect of hydrocolloids on the calcium equilibrium in skim milk, *Poster at the IFT Annual Meeting,* p. 189.
Thomas, W.R. (1992) Carrageenan, in *Thickening and Gelling Agents for Food*, ch. 2 (ed. A. Imeson), Blackie A & P, Glasgow, pp. 25–39.
Tilly, G. (1995) *Stabilizer Composition enabling the Production of a Pourable Aerated Dairy Dessert.* European Patent Application, EP 0 649 599 A1.
Towle, G.A. (1973) in *Industrial Gums* (eds R.L. Whistler and J.N. BeMiller), Academic Press, New York, pp. 83–114.
Tsutsumi, A., Deng, Ya., Hiraoki, T. *et al.* (1993) ESR studies of Mn^{2+} binding to gellan and carrageenan gels. *Food Hydrocolloids,* **7**(5), 427–34.

4 Gellan gum

4.1 Introduction

Bacteria have been an important source of polysaccharides since the early 1970s. Examples include dextran (used as a blood extender and for chromatographic purposes) and xanthan gum (which has been intensively tested in the last few years). Of the many gums produced by bacteria, only a few have commercial value, because only a few offer significant advantages over known and well-used hydrocolloids. A few examples of these novel polysaccharides are rhamsan gum (a better suspending agent than xanthan), biozan welan gum (a thickener used in oil-field applications, the thermal stability of which is superior to xanthan gum) and gellan gum (with its unique gelling abilities). Gellan, a fermentation product of the bacterium *Pseudomonas elodea*, is an extracellular polysaccharide. It can be produced on demand, and its quality is less influenced by the different qualities of the raw materials. Gellan can be produced in two forms, substituted or non-substituted, and it is beneficial to industries with very low levels of use in a large variety of applications. A variety of gel textures (soft elastic gels from the substituted gum, and hard brittle gels from the non-substituted gum) can be created (Sanderson, 1989, 1990).

4.2 Structure and chemical composition

Gellan gum is a linear anionic heteropolysaccharide of $\sim 0.5 \times 10^6$ Da. It comprises the monosaccharide building units glucose, glucuronic acid and rhamnose in a molar ratio of 2:1:1. It has a tetrasaccharide (Fig. 4.1) repeating unit (Jansson *et al*., 1983; O'Neill *et al*., 1983). In its native form (the polymer secreted by the bacteria), there are approximately 1.5 *O*-acyl groups (acyl substituents) per repeating tetrasaccharide unit. The *O*-acyl substituent was initially thought to be *O*-acetal, until Kuo *et al.* (1986) suggested that gellan gum contains both *O*-acetyl and *O*-L-glyceryl substituents on the 3-linked glucose unit, the former tentatively assigned to the 6-position and the latter to the 2-position (Sanderson, 1990).

Chemical analysis proved that glycerate substitution predominates over that with acetate (Sanderson, 1990). Moreover, glycerate substitution dramatically influences gellan properties since its bulk hinders chain associ-

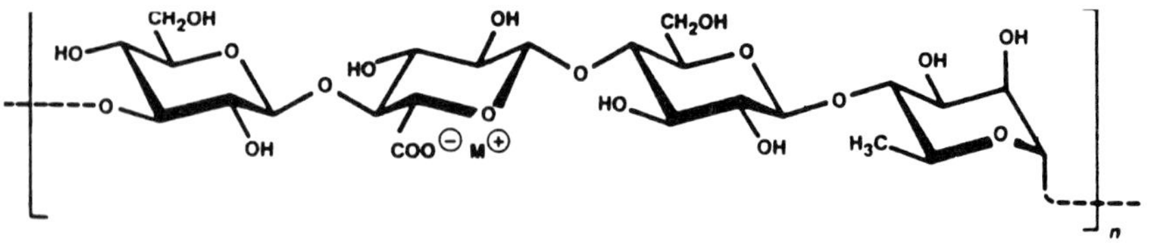

Figure 4.1 Gellan gum unsubstituted tetrasaccharide repeating unit. (Reproduced by permission of the Nutrasweet Kelco Co., a unit of Monsanto Co., San Diego, CA.)

ations and accounts for the change in gel texture brought about by de-esterification (Gibson, 1992). In comparison with welan and rhamsan, which have the same backbone but are substituted by different side-chains, no similarity in solution behavior is observed. The de-esterified gum is a polymer having non-substituted tetrasaccharide repeating units.

X-ray crystallography was used to propose the shape (conformation) of the gellan gum molecule. Early diffraction studies could not produce such a detailed structural analysis (Carroll *et al.*, 1982, 1983). A later effort failed to produce a structure consistent with the X-ray data (Upstill *et al.*, 1986). The crystal structure of gellan gum was clarified by the work of Chandrasekaran *et al.* (1988a) on the same data, indicating an extended, intertwined, threefold, left-handed, parallel double helix. Solid-state molecular shape is an indicator of molecular associations in solution, which presumably involve the ion-mediated aggregation of double helices (Chandrasekaran *et al.*, 1988b; Rinaudo, 1988; Sanderson, 1990). The structure includes uronic acid residues and can, therefore, be presented in the form of a variety of salts. The produced polysaccharide exists as a mixed salt, predominantly in the potassium form, but it can also contain sodium and calcium in small amounts, and even smaller amounts of magnesium.

4.3 Source, production, supply and regulatory status

4.3.1 Manufacture

Gellan gum is produced by a pure culture of the bacteria *P. elodea*. The fermentation medium for gum formation includes a carbon source such as glucose, a nitrogen source and the desired inorganic salts. For fermentation under sterile conditions to be successful, aeration, agitation and controlled temperature and pH need to be maintained (Kang *et al.*, 1982a). The fermentation broth increases in viscosity as the glucose is metabolized by the bacteria and the gum is secreted. Once fermentation is complete (when the carbon source is exhausted), the viable bacteria are killed by heat treatment before processing the broth to obtain the polysaccharide. Treatment of the pasteurized broth with alkali removes the acyl substituents on the gellan gum backbone. Then cellular debris is removed and the gum is recovered by precipitation with alcohol. Thus, a non-substituted form with a high degree of gellan gum purity is achieved. In fact two forms of gellan are produced: the fully acylated native form and the deacylated form.

Examples of products with low acyl content are Gelrite gellan gum for microbiological media and Kelcogel food-grade gellan gum ('low acyl gums'). Two kinds of gel with different textures can be produced. Cohesive elastic gels are produced from the native material and strong brittle gels from the deacylated material (Sutherland, 1992). Because the degree of acylation can be controlled by the deacylation step, many compositions with intermediate acyl content can be created, producing many different gel textures and products (Sanderson, 1990).

4.3.2 Nutritional aspects

Gellan gum is proprietary to Kelco (Kang *et al.*, 1980, 1982b,c; Kang and Veeder, 1983). Low acyl gellan gum is sold in the form of 60-mesh free-flowing powder. Kelcogel food-grade gellan gum is described for special uses in the food industry. Gellan gum safety studies in rats have shown that under a test of oral LD_{50}, the product is safe at > 5000 mg kg^{-1}. Following a 3-month diet, no signs of toxicity at up to 6% (w/w) of the diet were noted. In a two-generation reproduction study, no adverse effects were noted at up to 6% (w/w) of the diet. Teratology tests showed no dose-related effects at up to 5% (w/w) of the diet (Sanderson, 1990). Chronic dietary tests carried out with dogs for 52 weeks revealed no toxicological effects at concentrations of 3%, 4.5% or 5% (representing daily intakes of ~ 1.0, 1.5 or 2.0 g kg^{-1}, respectively) for either males or females. With monkeys subjected to 28 days of oral testing, no clinical signs or changes in blood chemistry were noted at doses of up to 3 g kg^{-1}. No skin or eye irritation was noted in rabbits. With humans, 23-day dietary tests yielded no adverse effects on plasma biochemistry, hematology or urinalysis at a daily dose of 200 mg kg^{-1}. Such results suggest that there is no cause for concern about gellan gum's use as a food product (Sanderson, 1990).

4.4 Functional properties

4.4.1 Hydration

Gellan contains divalent ions, inhibiting its hydration. Cold deionized water permits partial gellan hydration. Complete hydration is achieved by heating the dispersion to at least 70°C, or by mixing a pure monovalent salt form of the gum in cold deionized water. Because most water supplies contain divalent cations, hydration is restricted at ambient temperatures and the gellan gum can be easily dispersed (without hydration), avoiding problems related to lumping. The extent of hydration in cold water depends upon the cation concentration. In soft water, insufficient hydration of the gum can present problems of agglomeration and lumping. Under such circumstances, a dry blend of the gum with sugar and/or good agitation are used to improve dispersion (Gibson, 1992). In water containing > 150 ppm calcium carbonate at ambient temperature, some hydration can occur but its limited degree creates no dispersion problems. In general, the greater the cation concentration (the harder the water), the higher the hydration temperature. Sometimes (as is the case where water hardness is above ~ 200 ppm calcium carbonate), very high temperatures, achieved by ultra-heat treatment, are necessary for full hydration (Gibson, 1992). Discussions of hydration generally relate to the presence of divalent ions because the level of monovalent ions needed to inhibit hydration is much higher. Divalent ions can be sequestered by sodium or potassium citrate, or by phosphate-sequestering agents, enabling the gum's dissolution at low temperatures.

4.4.2 Solution properties

Once the gellan gum has been dissolved in cold water, highly viscous solutions can be achieved, with the aid of a sequestering agent. At a concentration of 1% (w/v), gellan gum is less pseudoplastic or shear-thinning than xanthan gum, but more pseudoplastic than high-molecular-weight sodium alginate (Gibson, 1992). A dramatic drop in gellan gum solution viscosity is observed in the range 25–50°C. This reflects a conformational change from some form of a relatively ordered, non-aggregated double helix to a random coil (Robinson *et al.*, 1987), a change that is reversible. The temperature dependence of viscosity enables the preparation of solutions with a high gum concentration, while avoiding solutions that are too viscous and difficult to manipulate. High temperatures combined with acidic conditions can result in hydrolytic degradation, and if gels are to be produced, a reduction in gel strength. Under neutral conditions, gellan gum is remarkably stable and gel quality is not affected, even when left at 80°C for several hours (Sanderson, 1990).

The influence of side-chains and substituents on the polyelectrolytic behavior of aqueous solutions of bacterial polysaccharides of the gellan family has been described (Campana *et al.*, 1992). Results of conductimetric and potentiometric titrations suggest that each of these polysaccharides adopts a double-helix conformation. Deacetylation destabilizes rhamsan and gellan helices, and topical rotation data confirm these results (Campana *et al.*, 1992). Topological features of several polysaccharides, including gellan gum, were studied by electron microscopy (Stokke *et al.*, 1993). The addition of potassium chloride to aqueous gellan gum changed its appearance from a dispersed polymer to superstrands with several associated chains. Macrocyclic species were also observed in gellan gum after the addition of a gel-promoting salt. The tendency to form macrocyclic structures in competition with intermolecular aggregates was determined by three factors: chain stiffness relative to overall length, parallel or antiparallel alignment of interacting chain segments and polymer concentration (Stokke *et al.*, 1993). The structural characteristics of gellan in aqueous solution were also studied by another group (Yuguchi *et al.*, 1993). Small-angle X-ray scattering from aqueous solutions of gellan in the sol and gel states was studied. Results were analysed in terms of the broken rod-like chain model, where rod thickness is taken into account. Gellan chains appeared to form branched clusters with a broad size distribution through interchain double helices in the sol state. Further growth of double-helix regions caused the alignment of rod-like domains and eventually gelation occurred. These results were in agreement with DSC measurements, which suggested that the cooperative dissociation of gellan chains is caused by the conformational transition at ~20°C (Yuguchi *et al.*, 1993).

Osmotic pressure measurements were carried out to determine the value of M_n for gellan gum and the osmotic second-viral coefficient using a high-speed membrane osmometer (Ogawa, 1993). Osmometry of the polyelectrolyte of gellan gum was carried out at 40°C in a solution of tetramethyl ammonium nitrate. The M_n values of the sample obtained were 5.4×10^{-4} and 5.5×10^{-4} in 0.05 and 0.075 mol dm^{-3} tetramethyl ammonium nitrate, respectively. The osmotic second-viral coefficients were rather large, as a result of the Donnan effect (Ogawa, 1993). Other physical measurements, such as ultrasonic velocities in aqueous gellan solutions and gels with and without sodium chloride, were measured over the temperature range 15–81°C. For an interpretation with respect to random coil-to-helix conformational changes in the gellan molecule followed by dehydration, see Tanaka *et al.* (1993).

Low-acyl content gellan gum materials are produced as mixed salts (potassium predominantly, but sometimes other divalent ions as well). In Gelrite by Kelco, 0.75% calcium, 0.25% magnesium, 0.70% sodium and 2.0% potassium ions are typically found. Low-acyl gum is soluble to some extent in cold water. Reducing the ionic content of the water and converting the gum to a pure monovalent salt form increases its solubility. Gelrite can be completely dissolved in deionized water using monovalent salt forms. Heating to at least 70°C aids in the dissolution of aqueous dispersions of low-acyl gellan gum. The higher the ionic strength of the aqueous phase, the higher the temperature required for good dissolution. Cooling of hot solutions results in gel formation (Sanderson, 1990).

In practical utilizations of low-acyl gellan gum, solubility can be suppressed by including ions such that the gum is dispersed in water and later activated by heating. Gellan solutions can react in cold water with mono- and divalent ions to produce gels that are resistant to heating (no melting). Therefore, in uses where the preparation of a solution is unavoidable in the early production stages, gellan gum powder should be introduced into a heated solution ($> 70°C$). Depending on the desired product, the dissolution procedure, blending of the gum powder with other ingredients and time of heating or addition should be planned with the desired gel texture and properties in mind (Sanderson, 1990).

Native gellan gum dispersed in cold water yields extremely high viscosities. However, it is very sensitive to salt concentration. In solutions of xanthan or gellan at the same concentration containing increasing concentrations of sodium chloride, the viscosity of the native gellan gum is strongly dependent on salt concentration, whereas that of xanthan is not. Native gellan solutions seem to be highly thixotropic and apparently high viscosities are the outcome of gel-like network formation. Similar thixotropic behavior is observed with solutions containing a blend of xanthan and LBG.

The viscoelastic properties of aqueous gellan solutions (0.5 to 1.33%) were studied, particularly the effects of concentration on gelation (Nakamura *et al.*, 1993). Gelation temperature increased with increasing gellan concentration. Gelation curves plotted as a function of time at a particular gelation temperature were approximated by the first-order reaction kinetics for gels of concentrations greater than 0.81%, and the rate constant increased with increasing concentration. The first-order kinetics were not applicable with gels at concentrations lower than 0.75%. Two distinct processes were observed along the gelation curve for a 0.5% solution. Master curves of the storage and loss modulus reduced to gelation temperature for gels with gellan concentrations higher than 0.75% indicated Zener type relaxation behavior (Nakamura *et al.*, 1993). Gel–sol transitions in gellan gum solutions (1–3%) as influenced by salts (chlorides of sodium, potassium, calcium and magnesium) were studied (Miyoshi *et al.*, 1994). Temperature effects depended upon the gellan gum and salt concentrations. Storage shear modulus for 1% gellan gum increased upon the addition of salts at lower temperatures, where gellan molecules have helical conformations, and decreased upon the addition of salts at higher temperatures, where gellan molecules exhibit coil transformations. Salts promoted helix aggregation at lower temperatures, whereas they reduced the coil dimension at high temperatures. Upon adding a lower concentration of salts (except for potassium chloride), the shear storage modulus became slightly smaller than that without salt. The reasons for these observations are discussed as well as an analysis of the results indicating that the mechanism of gel formation in gellan gum with divalent cations is markedly different from that with monovalent cations (Miyoshi *et al.*, 1994).

4.5 Gellan gel properties

4.5.1 Mechanism of gelation

Many studies suggest that gellan gelation initially occurs via the formation of double helices, followed by their ion-induced association (Gibson, 1992). This particular gelation mechanism suggests that heating and cooling in the absence of gel-promoting cations favor the formation of fibrils via double-helix formation between the ends of neighboring molecules (Fig. 4.2) (Gunning and Morris, 1990). In the presence of gel-promoting cations, these fibrils associate, resulting in gel formation. Divalent ions (ionic strength) are effective in gel production, and 3–7% of the recommended level of monovalent cations is enough for such purposes. The controlled release of divalent ions into the gellan solution, as has been done during alginate gelation, can be utilized to form homogeneous gellan gels, with the limitation that these gels are unstable and exhibit syneresis (Gibson, 1992). Different, more coherent demoldable gellan gels are achieved by cooling a hot solution (~ 90°C) of gellan that includes the sequestering agents needed to fully hydrate the gum. To achieve optimum gel strength, more divalent cations need to be added to the gum solution before cooling. Gelation temperatures depend on the cation concentration, with the effect of an increase from 35 to 55°C paralleling the increase in cations (Sanderson, 1990). Sodium salts of gellan gum form gels in the presence of potassium or calcium ions. Microfibrils in these gels are considerably longer and/or wider than in sols (Harada *et al.*, 1991). Potassium ions increase the number of junction zones and make them more heat resistant (Watase and Nishinari,

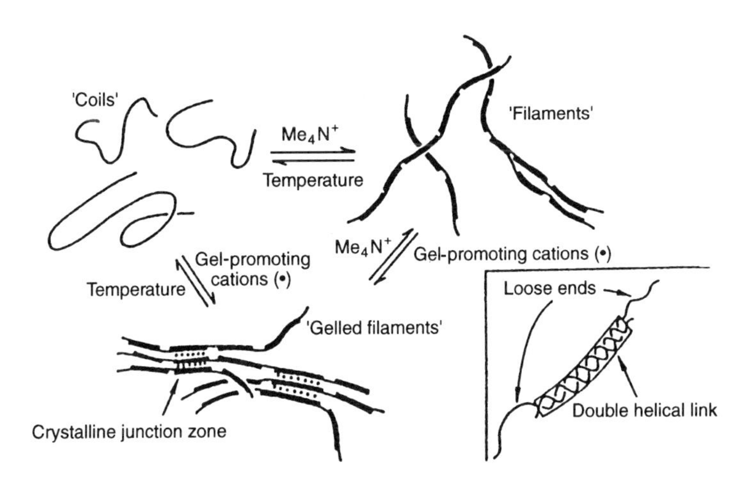

Figure 4.2 Model for the gelation of gellan gum. (From Gunning and Morris, 1990.)

1993). Junction zones are made up of various aggregations of double helices with different bonding energies and/or different degrees of rotational freedom of parallel links consisting of a single zipper (Watase and Nishinari, 1993). Gelling ability of the gellan gum decreased in the presence of urea. The dynamic Young's modulus of gellan gum decreased with increasing concentration of sucrose or glucose in the presence or absence of potassium chloride or calcium chloride. However, in the presence of urea and excessive amounts of these salts, the gelling ability of the gums was enhanced (Moritaka *et al.*, 1994).

4.5.2 Mechanical properties of gellan gels

Gellan gels' wide range of textural properties, from soft and elastic to hard and brittle, is dictated by the degree of esterification (Sanderson *et al.*, 1987b). The de-esterified product is commercially available and the properties of the hard and brittle gellan gum gels it produces are discussed here. The compressive strength (stress at failure) of 0.5–2.5% food-grade gellan gels was studied (Nussinovitch *et al.*, 1990a) and found to be the same as

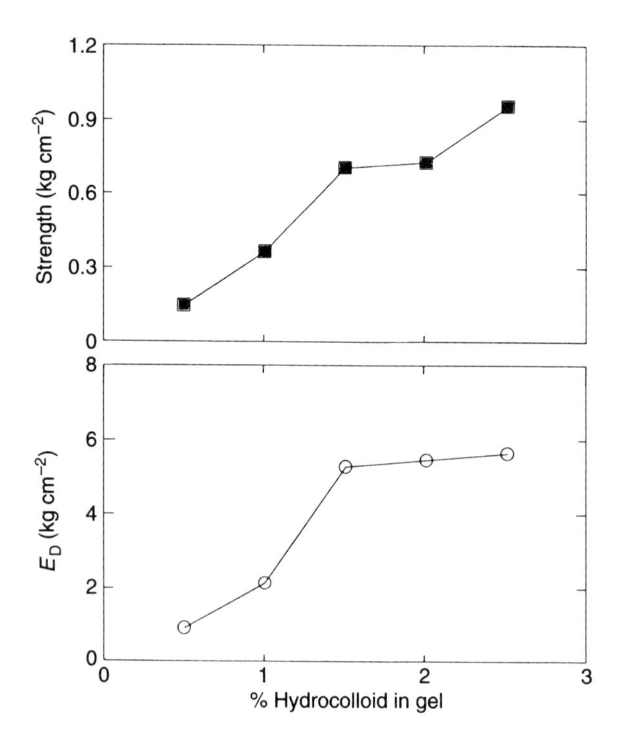

Figure 4.3 Compressive strength and deformability modulus of gellan gels of various concentrations. (From Nussinovitch *et al.*, 1990a.)

that of agar gels of similar concentration (Nussinovitch *et al.*, 1989). The fivefold increase in gum concentration resulted in an approximately tenfold increase in strength, from about 0.1 to 1 kg cm^{-2} (Fig. 4.3). The failure strain was usually in the range 15–20% but, unlike the strength, showed no clear dependency on gum concentration. Since almost up to failure the corrected (or true) stress versus Hencky's or 'true' strain relationship is highly linear, the deformability modulus can be calculated from the slope. The increase in deformability modulus values roughly corresponds to that of the gum concentration, i.e. roughly sixfold versus fivefold (Fig. 4.3). The overall modulus was in the range 1–6 kg cm^{-2}, about the same as that of agar gels at similar concentration (Nussinovitch *et al.*, 1990b). Gellan gels formed from 1% gellan gum solutions containing 7 mM calcium ions were tested to failure in tensile and torsional modes (Lelievre *et al.*, 1992). In the compression experiments, samples failed in shear at low deformation rates, and in a combination of compression and shear at higher strain rates. Tensile fracture was always evident in uniaxial tension measurements, and in the torsion experiments at low strain rates. At high deformation rates, the torsional samples underwent a combination of shear and tensile fracture (Lelievre *et al.*, 1992). Gellan gels (with or without added sucrose) have also been studied under large compressive deformation (Gao *et al.*, 1993; Willoughby and Kasapis, 1994). The stress–strain relationships of gellan and other gels were also tested by tensile test and data were fitted to a power law model (Hershko and Nussinovitch, 1995). The effect of polymer (0.6–1.8% (w/v)) and seven calcium ion concentrations (1.5–60 mM) on gellan gel strength and strain was tested (Juming *et al.*, 1994). Failure stresses and strains were measured in compressive, tensile and torsional modes on gellan gels. Shear stresses at failure were equal in all three testing modes and proportional to gellan content. Gels with low calcium levels increased linearly in strength with calcium ion concentration to a level of approximately 0.5 calcium ions per repeat tetrasaccharide unit of gellan gum polymer. Gel strength decreased linearly with calcium ion concentrations at higher concentrations. Gels with low calcium contents were extensible, with failure strains decreasing with the logarithm calcium ion concentration, whereas those with high calcium contents were brittle and failed at a constant strain, the value of which was twice as high in compression and torsion modes as in tension mode (Juming *et al.*, 1994, 1995). Gellan gels prepared at pH 2 were of no practical use, while those at pH 4 exhibited superior heat-resistance properties and were colorless and transparent. Gels formed at pH 6–10 could be used in foods that require reheating. Since gelation is affected by ions, it is suggested that pH effects on the gelation of gellan gums can be changed by the addition of salts (Moritaka *et al.*, 1995).

Gellan gels were also characterized by stress–relaxation studies. The force–relaxation curves of the gellan gels at 10% and 15% deformation were normalized and linearized by a method previously applied to other foods

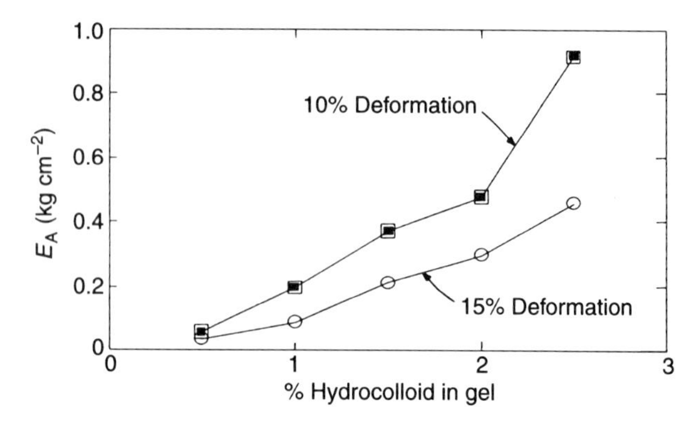

Figure 4.4 The asymptotic modulus of gellan gels at various concentrations determined from relaxation tests at two deformation levels. (From Nussinovitch *et al.*, 1990a.)

and gels (Peleg, 1980) and the asymptotic residual modulus (E_A), which serves as a measure of the gel's solidity, was calculated (Nussinovitch *et al.*, 1990b). The value of E_A, determined at two strains, was found to be dependent on gum concentration (Fig. 4.4). In general, and almost irrespective of concentration, the magnitude of E_A was of the order of a tenth or less of the corresponding deformability modulus. This is an indication that the gellan gels cannot sustain an appreciable amount of unrelaxed stress. Despite their resilience over short periods, they are not very elastic in the true sense of the term, compared, for example, with rubbery material. The magnitude of the asymptotic modulus at 10% deformation was significantly higher than that at 15%, indicating that gellan gels have a yielding structure and that gross failure is probably the culmination of a continuous process. A linear correlation was found between the magnitude of the asymptotic

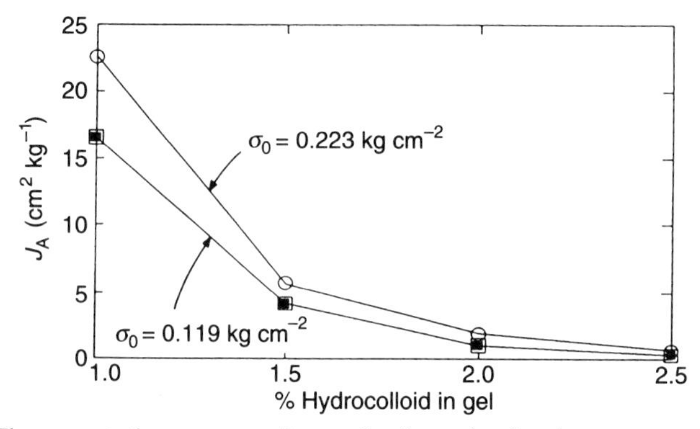

Figure 4.5 The asymptotic creep compliance of gellan gels of various concentrations determined in tests under two loads. (From Nussinovitch *et al.*, 1990a.)

modulus and the percentage recoverable work at a particular strain (Nussinovitch *et al.*, 1990b), a sign that both parameters are indicative of the gel's degree of elasticity and its strain dependency. Creep curves of gellan gels under two constant loads, corresponding to an initial stress of 0.12 or $0.22\,\mathrm{kg\,cm^{-2}}$, were normalized and linearized (Nussinovitch *et al.*, 1990a). Asymptotic compliance of gellan gels as a function of gum concentration is presented in Fig. 4.5. The plots demonstrate that, in agreement with the relaxation behavior, gellan gels are yielding materials, i.e. their creep compliance increases with imposed stress. The compliance of the more concentrated gels was much smaller than that of the studied gels. The concentrated gels, i.e. 1.5–2.5%, could reach creep strains of the order of 0.2–0.6, which were higher than the failure strain in uniaxial deformation. The more dilute gels, i.e. of 0.5 and 1%, never reached equilibrium or even quasi-equilibrium under the employed stresses, and their deformation continued indefinitely.

The stress or strain histories in stress–relaxation and creep tests are very different. Therefore, the behavior of materials, particularly non-linear, viscoelastic materials, in creep cannot be used to predict the behavior in stress relaxation, or vice versa. However, in some food materials the equilibrium conditions, as calculated from the asymptotic parameters, in creep and stress relaxation bear some similarity. The relationship between the asymptotic modulus and strain was compared with that between the reciprocal of the asymptotic compliance and the asymptotic strain (Nussinovitch *et al.*, 1990b). The most obvious feature of the two plots is that the strain levels are not the same, and that gels with 0.5 and 1% concentration did not reach equilibrium in creep.

Compressed gellan gels tend to exude liquid (Table 4.1). It is quite possible that this liquid release relieves hydrostatic pressure and serves as a mechanism for stress relaxation. In creep, syneresis acts in two ways. It reduces resistance to deformation by relieving the hydrostatic pressure, and at the same time causes gel strengthening by increasing the gum's effective concentration. This may be the reason why in creep the specimens sustained a much higher deformation than in uniaxial compression. The setting rate of gellan gels is dependent on their rate of cooling. When the setting temperature is reached, the gel reaches a strength that stays constant if stored under those conditions. This stability is a sign that gellan gum gels do not undergo syneresis (since syneresis is followed by an increase in gel strength). However, at low gum levels, free water can be expressed from the gel by pressure or the gel can undergo syneresis under its own weight. Therefore, higher gum concentrations are required for food products requiring the lowest potential syneresis.

The texture of gellan gels was also studied by texture-profile analysis (TPA), which measures the effects of a number of variables (Fig. 4.6). The technique uses two successive compressions of a gel specimen and the

Table 4.1 Syneresis of pure gellan gels during mechanical testing

Gellan (%)	Percentage exuded liquid				
	In compression to failure	After 5 min in relaxation		After 5 min in creep	
		$\varepsilon = 0.11$	$\varepsilon = 0.16$	$\sigma_0 = 12\,\mathrm{kg\,cm^{-2}}$	$\sigma_0 = 0.22\,\mathrm{kg\,cm^{-2}}$
0.5	15	8	15	–[a]	–[a]
1	10	5	10	20	38
1.5	9	5	13	5	13
2	8	3	7	3	8
2.5	7	5	9	1.5	4

Data are mean values of 3–4 replicates.
[a]Specimens disintegrated under these loads.
ε is strain.

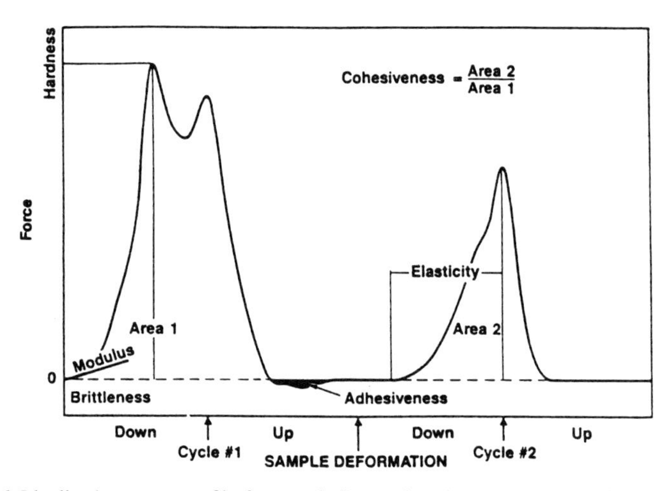

Figure 4.6 Idealized texture profile for a gel. (Reproduced by permission of the Nutrasweet Kelco Co., unit of Monsanto Co., San Diego, CA.)

evaluation of a few values that are calculated from the force–deformation curves generated, with the help of a computer attached to a universal testing machine. In the absence of ions, a very weak gel is formed. Hardness (maximum force occurring at any time during the first cycle of compression, correlated to the rupture strength of the specimen) rapidly increases to a maximum at low calcium levels and then gradually decreases as ionic concentration increases. The modulus (initial slope of the force–deformation curve, usually correlated with sensory perception of the sample's firmness) shows a maximum between 0.016% and 0.05% added calcium (400 and 1250 ppm calcium carbonate, respectively). Gel brittleness is related to ion concentration such that a reduction in brittleness is observed with decreasing ion concentration. In contrast, elasticity drops rapidly as the ion level increases to an almost constant value of around 10%. Divalent magnesium has a similar effect on gel texture. The same trends are also observed with monovalent ions such as potassium and sodium, albeit at much higher concentrations (Sanderson, 1990). Maximum hardness with sodium or potassium ions is achieved at an ionic concentration which is ~25 times greater than the molar concentration of calcium or magnesium ions. If low concentrations of gellan gum are to be used in a food formulation, sodium ions will have practically the same effect as potassium ions on gel strength. At high gum concentrations, however (Grasdalen and Smidsrod, 1987), potassium ions at the same concentration produce stronger gels. The hardness of agar, κ-carrageenan and gellan gels, all prepared at optimum ion concentrations, was compared by TPA. The gellan gum gels were stronger at equal gum concentrations. The same tendency was observed for the modulus (firmness). With brittleness, a small but insignificant increase of about 6% was observed, with an over fourfold increase in gum concentration. The elasticity of gellan gels doubled over the same gum concentration range (Gibson, 1992), but a much larger increase was required to produce a different texture. The texture of gellan gum gels is supposed to remain constant over a wide range of pH values. Sugars have a remarkable influence on texture, at concentrations higher than 40%, when firmness decreases and gels become less brittle and more elastic. At higher sugar concentrations, less interference (as reflected by higher firmness) was observed when smaller sugar molecules were used in the formulation (monosaccharides instead of disaccharides).

4.5.3 Melting and setting points

Melting and setting points, which are important gel properties, depend on ion concentration and type, and to a lesser extent on gum concentration (Sanderson, 1990). Gels differ in their setting temperatures. For example, those with calcium set at between 25 and 40°C, and those with sodium set at 40–50°C. At lower ion levels, gels remelt upon heating; at higher ion levels, they do not melt below 100°C. Gellan gels are clear, firm to the touch and do not melt, but the release of liquid during mastication gives a melt-in-the-mouth sensation. This is important to flavor release and can be useful in, for example, dessert products (Owen, 1989). Other investigations have suggested that overall flavor increases as gel hardness decreases and gel brittleness increases (Clark, 1990). Therefore, very brittle gellan gels are important as good flavor releasers.

4.5.4 Comparison with other hydrocolloids

In a comparison of gellan with other gums (such as κ-carrageenan and agar) a large hysteresis in setting and melting was notable for agar. The properties of both gellan and carrageenan are ion dependent. In certain applications, it is also possible to produce gellan gels via methods that are appropriate for alginates. Alginate precipitation can be induced by the addition of acid, and this is also an effective alternative means of isolating gellan gum. If gellan gels are produced as a result of hydrogen ion addition, the resultant gels are extremely strong.

Blends with other hydrocolloids are sometimes necessary in the manufacture of different food products. The blending of hydrocolloids can be advantageous because of synergism. In the field of hydrocolloids, non-gelling agents (e.g. xanthan and guar gum), or gelling and non-gelling agents (e.g. carrageenan and LBG) are commonly combined to achieve increased viscosity or superior properties of the formed gels, such as higher elasticity. When viscosity-forming agents such as guar gum, LBG, xanthan gum, CMC or tamarind gum were added to gellan gum, and their amount was increased while the total gum concentration was held constant, a progressive reduction in hardness and modulus was noted, whereas brittleness remained essentially constant and elasticity increased slightly (Sanderson, 1990). The addition of thickeners to gelling agents also helps reduce syneresis, improve freeze–thaw stability and in some cases eliminate undesirable interactions between the components. When blends of low-acyl gellan gum and agar (0.50% and 0.25% total gum concentration, respectively, in 4 mM calcium ions) were tested, hardness and modulus decreased as the blend became richer in the agar component. The higher the gum concentration, the more pronounced the decrease. In this case, gel brittleness and elasticity remained almost constant. When blends of κ-carrageenan and low-acyl gellan gum (in 0.16 M potassium ions) were prepared, a rapid drop in hardness was observed (from 4.5 to 2) in going from a 0.5% low-acyl gellan gum to a 0.5% 80:20 blend containing 80% gellan gum (Sanderson, 1990). When low-acyl gellan gum and xanthan–LBG were combined, the gels became less brittle, less hard and stiff, and elasticity increased in parallel to an increase in the xanthan–LBG fraction. Similar textural changes are induced when LBG is replaced with other gums, such as cassia gum and konjak mannan. Even without the addition of other gums, various textures can be achieved by using different concentrations of low-acyl gellan gum by itself or in a blend with different proportions of the high-acyl form. Starches give a thick, pasty consistency to foods. The addition of gellan gum does not markedly influence starch viscosity. Upon cooling of the former, the addition of $>0.1\%$ gellan gum resulted in a firmer texture (Sanderson *et al.*, 1987a). This property is important when incorporating gellan into starch-based products such as puddings and pie fillings, where better flavor is of great importance. Blends of high-acyl gellan gum and starch have textural and functional properties comparable to those of starch alone (Clark and Burgum, 1989). The resultant food products prepared with this blend (e.g. fillings, toppings, mousses, soups and sauces), exhibit high shear stability and low rigidity (Clark and Burgum, 1989).

The addition of gellan to gelatin results in a range of textures that depend on the relative proportion of the gum in the mixture (Wolf *et al.*, 1989). The properties of the gellan–gelatin blend are dependent upon pH, temperature, ionic strength, time, total and relative hydrocolloid concentrations and gelatin type. If precipitation is induced under special conditions, coacervation occurs and designed microcapsules can be formed (Chilvers and Morris, 1987). The inclusion of gellan in gelatin is capable of elevating the latter's setting temperatures, a phenomenon that is important in the preparation of multi-layered desserts. The mixed gel is assumed to be built as a gelatin gel entrapped within a gellan gel structure, because two distinct setting temperatures are observed, coinciding with the setting of the two components. Manipulation of polymer concentration and ionic strength promotes phase inversion from a gellan continuous phase to a system where gelatin forms the supporting matrix. Knowledge of these interactions might assist in developing food products with novel textures (Papageorgiou *et al.*, 1994).

4.6 Food and other applications

Gellan has been tested for use in many applications, e.g. in microbiological media, tissue-culture media, foods and pet foods, deodorant gels, films and coatings, capsules, cereals and bakery products, photographic emulsions and microcapsules. It is also used as a model in studying gel swelling and shrinkage in gels (Sanderson, 1990; Chalupa *et al.*, 1994; Nussinovitch *et al.*, 1995). Gellan gum can be used as a fining agent for alcoholic beverages including beers, wines and fortified wines (Dartey, 1993). It can be added prior to fermentation or for fining mature beer or wine during the final packaging stage. Typically, a 0.1% gellan gum solution is prepared in a 0.05% sodium citrate solution and added to the beer or wine at 1–100 ppm. Gellan gum can be used to extend or completely replace isinglass in beer production (Dartey, 1993). The addition of up to 0.5% gellan gum to corn grits extrusion resulted in increase in water absorption as well as reduced extruder torque (Maga *et al.*, 1991).

Low-acyl gellan gum can replace carrageenan–LBG at around 33–50% of the level used today. Since such gels are more brittle, the addition of xanthan–LBG helps simulate the texture of the replaced gels. The addition of propylene glycol should be considered if freeze–thaw properties are important. In soft gelatin capsules and photographic emulsions, blends of gelatin and low-acyl gellan gum have been found to be a reasonable replacement for gelatin alone. The inclusion of gellan in the capsule changes its solubility in the gut, and in fact the relative insolubility could be advantageous for slow-release preparations. Low-acyl gellan can also serve as a component of microcapsules, which are normally based on gelatin and gum arabic and produced by coacervation (Sanderson, 1990).

The use of Gelrite as a substitute for agar has been studied. The advantages of such gels relate to their thermostability, enabling long incubations at high temperatures. In addition, such gels are stronger than agar gels at equivalent concentrations. Gellan culture medium is advantageous in that it reduces the time required for plate preparation, it produces a drier medium and, in the case of some mesophilic species, it reduces the required incubation time. The transparency of the gel is also an advantage. Limitations of such culture media include difficulty remelting some media and high setting temperatures, which could cause, for example, blood hemolysis on blood plates. The presence of sulfur or other impurities in the agar can affect plant-tissue-culture growth. Plant tissue culture based on gellan could, therefore, benefit from the latter's purity. The level of gellan needed for such applications is much lower than that of agar. Moreover, gellan exhibits good resistance to contamination by molds, easy washing from the plant tissue for transplantation and the ability to observe stages in culture development. The extensive use of gellan for plant tissue culture will depend on comprehensive research in the near future (Sanderson, 1990). Gellan was used to immobilize living bifidobacteria (Chapter 15) (Camelin *et al.*, 1993). Commercial gellan needs to be purified by cation- and anion-exchange resins to obtain beads by dropping gellan solution into solutions of divalent cations (Doner and Douds, 1995).

Pet food may include meat-like chunks that are produced by alginate–calcium reaction. Alginate can also be used to build a block containing comminuted meat. Alginate remains the vehicle for chunk production, whereas low grades of κ-carrageenan are used for binding in a gelled block. The carrageenan is gelled via potassium ions and LBG is added to control its elastic properties. Low-acyl gellan gum has been proposed as a component of pet food formulations. Low-acyl gellan can replace the carrageenan at about half of the latter's concentration (Sanderson, 1990). Gellan gum can be used alone or in combination with other hydrocolloids in low-fat (0–20%) ground meat, poultry or seafood products (Laaman and Tye, 1991), where the micelle structure of the gellan dispersion makes it a suitable fat substitute. Other applications include brine for cured meat and sea-food products, where gellan gum acts by binding water followed by gelation during cooking. Advantages of gellan gum include its ability to function at low concentrations (0.01–3% of the total system's weight) and its good pH and thermal stability (Laaman and Tye, 1991). Gellan gum can be used for breadings and batters, and with cheese, chicken, fish, potatoes and vegetables for fat reduction (Duxbury, 1993). Foods coated with gellan gum batter coatings contain low levels of fat and have the desirable qualities of fried foods, such as crispiness and juiciness (Chalupa and Sanderson, 1994).

Gellan gums are used mainly in confectioneries, jams and jellies, fabricated foods, water-based gels, pie fillings and puddings, pet foods, icings and frostings, and dairy products. Low-acyl gellan gum is of commercial value at very low levels of usage. A further reduction of ~33% in its concentration can be achieved by using clarified gellan gum rather than the non-clarified counterpart. Clarified Kelcogel, for example, is used at concentrations of about 50% to less than 33% of the agar or carrageenan it replaces to yield a similar texture. Gellan gum and polydextrose were used to produce low-calorie or sugar-free jelly sweets, where the final total solids content exceeds 80% (Gibson *et al.*, 1994). This is not believed to be possible with other confectionery systems, e.g. gelatin, pectin or agar. In comparison with other gums, gellan gum does not appreciably add to the viscosity of the depositing mix and is hence an ideal gelling agent for these low-calorie systems (Gibson *et al.*, 1994). Gellan gum is useful in cocoa and chocolate products (Anon., 1995) and in the development of alcoholic and non-alcoholic beverages (Giese, 1995).

In selected Japanese foods, such as Mitsumame jelly cubes, hard red-bean jelly, soft red-bean jelly and Tokoroten noodles, gellan can replace agar to yield similar or better textures at a concentration of ~33–50% of the agar concentration traditionally used. These products are firmer than their agar counterparts (Sanderson, 1990). Gellan gum can be used in bakery fillings (Anon., 1991). Its inclusion in fruit bakery fillings results in a smooth texture and better mouthfeel than traditional fillings made only with modified starch. Gellan gum's unique water-binding mechanisms enable clean, quick release of the fruit flavor. Its use in bakery fillings also results in improved storage and heat stability, and reduced boil-out and moisture-loss rates. The lower viscosity of a starch–gellan gum combination at elevated temperatures improves heat transfer and allows the product to cook faster: processing times are reduced, flavor abuse is minimized and gelation temperatures are more rapidly reduced. Gellan gum also maintains its gel strength over a wide pH range: 3.5–8.0 (Anon., 1991). The amount of gellan in different foods can be determined by precipitation with sodium chloride, followed by trapping on glass wool, washing with salt solution, elution with boiling acid or boiling base and determination by the carbazole test. Recoveries from milk, beverages, salad dressings and other foods were 75–94%, with a reproducibility of 2 to 4% (Graham, 1991). Other proposed analytical procedures for determining the level of gellan gum in foods are based on solubilization and precipitation (Baird and Smith, 1989). Gellan and other gelling agents have been reported to stabilize water-in-oil dispersions (Norton, 1992). Gellan gum can be used to prepare gels that can be dried and subsequently rehydrated to their original shape (Bell *et al.*, 1993). Potential uses for gellan pieces prepared in such a manner include dried rehydratable particulates for instant foods.

Other hydrocolloids can be added to gellan to achieve better performance in the sense that a broader range of rheological properties becomes available. In addition, the clarity of the gellan gels and their slow-release properties can be used to improve currently used products, in parallel to developing outstanding new ones.

References

Anon. (1991) Gellan gum: a stabiliser of many means. *Prepared Foods*, **160**(6), 125.
Anon. (1995) A user's guide to Kelcogel gellan gum. *Confect. Product.*, **61**(10), 750–51.
Baird, J.K. and Smith, W.W. (1989) An analytical procedure for gellan gum in food gels. *Food Hydrocolloids*, **3**(5), 407–11.
Bell, V.L., Giampetro, D.A., Ortega, D. *et al.* (1993) Use of gellan gum to prepare gels that can be dried and subsequently rehydrated to their original shape. *Res. Disclosure*, No. 345, 42.
Camelin, I., Lacroix, C., Paquin, C. *et al.* (1993) Effect of chelatants on gellan rheological properties and setting temperature for immobilization of living bifidobacteria. *Biotechnol. Progr.*, **9**(3), 291–7.
Campana, S., Ganter, J., Milas, M. *et al.* (1992) On the solution properties of bacterial polysaccharides of the gellan family. *Carbohydr. Res.*, **231**, 31–8.
Carroll, V., Miles, M.J. and Morris, V.J. (1982) Fibre-diffraction studies of the extracellular polysaccharide from *Pseudomonas elodea. Int. J. Biol. Macromol.*, **4**, 432.
Carroll, V., Chilvers, G.R., Franklin, D. *et al.* (1983) Rheology and microstructure of solutions of the microbial polysaccharide from *Pseudomonas elodea. Carbohydr. Res.*, **114**, 181.
Chalupa, W.F., Colegrove, G.T., Sanderson, G.R. *et al.* (1994) Simple films and coatings made with gellan gum. *Res. Disclosure*, No. 361, 244.
Chalupa, W.F. and Sanderson, G.R. (1994) *Process for Preparing Low-fat Fried Foods*. US Patent No. 6027 (930115).
Chandrasekaran, R., Millane, R.P., Arnott, S. *et al.* (1988a) The crystal structure of gellan gum. *Carbohydr. Res.*, **175**, 1–15.
Chandrasekaran, R., Puigjaner, L.C., Joyce, K.L. *et al.* (1988b) Cation interactions in gellan: an X-ray study of the potassium salt. *Carbohydr. Res.*, **181**, 23–40.
Chilvers, G.R. and Morris, V.J. (1987) Coacervation of gelatin–gellan gum mixtures and their use in micro-encapsulation. *Carbohydr. Polym.*, **7**, 111–20.
Clark, R.C. (1990) Flavour and texture factors in model gel systems, in *Food Technology International, Europe* (ed. A. Turner), Sterling Publications International, London, pp. 272–7.
Clark, R.C. and Burgum, D.R. (1989) *Blends of Acyl Gellan Gum with Starch*. US Patent No. 4,869,916.
Dartey, C.K. (1993) Applications of gellan gum as a fining agent in alcoholic beverages. *Res. Disclosure*, No. 348, 256.
Doner, L.W. and Douds, D.D. (1995) Purification of commercial gellan to monovalent cation salts results in acute modification of solution and gel-forming properties. *Carbohydr. Res.*, **273**(2), 225–33.
Duxbury, D.D. (1993) Fat reduction without adding fat replacers. *Food Processing, USA*, **54**(5), 68, 70.
Gao, Y.C., Lelievre, J. and Tang, J. (1993) AQ constitutive relationship for gels under large compressive deformation. *J. Text. Studies*, **24**(3), 239–51.
Gibson, W. (1992) Gellan gum, in *Thickening and Gelling Agents for Food* (ed. A. Imeson), Chapman & Hall, London, pp. 227–49.
Gibson, W., Rolley, N.A., Akintokumbo-Sofuyi, O.Y. *et al.* (1994) Production of low calorie (low joule) or sugar-free jelly sweets using polydextrose and gellan gum, where final total solids exceeds 80%. *Res. Disclosure*, No. 361, 276–7.
Giese, J. (1995) Developments in beverage additives. *Food Technol.*, **49**(9), 63–5, 68–70, 72.
Graham, H.D. (1991) Isolation of gellan gum from foods by use of monovalent cations. *J. Food Sci.*, **56**(5), 1342–6.
Grasdalen, H. and Smidsrod, O. (1987) Gelatin of gellan gum. *Carbohydr. Res.*, 371–93.
Gunning, A.P. and Morris, V.J. (1990) Light-scattering studies of tetramethyl ammonium gellan. *Int. J. Biol. Macromol.*, **12**, 338–41.
Harada, T., Kanzawa, Y., Kanenaga, K. *et al.* (1991) Electron microscopic studies on the ultrastructure of curdlan and other polysaccharides in gels used in foods. *Food Struct.*, **10**(1), 1–18.
Hershko, V. and Nussinovitch, A. (1995) An empirical model for the stress–strain relationships of hydrocolloid gels in tension mode. *J. Text. Studies*, **26**, 675–84.

Jansson, P.E., Lindberg, B. and Sandford, P.A. (1983) Structural studies of gellan gum, an extracellular polysaccharide elaborated by *Pseudomonas elodea*. *Carbohydr. Res.*, **124**, 135.
Juming, T., Lelievre, J., Tung, M.A. *et al.* (1994) Polymer and ion concentration effects of gellan gel strength and strain. *J. Food Sci.*, **59**(1), 216–20.
Juming, T., Tung, M.A. and Yanyin, Z. (1995) Mechanical properties of gellan gels in relation to divalent cations. *J. Food Sci.*, **60**(4), 748–52.
Kang, K.S., Colegrov, G.T. and Veeder, G.T. (1980) *Heteropolysaccharide Produced by Bacteria and Derived Products.* European Patent No. 0 012 552.
Kang, K.S., Colegrov, G.T. and Veeder, G.T. (1982a) *Polysaccharide S-60 and Bacterial Fermentation Process for its Preparation.* US Patent No. 4,326,053.
Kang, K.S., Colegrov, G.T. and Veeder, G.T. (1982b) *Deacetylated Polysaccharide S-60.* US Patent No. 4,326,052.
Kang, K.S., Veeder, G.T., Mirrasoul, P.J. *et al.* (1982c) Agar-like polysaccharide produced by a *Pseudomonas* species: production and basic properties. *Appl. Environ. Microbiol.*, **4**(5), 1086–91.
Kang, K.S. and Veeder, G.T. (1983) *Fermentation Process for Preparation of Polysaccharide S-60.* US Patent 4,377,636.
Kuo, M.S., Dell, A. and Mort, A.J. (1986) Identification and location of L-glycerate, an unusual acyl substitution in gellan gum. *Carbohydr. Res.*, **156**, 173–87.
Laaman, T.R. and Tye, R.J. (1991) Application of gellan gum to meat systems. *Res. Disclosure*, No. 323, 212.
Lelievre, J., Mirza, I.A. and Tung, M.A. (1992) Failure testing of gellan gels. *J. Food Eng.*, **16**(1/2), 25–37.
Maga, J.A., Kim, C.H. and Wolf, C.L. (1991) The effect of gellan gum addition on corn grits extrusion. *Food Hydrocolloids*, **5**(5), 435–41.
Miyoshi, E., Takaya, T. and Nishinari, K. (1994) Gel–sol transition in gellan gum solutions. I. Rheological studies on the effects of salts. *Food Hydrocolloids*, **8**(6), 505–27.
Moritaka, H., Nishinari, K., Nakahama, N. *et al.* (1994) Effects of sucrose, glucose, urea and guanidine hydrochloride on the rheological properties of gellan gum gels. *J. Jap. Soc. Food Technol.*, **41**(1), 9–16.
Moritaka, H., Nishinari, K., Taki, M. *et al.* (1995) Effects of pH, potassium chloride and sodium chloride on the thermal and rheological properties of gellan gum gels. *J. Agric. Food Chem.*, **43**(6), 1685–9.
Nakamura, K., Harada, K. and Tanaka, Y. (1993) Viscoelastic properties of aqueous gellan solutions: the effects of concentration on gelation. *Food Hydrocolloids*, **7**(5), 435–47.
Norton, I.T. (1992) *Water in Oil Dispersion.* European Patent Application 0 473 854 A1.
Nussinovitch, A., Ak, M.M., Normand, M.D. *et al.* (1990a) Characterization of gellan gels by uniaxial compression, stress relaxation and creep. *J. Text. Studies*, **21**(1), 37–49.
Nussinovitch, A., Kaletunc, G., Normand, M. D. *et al.* (1990b) Recoverable work versus asymptotic relaxation modulus in agar, carrageenan and gellan gels. *J. Text. Studies*, **21**(4), 427–38.
Nussinovitch, A., Peleg, N. and Mey-Tal, E. (1995) Continuous monitoring of changes in shrinking gels. *Lebensm. Wissen. Tech.*, **28**(3), 347–9.
Nussinovitch, A., Peleg, M. and Normand, M.D. (1989) A modified Maxwell and a non-exponential model for characterization of the stress-relaxation of agar and alginate gels. *J. Food Sci.*, **54**, 1013–16.
Ogawa, E. (1993) Osmotic pressure measurements for gellan gum aqueous solutions. *Food Hydrocolloids*, **7**(5), 397–405.
O'Neill, M.A., Selvendran, R.R. and Morris, V.J. (1983) Structure of the acidic extracellular gelling polysaccharide produced by *Pseudomonas elodea*. *Carbohydr. Res.*, **124**, 123.
Owen, G. (1989) Gellan gum-quick setting gelling systems, in *Gums and Stabilisers for the Food Industry*, vol. 5 (eds G.O. Phillips, D.J. Wedlock and P.A. Williams), IRL Press, Oxford, pp. 345–9.
Papageorgiou, M., Kasapis, S. and Richardson, R.K. (1994) Steric exclusion phenomena in gellan/gelatin systems. I. Physical properties of single and binary gels. *Food Hydrocolloids*, **8**(2), 97–112.
Peleg, M. (1980) Linearization of relaxation and creep curves of solid biological materials. *J. Rheol.*, **24**, 451–63.

Rinaudo, M. (1988) Gelation of ionic polysaccharides, in *Gums and Stabilisers for the Food Industry*, vol. 4 (eds G.O. Phillips, D.J. Wedlock and P.A. Williams), IRL Press, Oxford, p. 119.
Robinson, G., Manning, C.E., Morris, E.R. *et al.* (1987) Sidechain and mainchain interactions in bacterial polysaccharides, in *Gums and Stabilisers for the Food Industry*, vol. 4 (eds G.O. Phillips, D.J. Wedlock and P.A. Williams), IRL Press, Oxford, pp. 173–81.
Sanderson, G.R. (1989) The functional properties and applications of microbial polysaccharides – a supplier's view, in *Gums and Stabilisers for the Food Industry*, vol 5 (eds G.O. Phillips, D.J. Wedlock and P.A. Williams), IRL Press, Oxford, pp. 333–44.
Sanderson, G.R. (1990) Gellan gum, in *Food Gels* (ed. P. Harris), Elsevier Applied Science, London, pp. 210–32.
Sanderson, G.R., Bell, V.L., Burgum, D.R. *et al.* (1987a) Gellan gum in combinations with other hydrocolloids, in *Gums and Stabilisers for the Food Industry*, vol. 4 (eds. G.O. Phillips, D.J. Wedlock and P.A. Williams), IRL Press, Oxford, pp. 301–8.
Sanderson, G.R., Bell, V.L., Clark, R.C. *et al.* (1987b) The texture of gellan gum gels, in *Gums and Stabilisers for the Food Industry*, vol. 4 (eds G.O. Phillips, D.J. Wedlock and P.A. Williams), IRL Press, Oxford, pp. 219–29.
Stokke, B.T., Elgsaeter, A. and Kitamura, S. (1993) Macrocyclization of polysaccharides visualized by electron microscopy. *Int. J. Biol. Macromol.*, **15**(1), 63–8.
Sutherland, I.W. (1992) The role of acylation in exopolysaccharides including those for food use. *Food Biotechnol.*, **6**(1), 75–86.
Tanaka, Y., Sakurai, M. and Nakamura, K. (1993) Ultransonic velocities in aqueous gellan solutions. *Food Hydrocolloids*, **7**(5), 407–15.
Upstill, C., Atkins, E.D.T. and Atwool, P.T. (1986) Helical conformations of gellan gum. *Int. J. Biol. Macromol.*, **8**, 275.
Watase, M. and Nishinari, K. (1993) Effect of potassium ions on the rheological and thermal properties of gellan gum gels. *Food Hydrocolloids*, **7**(5), 449–56.
Willoughby, L. and Kasapis, S. (1994) The influence of sucrose upon gelation of gellan gum in large deformation compression analysis. *Food Sci. Technol. Today*, **8**(4), 227–33.
Wolf, C.L., LaVelle, W.M. and Clark, R.C. (1989) *Gellan Gum/gelatin Blends.* US Patent No. 4,876,105.
Yuguchi, Y., Mimura, M., Kitamura, S. *et al.* (1993) Structural characteristics of gellan in aqueous solution. *Food Hydrocolloids*, **7**(5), 373–85.

5 Pectins

5.1 Introduction

The natural polymers (structural materials) found in all land plants are termed pectins (Braconnot, 1825). Like starch and cellulose, pectin is a structural carbohydrate (Christensen, 1986). It was discovered in the 18th century (Vauquelin, 1790), and Braconnot was the first to characterize it as the active fruit component responsible for gel formation. He also suggested the word 'pectin', which originates from a Greek word meaning 'to congeal or solidify'.

Commercially, pectin is extracted from citrus peel or apple pomace. The commercial isolation of pectins from suitable plant material began early in the 20th century and has been developing ever since. Pectic substances are integral structural components of the cell and play an important role as cementing material in the middle lamellae (Fig. 5.1) of primary cell walls (Christensen, 1986). The many reviews and comprehensive texts on pectin are valuable sources for the reader (Kertesz, 1951; Doseburg, 1965; Pilnik and Zwiker, 1970; Christensen and Towles, 1973; Pedersen, 1980; May, 1992; Sakai *et al.*, 1993). The release of pectin involves acidic extraction and isolation by precipitation, followed by drying to obtain a powder with standard properties. Ultrasound has been suggested to intensify pectin de-esterification (Panchev *et al.*, 1994). Pectins are normally dried to less than 10% water content. The product is kept in a vapor-tight package under cool, dry conditions. Commercial pectins usually have particle sizes of $\sim$0.25 mm and a low density, $\sim$0.7 g cm^{-3}. Commercial pectins include mainly polymerized galacturonic acid that has been partly esterified with methanol (Rolin and De Vries, 1990). The percentage of the partially esterified portion of polymerized galacturonic acid strongly influences the functional properties of the pectin, and pectins with both low and high ester contents are sold. At low pHs pectins with high ester contents with the addition of enough sugar create fruit-system gels (Rolin and De Vries, 1990).

Pectin is used as a gelling agent in traditionally manufactured fruit-based products, especially jams and jellies. The heat stability of pectin under acidic conditions makes it an ideal candidate for the conditions occurring when texturization or stabilization are required in acidic food systems. Home-made jam-making is based on the fruit pulp's ability to form gels when boiled with sugar: the natural pectin content in the pulp is responsible for the gelation. Commercial jam processing adds already produced pulp,

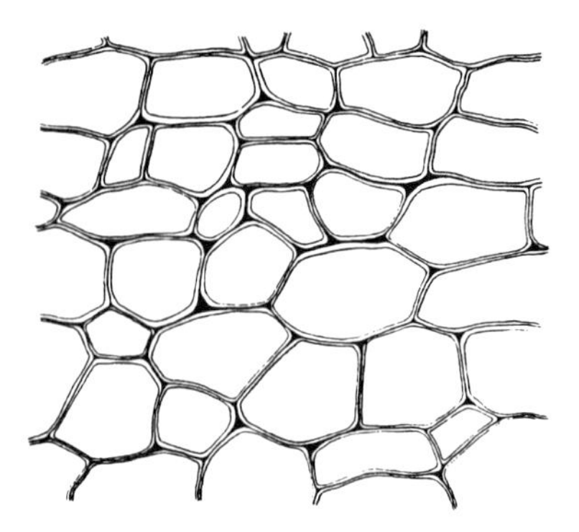

Figure 5.1 The middle lamella seen as a solid black mass between the cells of an unripe apple (× 1350), adapted from *The Pectic Substances* by Z.I. Kertesz (1951), Interscience Publishers, New York.

yielding more uniform preparations. High-ester pectins can form gels at low pH when sufficient amounts of sugar are added (thereby reducing water activity in the system). Low-ester pectins create gels in the presence of calcium ions via a different mechanism. The increasing amounts of pectins produced are now utilized outside their traditional industry as part of the confectionery industry, as stabilizers in the milk industry, and for pharmaceutical purposes.

5.2 Nomenclature

Pectin and pectic substances are heteropolysaccharides consisting mainly of galacturonic acid and galacturonic acid methyl ester residues (Christensen, 1986). To obtain uniform definitions in this area, the American Chemical Society adopted a revised nomenclature for pectic substances (Baker *et al.*, 1944) as follows.

Pectic substances are those complex colloidal carbohydrate derivatives that occur in or are prepared from plants and contain a large proportion of anhydrogalacturonic acid units, which are thought to exist in a chain-like combination. The carboxyl groups of polygalacturonic acid may be partly esterified by methyl groups and partly or completely neutralized by one or more bases (Christensen, 1986).

Protopectin is the water-insoluble parent pectin substance that occurs in plants and which, with restricted hydrolysis, yields pectin or pectinic acid.

Pectinic acids are the colloidal polygalacturonic acids containing more than a negligible proportion of methyl ester groups. Pectinic acids, under suitable conditions, are capable of forming gels in water with sugars and acid, or if suitably low in methoxyl content, with certain ions. The salts of pectinic acids are either normal or acid pectinates.

Pectin (or pectins) are those water-soluble pectinic acids of varying methyl ester content and degree of neutralization that are capable of forming gels with sugar and acid under suitable conditions.

Pectic acid is a term applied to pectic substances composed mostly of colloidal polygalacturonic acids and essentially free of methyl ester groups.

Protopectinase is the enzyme that converts protopectin into a soluble product. It is also called pectosinase or propectinase (Christensen, 1986).

Pectinesterase (PE) or pectinmethylesterase is the enzyme that catalyses the hydrolysis of the ester bonds of pectic substances to yield methanol and pectic acid. The name *pectase* does not indicate the nature of the enzyme action and has, therefore, been replaced by these more specific names.

Polygalacturonase (PG) or pectin polygalacturonase is the enzyme that catalyses the hydrolysis of glycosidic bonds between de-esterified galacturonide residues in pectic substances.

Pectinase is frequently used to designate the glycosidase as well as pectic-enzyme mixtures (Baker *et al.*, 1944; Christensen, 1986).

A modern definition of pectin takes into consideration the low methyl ester content and the amidated pectinic acids as follows: pectin is a complex, high-molecular-weight polysaccharide consisting mainly of the partial methyl esters of polygalacturonic acid and their sodium, potassium and ammonium salts. In some types (amidated pectins), galacturonamide units further occur in the polysaccharide chain. The product is obtained by aqueous extraction of appropriate edible plant material, usually citrus fruit and applies (Christensen, 1986).

5.3 Structure

Commercial pectins are composed mainly of polymerized, partly methanol-esterified (1–4)-linked α-D-galacturonic acid. The pectin molecule can contain 200–1000 linked galacturonic acid units. In some pectins, the methyl ester groups are partially replaced by amide groups, to a maximum of 80% (Fig. 5.2). During extraction, only part of the pectin molecules can be extracted by non-degradative means, whereas dilute acids are generally used

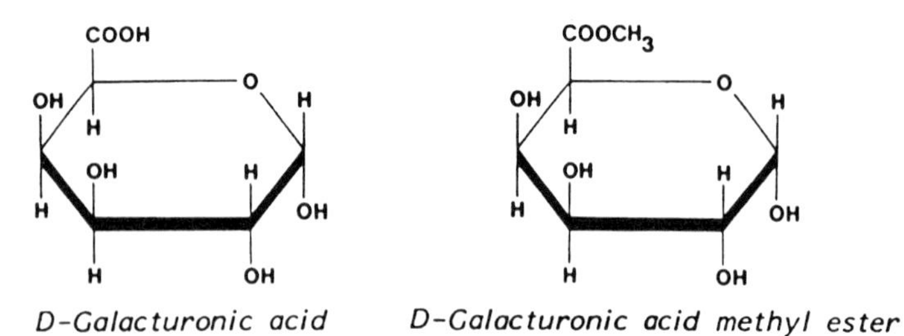

Figure 5.2 Principal units in the pectin molecule.

(Rolin and De Vries, 1990). Therefore, the structure of the resultant pectin differs greatly. About 5–10% of the galacturonic acids are neutral sugars such as galactose, glucose, rhamnose, arabinose and xylose. They can be bound to the galacturonate main-chain, be inserted into the main-chain (rhamnose) or be a part of contaminating polysaccharides (glucans and xyloglucans). Pectins from apple, citrus, cherry, strawberry, carrot, pumpkin, sugar beet, potato, onion and cabbage have the same neutral sugar composition (Amado and Neukom, 1984; Guillon *et al.*, 1986; Rolin and De Vries, 1990), in contrast to pectins from mountain pine pollen, Japanese kidney beans and duckweed, which contain large amounts of xylose or apiose (Mascaro and Kindell, 1977; Matsuura, 1984). Bacterial enzymes can be used to extract pectin from pumpkin and sugar beet (Matora *et al.*, 1995). Information on the characterization of pectic substances from selected tropical fruits such as orange, lime, banana, mango, avocado, pawpaw, cashew apple, star apple, tomato and guava, in terms of their gelation properties, can be found elsewhere (Nwanekezi *et al.*, 1994).

X-ray diffraction studies performed on dried fibers to study the structure of pectin indicated that the galacturonan backbone forms a right-handed helix, with three galacturonic acid units in C_1 conformation as the repeating sequence, corresponding to a repeat distance of 1.34 nm (Palmer and Hartzog, 1945; Walkinshaw and Arnott, 1981a). Morris *et al.* (1982) suggested that gel formation with calcium involves polygalacturonic acid sequences with a 2_1, ribbon-like symmetry. Upon drying the gel, however, the 3_1 helical symmetry is restored via polymorphic phase transition. Commercial pectins contain lower amounts of neutral sugars relative to pectin extracted under mild conditions. A large proportion of these sugars is 1,2-bound rhamnose present in the galacturonan backbone. However, rhamnose's distribution along the pectin chains has not yet been fully elucidated (Christensen, 1986). The length of polygalacturonate sequences between rhamnose interruptions has been suggested to be fairly constant, corresponding to ~25 residues (Powell *et al.*, 1982). Analysis of similar sequences with 20 to 30 degrees of polymerization (Neukom *et al.*, 1980) showed both oligomers to be almost fully made up of galacturonic acid units with only traces of rhamnose, and galacturonan segments containing rhamnose from apple tissue. Therefore, it was concluded that cell walls can contain both a pure galacturonan-type pectin and a rhamnogalacturonan-type pectin in different regions. A molecular model of their blockwise occurrence in a few hairy regions was suggested, based on results obtained by the specific enzymatic degradation of apple pectins (De Vries *et al.*, 1982).

5.4 Sources and properties

Pectins can differ as a result of ripening and these differences can influence the efficiency of the extraction process (De Vries *et al.*, 1984; Huber, 1984; Boothby, 1983). Pectin extracted from the primary cell wall may have more branches of neutral sugars than that extracted from the middle lamella (Redgwell and Selvendran, 1986). Side-chains (neutral sugar side-chains) are distributed unevenly along the main-chain. Therefore, models describing smooth and hairy regions within pectin that has been extracted by a mild process can be deduced for pectic substances from citrus, sugar beet, cherry and carrot (Rolin and De Vries, 1990).

A description of pectic fractions from different sources can be found elsewhere (Rolin and De Vries, 1990). Substituents such as acetyl groups (in potato and sugar beet pectins) can prevent gelation. In apple and citrus, only a very low degree of acetylation is measured, and the acetyl groups may be located in the hairy regions (Vorgen *et al.*, 1986). Active pectin oligomers have also been detected in ripening tomato fruits (Melotto *et al.*, 1994). Moreover, associations of pectin with boron in the cell walls of squash and tobacco have been reported (Hu-Hi and Brown, 1994). Since pectin can come from different agricultural sources, it is not surprising that different pectins have different substituents located in different positions. Recently, pectin has been extracted from Galgal (*Citrus Pseudolimon Tan*) (Attri and Maini, 1996). The process was standardized for maximum recovery of pectin from these peels using various extractants and varying the extractant, peel ratio, extraction time, number of extractions and peel particle size (Attri and Maini, 1996). Pectin extraction from citrus peel using PG produced on whey has been reported (Donaghy and Mckay, 1994). Dried sweet whey was used as a complete medium for the production of the enzyme by the yeast *Kluveromyces fragilis*. The concentrated enzyme was then used to release pectin from the peels and apple pomace but was unable to release pectin from sugar beet pulp. Conditions for pectin extraction from orange peel were optimized with regards to enzyme concentration, water:peel ratio, temperature and duration of treatment (Donaghy and Mckay, 1994).

The most abundant substituent is the methanol ester of galacturonate residues. If apple or citrus pectins are not subjected to de-esterification, their

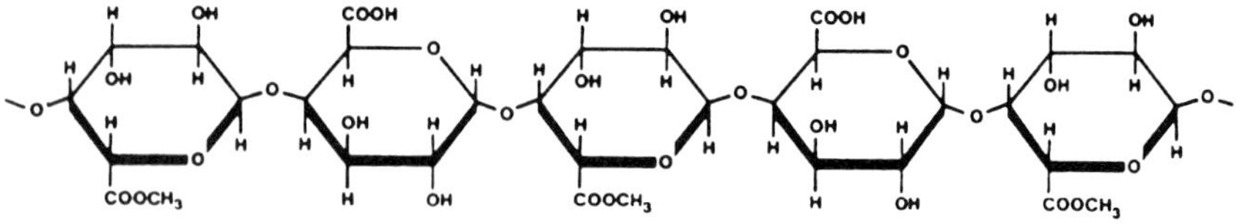

Figure 5.3 Section of a high ester content pectin molecule, with a DE ~60%.

degree of esterification (DE) is high (~70%) compared with the low DE values for pectins extracted from sunflower heads, potato, tobacco and pear (Vorgen *et al.*, 1986; Turmucin *et al.*, 1983; Pathak and Shukla, 1981). The DE is defined as the ratio of esterified galacturonic acid units to the total number of galacturonic acid units in the molecule (Fig. 5.3). These values can also be influenced by the degree of ripening of the raw material and changes in the extraction procedure. Ester group distribution depends on the source. There is evidence of random intramolecular distribution in mildly extracted apple pectins, contradicting a work which reported some regularity (De Vries *et al.*, 1983 and De Vries *et al.*, 1986, respectively). Non-random distribution has been reported in commercial pectins (Anger and Dongowsky, 1984, 1985; De Vries *et al.*, 1986). Information on de-esterification by fungal enzymes, pectin structure, conformation in solution and gels, and other properties can be found elsewhere (Kohn *et al.*, 1983, 1985; Markovic and Kohn, 1984).

The main sources of commercial pectins are citrus (lemon, lime, orange and grapefruit) peel and apple pomace. Peels are supplied for pectin production after the juice has been squeezed and the essential oils extracted. After conveying the peels to the extraction site, a water wash is used to remove as much water-soluble material as possible, other than pectin, and then extraction is begun or the peel is dried for future processing. It is not surprising to find pectin plants near plants that can supply them with the raw material directly, such as those producing apple or citrus juice, or cider. Apple pomace, which once served as the major raw material, has been replaced to a large extent by citrus peel because the latter contains 15–20% more pectin on a dry weight basis. During the Second World War, sugar beet waste (from sugar production) served as a source of pectin production. Since this pectin contains acetyl ester, other better sources are preferred. As mentioned earlier other raw materials exist (Pathak and Shukla, 1978).

5.5 Pectin manufacture

Pectin manufacturing processes are generally known (Fig. 5.4). However, variations in, or fine-tuning of, the processes, i.e. the specific conditions used, are kept confidential by the manufacturers who consider them trade secrets.

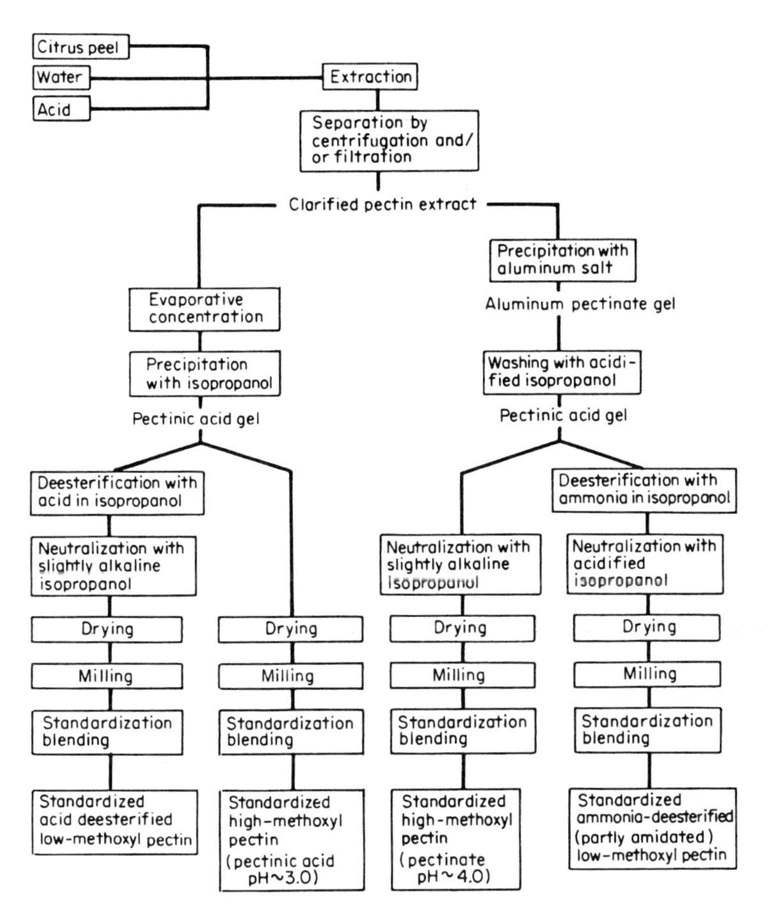

Figure 5.4 Various process routes for the manufacture of pectins. (Adapted from *Handbook of Water-Soluble Gums and Resins*, McGraw-Hill, NY, 1980, R.L. Davidson, ed.)

In general, the fresh or dried raw material (apple pomace, citrus peel and a number of other surplus materials such as sunflower bottoms and sugar beet waste) (Karpovich *et al*., 1981) is extracted in demineralized water that has been acidified with mineral acid to give a pH of 1.5 to 3.0 (hydrochloric or nitric acid are most often used) at 70°C and for ~3 h. In the case of citrus peel, pretreatment of the peel by blanching and washing to eliminate PE activity and to remove glucosides, sugars and citric acid is common. Dried peel is stable under storage conditions, making its transport over great distances feasible. Dried citrus peel contains ~20–30% pectin. Dried apple pomace yields 10–15% pectin.

Since a certain degree of pectin de-esterification takes place during the extraction, conditions should be chosen to fit the desired product (Rolin and De Vries, 1990). Temperature, pH and time need to be carefully controlled.

Rapid-set high-ester pectins are generally extracted at temperatures close to boiling. At these high temperatures, hydrolysis of the parent pectic substances accelerates, viscosity is lowered and diffusion facilitated. This process may take less than 1 h with only minor de-esterification, whereas lower extraction temperatures and longer extraction times favor de-esterification to yield slow-set high-ester pectins or even low-ester pectins. After separating out the extract, the peels can be used as cattle feed, while the extracted liquid (viscous liquid containing 0.3–1.5% dissolved pectin) is clarified by filtration and centrifugation. At this stage, the clear pectin extract can be further de-esterified by maintaining controlled pH and temperature. The extract can be concentrated and, after preservation with sulfur dioxide, sold as 'liquid pectin'.

Pectin can be isolated by alcohol precipitation or by precipitation as an insoluble salt. With the former procedure, the pectin (unlike the water-soluble materials in the extract) precipitates out, and the alcohol is recovered by distillation. Alcohol precipitation is effected by mixing the extract with methanol, ethanol or 2-propanol. In some processes, the extract is concentrated by evaporation prior to precipitation to minimize distillation costs (Rolin and De Vries, 1990). Other alternative procedures involve pectin separation by aluminum or copper ions as an insoluble salt (Kausar and Nomura, 1982; Michel *et al.*, 1981). These metal ions can later be removed by acidified alcohol washes then a wash in alkaline alcohol to neutralize the product. The resultant alcohol-wetted pectin is pressed, dried and milled, or is de-esterified in the alcohol suspension (Rolin and De Vries, 1990).

De-esterification can be achieved with an acid or base. If ammonia is used, then some of the methyl ester groups are replaced by amide groups and the product is termed 'amidated pectin'. For pectin manufacture, as for other hydrocolloids, a blending and standardization stage is important. With the inclusion of this stage, the marketed blends exhibit very similar performance with respect to firmness of the resultant gel and the time necessary to gel high-ester pectins under predetermined constant conditions. In a similar manner, low-ester pectins are standardized in terms of their calcium reactivity.

Pectin manufacturers are located in many places worldwide. A few examples are Hercules (factories in Denmark, Germany and Florida, USA), Unipectine (France), Pektin-Fabrik (Germany), General Foods Corp. (USA) and Pectina de Mexico. Smaller pectin manufacturers are found in Switzerland, Brazil, Israel, Argentina and a few other European countries.

5.6 Commercial availability, specifications and regulatory status

Descriptions and terminology of commercial pectins exist (Doesburg, 1965). Commercial pectin is defined as the partial methyl esters of polygalacturonic acids and their sodium, potassium, calcium and ammonium salts. The pectin is extracted from edible plant organs and no organic precipitants other than methanol, ethanol and isopropanol are used. Amidated pectins can be produced by ammonia treatment. Standardization can be achieved by dilution with sugars. Buffer salts are permitted to yield desirable setting conditions.

Commercial pectins are divided into high- and low-ester pectins in accordance with their DE values: a value over 50% is considered a high-ester pectin, values from 50% to negligible amounts define a low-ester pectin (Rolin and De Vries, 1990).

Pectate is a polymerized galacturonic acid with no or only negligible esterification. The degree of amidation (DA) is the percentage of galacturonic acid subunits that are amidated. High-ester pectins used for gel-making can be divided into rapid-set, medium-set and slow-set pectins, depending on the time necessary for solidification. The higher the DE (in high-ester pectins), the shorter the setting time. High-ester pectins are regularly standardized to 150 grade of USA-SAG, meaning that 1 part pectin can solidify 150 parts of sucrose into a jelly with the standard properties of 65° Brix (soluble solids), pH 2.2–2.4 and 23.5% SAG (indication of gel strength, see section 5.8.1).

In addition to the definitions of commercial pectins, purity is also defined by several requirements: galacturonic acid content >65%, DA <25%, loss on drying not more than 12%, acid-insoluble ash not more than 1%, alcohol residues of all kinds not more than 1%, nitrogen not more than 2.5% and sulfur dioxide not more than $50\,mg\,kg^{-1}$ (Food Chemicals Codex, 1981; Anon., 1978; Anon., 1981a,b). Since pectin is an important constituent of land plants, it is consumed in significant quantities. Pectin passes unchanged (no enzymatic degradation) to the large intestine, where bacteria use it as a carbon source. However, its hydrolysation in the intestinal tract produces next to no calories (Cambell and Palmer, 1978). From a toxicological point of view, there are no limitations on its use (Anon., 1981a,b). Pectins are GRAS for use in human foods. The FDA has not issued any specific limitations or guidelines for their use in any food (Anon., 1981c). The potential dietary benefits of citrus pectin and fiber have been reviewed by Baker (1994). Other health aspects of pectin are important and have been studied by many researchers. Examples include the role of pectin in cholesterol regulation (Cerda, 1994), citrus pectin and cholesterol interactions in the regulation of hepatic cholesterol homeostasis and lipoprotein metabolism in the guinea pig (Fernandez *et al.*, 1994), the use of pectin as a fat replacer (Hoefler, 1994), oral administration of modified citrus pectin as an inhibitor of spontaneous prostate cancer metastasis in rats by inhibiting carbohydrate-mediated cell–cell interactions (Pienta and Raz, 1994), the dose response of colonic carcinogenesis to pectin and guar gum (Klurfeld *et al.*, 1994), and the effects of structural parameters of pectin on its interaction with drugs *in vitro* (Fritzsch *et al.*, 1994). Recently, a pectin-supplemented enteral diet was reported to reduce the severity of methotrexate-induced enterocolitis in rats (Mao *et al.*, 1996). Information on the preparation and physicochemical properties of polymer complexes of benzimidazolyl-2-methylcarbamate and apple pectin can be found (Khalikov *et al.*, 1995), and a new pectin-based material for selective low density lipoprotein-cholesterol removal has been reported (Lewinska *et al.*, 1994). The physiological effect of low-molecular-weight pectin is discussed by Yamaguchi *et al.* (1994). Such pectins that can retain their activities are important since high viscosity reduces their usability. This preparation exhibited high solubility and a repressive effect on lipid accumulation in the liver (Yamaguchi *et al.*, 1994). Pectin formulations have also been used for colonic drug delivery (Ashford *et al.*, 1994).

High-methoxy pectin loses about 5% of its USA-SAG grading when stored at 20°C in a dry atmosphere, whereas low-ester pectin is more stable and under favorable conditions loss is undetectable (Food Chemicals Codex, 1972; Anon., 1978; Anon., 1981c). The microbiological purity of pectins is specified in many cases by the manufacturer, since it is used mainly in acidic media and, therefore, yeast and mold counts are relevant. Typical specifications may include a total plate count at 37°C of less than 500 cells g^{-1}; a yeast and mold count at 25°C of less than 10 cells g^{-1} and *Escherichia coli*, salmonella and staphylococcus test results being negative.

5.7 Solution properties

Good pectin solubility can be achieved by following recommended dissolution procedures. In general, pectin is not soluble under conditions in which it forms a gel. The powder needs to be dispersed in warm water (not less than 60°C), at reduced mixing rates and then at full speed. Ignoring manufacturer recommendations could result in the formation of lumps that are difficult to dissolve. Good dissolution is achieved by mixing pectin with five times its own weight of sugar. Other blending media, such as a 65% sugar solution or alcohol to wet the pectin for small-scale laboratory use, are recommended. If a high-shear mixer is not used, then boiling for 1 min is necessary to guarantee full dissolution (Rolin and De Vries, 1990).

5.7.1 Viscosity

The viscosity of pectin solutions is dependent on their concentration, presence of calcium or similar non-alkali metals, pH, the chemical properties of the pectin, the DE and the average molecular weight. Dilute pectin solutions (up to approximately 0.5%) are Newtonian and only slightly affected by calcium ions. Increased pH results in increased viscosity. Salts of monovalent cations reduce pectin solution viscosity, because of reductions at high ionic strength. The higher the average molecular weight, the higher the solution viscosity. The molecular weight of pectin can be estimated by using intrinsic viscosity methods. Pseudoplastic solutions can be achieved with concentrations higher than 1%. In contrast to dilute solutions and in the absence of calcium, such solutions increase in viscosity if the pH is reduced within the typical application range of 2.5–5.5. Pectins in the presence of calcium form thixotropic solutions, the viscosity of which increases with increasing pH within the aforementioned range. In fact, different textures can easily be achieved by combining pectin types and concentrations, ion concentrations and pH (Michel *et al.*, 1982; Christensen, 1954; Berth *et al.*, 1982). Solution properties of pectins are changed by hydrolysis of side-chains. Hydrolysis did not affect the specific viscosity of dilute (0.5%) pectin solutions; however, viscosity significantly decreased in concentrated 2.0–6.0% pectin solutions. Results suggest that pectin side-chains exist in an entangled state in concentrated solutions. In these latter solutions, the extent of viscosity reduction was dependent on pectin concentration (Hwang and Kokini, 1995). Based on viscometry measurements, the average molecular weight of commercial pectin normally falls between 50×10^3 and 150×10^3. It is important to note that by using other techniques such as light-scattering, other results ($\sim 1 \times 10^6$ or higher) have been found owing to intermolecular associations and aggregation of pectin molecules.

5.7.2 Chemistry and properties

Pectin is in fact a polyacid. The negative charge on dissolved pectin is smaller at low pH than at high pH. This charge attracts protons and, therefore, dissociations of individual acid groups are not independent (Rinaudo, 1974). Via a mechanism known as 'membraneless osmosis', pectin can concentrate solutions of proteins such as milk proteins. Since pectin and the protein cannot exist in the same solution, two phases develop, one rich in pectin and the other in protein. Pectin has a higher affinity for water and the protein phase is thus concentrated by a factor of 5–12. The addition of metal ions to pectin solutions causes an increase in viscosity, or gel formation or pectin precipitation. Reactions of polyanions (pectin) with polycations (other macromolecules) form insoluble products. Dissolved pectin exhibits good stability at pH 4. Far from this optimum, depolymerization occurs at low pH, whereas at high pH (any pH > 5) degradation occurs owing to β-elimination. High-ester pectins are more vulnerable to such degradation than their low-ester counterparts. In the juice industry, pectin-degrading enzymes are often used to obtain a clarified product.

A study of synergistic interactions in dilute polysaccharide solutions was recently performed (Goycoolea *et al.*, 1995). A simple viscometric approach was used in cases in which exclusion effects should be negligible. There were no viscosity changes for alginate and pectin with sufficient calcium ions to induce almost complete conversion to the dimeric 'egg box' form, demonstrating that conformational rigidity is not, in and of itself, sufficient for other polysaccharides to form heterotypic junctions with mannan or glucomannan chains.

Atomic force microscopy (AFM) was used for imaging polysaccharides such as pectin, ι-carrageenan, xanthan and acetan (Kirby *et al.*, 1996). The polysaccharides were deposited from an aqueous solution onto the surface of freshly cleaved mica, air-dried and then imaged under alcohols. Improved resolution was obtained relative to the more traditional metal-coated samples or replicas (Kirby *et al.*, 1996).

5.8 Pectin gels

High-ester pectin gels can be successfully prepared following good dissolution. Jam preparation procedures can be found elsewhere. Briefly, they include heating the sugar and fruit fraction in amounts that will yield 65% soluble solids in the final batch. The pectin is added in solution form and stirring and boiling are carried out under vacuum to achieve the desired soluble-solids content. The vacuum needs to be broken before heating to pasteurization. Then citric acid is added to reduce the pH to 3.0–3.1. The mixture is cooled to filling temperatures and gelation occurs in the container itself. For high-ester pectin gelation, low pH, a high soluble-solids concentration and appropriate temperatures are needed to fulfill the desired requests. A high-ester pectin gel cannot be melted after solidification. Pregelation phenomena (stirring while gelation is in progress) result in lower-strength gels, or the absence of gelation with continued interference. For gel formation, a three-dimensional network is necessary to hold water, sugar and other solutes (Fig. 5.5). The junction zones in the high-ester pectin gel network have been described by a model suggested by Walkinshaw and Arnott (1981b). According to this model, three to ten polymer-chain segments with a helical structure form aggregates of parallel chains that are limited in size because of steric barriers, entropic factors and possibly rhamnose insertions (Christensen, 1986). Local crystallization is sustained by intermolecular hydrogen bonds and is probably reinforced by hydrogen bonding with water molecules in one set of triangular channels, and hydrophobic attractions between methyl groups forming columns in a second set of triangular channels. In molecular gel networks, at least two types of bonding are involved. One is strong and responsible for the elastic properties of the gel and the other is weaker and capable of reforming after disruption. Sugars play an active role in the formation of the pectin gel network by associating with pectin molecules via hydrogen bonding to form secondary links that reinforce the molecular network structure. The aging

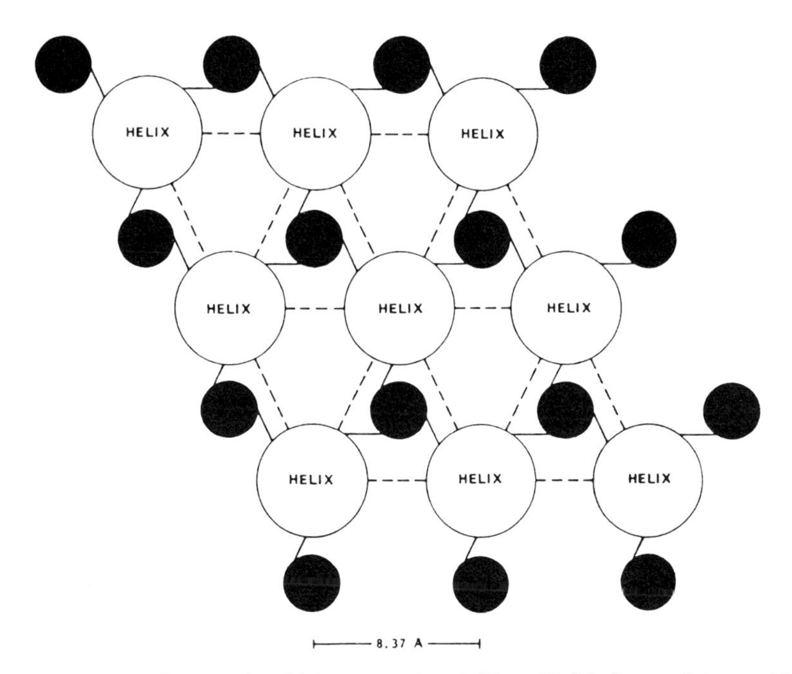

Figure 5.5 Junction zone in a high-ester pectin gel. (From Walkinshaw and Arnott, 1981b.)

process of high-methoxy pectin (HMP)–sucrose aqueous gels can be followed by low amplitude oscillation (Dasilva and Goncalves, 1994). Dynamic mechanical measurements enabled a determination of the point at which the system undergoes the sol–gel transition. The HMP–sucrose system is extremely sensitive to temperature variations during aging, especially in the lower temperature range. The gel's viscoelastic behavior indicates changes with aging temperature, probably because of variations in the mobility of the pectin chains and, consequently, in the lifetime of the junction zones (Dasilva and Goncalves, 1994). HMP–sugar gels are formed by a combination of hydrogen bonding and hydrophobic effects. Because magnitudes of the latter are affected by the solute used and temperature, gel strength and rate of structure development are also affected. Technologically important events that take place during the sol–gel transition have been considered, including a profile of complex viscosity during gelation, and the effects of rate of cooling and pectin concentration (Dasilva and Rao, 1995). Structure developments in HMP–fructose gels were also characterized by Rao and Cooley (1993, 1994). Weaker pectin networks are formed under thermal conditions unfavorable to the development of hydrophobic interactions. Gelling time and elastic modulus have a complex dependence on temperature, which can be attributed to the different thermal behaviors of the intermolecular interactions that stabilize the non-permanent cross-links of these physical networks. The influence of temperature on the dynamic and steady-shear rheology of pectin dispersions was studied by Dasilva *et al.* (1994). The authors used the time–temperature superposition principle to calculate activation energies, and their dependence on temperature and shear rate was analysed.

Low-ester pectin gels do not require a high solids content or low pH, but they do need the presence of calcium, which can be provided by the fruit pulp if a fruit product is desired. Calcium binding to low-ester pectin cannot be explained as a simple electrostatic interaction: it involves intermolecular chelate binding of the cation leading to the formation of macromolecular aggregates (Kohn and Luknar, 1977). An 'egg-box' model has been suggested for primary junction zones in the low-ester pectin molecular gel network (Rees, 1982). Chain segments with 14 or more residues having a ribbon-like symmetry are believed to form parallel-oriented aggregates. Chelate bonds with oxygen atoms from both galacturonan chains formed by calcium ions are formed when calcium ions fit into 'cavities' in the structure (Fig. 5.6). Although they differ from those involving high-ester gels, the concepts of good manufacturing practice need to be maintained, and the ingredients appropriately selected. Main differences between high- and low-ester systems are the ability to melt a low-ester pectin gel and the immediacy with which solidification occurs in the low-ester pectin system, relative to the slow rate of the high-ester pectin gel. Amidated low-ester pectins are usually able to jellify preserves, jams and jellies with calcium ions originating from fruit and water (Broomfield, 1988). Non-amidated low-ester pectins generally require a higher calcium level and the addition of extra calcium is very often necessary to obtain proper gel formation (Christensen, 1986). The degree of amidation and esterification controls the readiness of low-ester pectin reactions with calcium to induce gel formation. Low-ester pectins with a DE of 25–35% (non-amidated), and pectins with 20–30% DE and 18–25% DA are highly reactive with calcium and are, therefore, used in low-calcium and low-soluble-solids content systems. Pectins with a low ester content of 35 to 45% (non-amidated) and those with 30–40% DE and

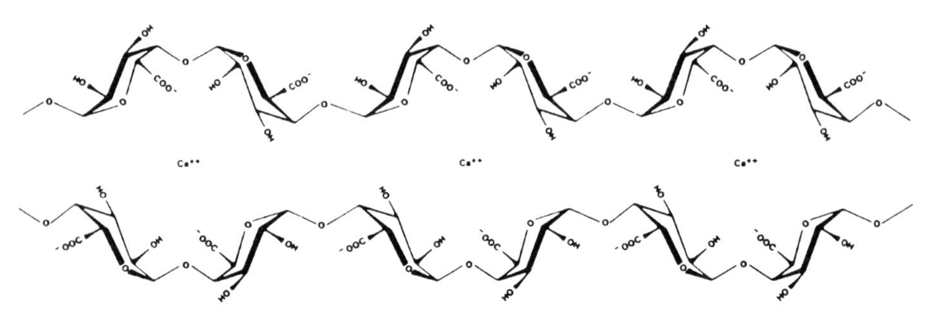

Figure 5.6 Junction zone in a low-ester pectin gel.

10–18% DA, because of their lower calcium reactivity, can serve in high-calcium or high-soluble-solids content systems (Buhl, 1990).

5.8.1 Gel properties

Results of gel-strength measurements yielded by different methodologies are not necessarily correlated. This is partly the result of the different properties defined and of the different measures such as deformation within the elastic limits and the force needed to break the gel. Traditional methods are the SAG method (this traditional method, based on Cox and Higby's work (1944) and adopted by the Institute of Food Technologists (1959), defines a jelly grade designation of 150° USA-SAG as the transformation of 150 parts sucrose by 1 part pectin into a gel of 65° Brix, pH 2.2–2.4 and a gel strength of 23.5% SAG after being cast for 2 min and removed from a standard glass with exactly specified inner dimensions), the use of the LFRA texture analyser, the Boucher Electronic Jelly Tester, the FIRA tester, the family of universal testing machines, the Herbstreith pectinometer, a spreadameter and the Bostwick Consistometer. In addition to the variety and plurality of gel texture measurements, other measurements of time–temperature relationships have also been developed within the industry. Gelation is influenced by the degree of methyl esterification, amidation, pectin concentration, water activity, the presence of calcium ions and pH.

The profile of shear modulus temperature meltdown of pectin gels was described by Clark *et al.* (1994). In this study, a cascade-theory approach to biopolymer gelation was developed to describe variations in the shear modulus with temperature for thermoreversible gels. The broadness of this 'melting transition' is seen to depend critically on the enthalpy of cross-linking, whereas the critical gel-melting temperature is determined by additional factors, such as the entropy of cross-linking, polymer concentration, molecular weight and the number of cross-linking sites. When the model was used to fit experimental data from a pectin system, a broad melting transition and the high melting point of the pectin system were consistent with much smaller negative values for these parameters (Clark *et al.*, 1994).

The mechanisms of gelation hint at a combination of hydrogen bonding and hydrophobic interactions in the case of high-ester pectin gel formation. The hydrophobic parts of the high-ester pectin molecule are the ester groups. Contact between these hydrophobic areas is associated with energy contribution. Hydrogen bonds formed between adjacent galacturonan chains contribute even more to decreases in the energy of junction-zone formation. However, the energy contribution of the hydrophobic interaction is necessary in order to make the sum of the energy contributions favoring gelation large enough to exceed energy contributions that resist gelation. The proposed mechanism suggests that gel formation with high-ester pectins relates to the rigidity of its molecule, its correlation to DE level and the presence of sugars in the system. For low-ester pectin, the 'egg-box' model used to explain alginate gelation is proposed. Twofold helices are bridged by calcium ions of opposing carboxyl groups. In dried pectins, helices with three subunits per turn are detected, hinting that helix structure changes from twofold to threefold when the gel is dried to a powder. Support for the 'egg-box' model also comes from the direction of equilibrium dialysis: ~50% of the calcium ions cannot be removed by exposure to a very large concentration of univalent cations.

5.9 Applications

The most common use of pectins is in the preparation of jams, jellies, or similar gels. Detailed information on jam production can be found elsewhere (Kertesz, 1951). Ordinary jam is generally made from high-ester pectin, whereas the low-ester pectins are used when a softer, more spreadable texture is desired. If fruit particles (pulp) are to be contained in the jam, a high gelation temperature is used and solidification begins almost immediately after filling the containers, with almost no floatation of the particles being observed. If very large containers are used for jam-filling, then pectins with lower filling temperatures should be considered, to minimize flavor and color destruction, especially in the center of the container. The rheological indices of fruit content in jams and the effect of formulation on flow plasticity of sheared strawberry and peach jams has been studied (Costel *et al.*, 1993). The effect of formulation factors on Casson yield values measured at low and medium shear rates are reported.

To prepare low-sugar (less sweet) gums, low-ester pectins are used in combination with calcium in an amount related to gelation temperature and the quality of the formed texture. When jellies that contain no particles are produced, slow-setting pectins that solidify a long time after filling, allowing air bubbles to float and escape from the product, are preferred. For confections, a slow-setting high-ester pectin is used. The solid content of such preparations is high, ~78%, in contrast to ordinary jams at ~65% or low-sugar jams at ~30–55% (Rolin and De Vries, 1990). For baked goods, a heat-resistant gel is usually produced with a soluble-solids content of 45–75% and, depending on the type of pectin used, a typical dosage yields pH values of 3.3–3.6; if a cold-setting gel is used, a product with ~61% soluble solids and pH 4.0 is produced using rapid-set high-methyl-ester pectin (0.7%). Heat-resistant gels are generally prepared from high-ester pectins but can be produced from low-ester pectins if calcium citrate is used in the formulation to elevate the gelation temperature subsequent to the setting of the system. Fruit preparations for dairy products are often sold as semi-gel/thixotropic products with a typical soluble-solids content of 30–

65% and a pH of 3.6–4.0, usually prepared with 0.3–0.6% low-methyl-ester pectin. These should be prepared in such a way that the big fruit chunks or berries are distributed uniformly even after storage and pumping or transport (Rolin and De Vries, 1990).

Pectins are used to prepare bakery fillings and glazes. Oven-resistant high-sugar jams are produced at a solids content of ~70% using rapid-set pectin. Another demand of such products is mechanical stability. The less the gel is ruptured, the lower the syneresis at elevated heating temperatures. Non-amidated low-ester pectins are recommended for the production of bakery jams with satisfactory stability. Low-ester pectin gels are produced with ~65% soluble solids and a relatively high dosage of calcium-reactive, low-ester pectin. Before being applied to the baked goods, water is added, then the gel is heated to ~85°C to induce melting, and hot coating of the product follows. Upon cooling, a glossy coverage is formed. Other studies on the physical and mechanical properties of highly plasticized pectin–starch films can be found (Coffin and Fishman, 1994).

The thermomechanical properties of pectin and polyvinyl alcohol (PVA) blends have been recently studied (Coffin *et al.*, 1996). Increasing the amount of PVA in the blends reduced the storage and loss modulus of the films above the glass transition temperature. Changes in the molecular weight and degree of ester hydrolysis of PVA exerted a rather small effect on the blends. The composition should be targeted to the specific aim (Coffin *et al.*, 1996).

Stabilization of pasteurized or sterilized, acidified milk products (pH values of ~3.5–4.2) can be achieved by using high-ester pectins with DE greater than ~70. Acidification can be produced by either fermentation or the addition of fruit juice. If casein stabilization is not achieved, an undesirable grain-like texture is obtained. The pectin, added before homogenization, is absorbed onto the casein particles, which have a positive charge in the unstabilized milk. If the amount of added pectin is small, then the charge is neutralized and the system tends to collapse owing to the removal of repulsive forces. If pectin addition is continued however, a new repulsive force builds up, resulting in stabilization of the acidified milk system. Hydrophobic as well as electrostatic interactions are important in stabilizing pectin–casein dispersions (Pereyra *et al.*, 1995). The shear rate and time dependency of stirred yoghurt rheology were evaluated as influenced by added pectin and strawberry concentrate (Basak and Ramaswamy, 1994). The rheology of the flavored yoghurt was influenced by both pectin (0–0.5%) and the concentrate, and the desired product viscosity could be obtained by postfermentation mixing of stirred yoghurt with the pectin and fruit concentrate. Quality requirements for yoghurt–fruit preparations and the rheological parameters used to assess their properties are discussed in Kratz and Dengler (1995). Three different types of pectin were compared in yoghurt preparations. The possibility of using yoghurt–fruit preparations instead of increasing the dry matter content in order to improve the consistency of fruit yoghurt was considered. In fat-free yoghurt, good mouthfeel and stability were achieved by increasing the percentage of milk solids and adding a mixture of gelatin, starch and pectin (Moller, 1995). In order to eliminate unacceptable viscosity, poor mouthfeel and syneresis in pasteurized yoghurt, processing conditions need to be adjusted and starch, gelatin and pectin added (Moller, 1995). The use of non-traditional additives (dried fruit and vegetable powders) in the manufacture of cultured milk products for therapeutic and prophylactic uses was reported by Arkhipova and Krasnikova (1994). Results of clinical trials showed that in less than 2 weeks patients receiving these cultured milks shared increased appetites, improved intestinal microflora and 20–25% decreases in blood cholesterol level (Arkhipova and Krasnikova, 1994).

Pectin is also used to stabilize clouding in beverages. Such stability is dependent on the nature and amount of the pectin present. Natural clouding agents can be produced (Elshamei and Elzoghbi, 1994) from orange and lemon peels using enzyme preparations to hydrolyse the pectin in the peel. The chemical and physical properties of the clouds were evaluated in parallel to the drink's properties, taste and stability. The cloudiness of the produced drinks stabilized after 42 days of storage at 25°C. Other reports on the physicochemical nature of pectin associated with commercial orange juice clouding can be found elsewhere (Klavons *et al.*, 1994).

The confectionery industry makes use of slow-set high-ester pectins to prepare fruit jellies and jelly centers. Low-ester pectins are also used to impart thixotropic behavior at low concentrations, or to achieve a cold-set type of gelation if diffusion of calcium ions occurs. The use of pectin in confections permits the manufacture of products with tailor-made textural properties, good flavor release and compatibility with continuous processing. High-methoxy pectin was also reported to produce coatings that inhibit lipid migration in a confectionery product (Brake and Fennema, 1993).

Low-ester pectins are used as gelling agents and texturizers in many very different food products such as artificial caviar, meat products and dessert jellies. Pectin–alginate combinations have a synergistic effect in terms of gel-formation properties. Xanthan and pectin together can serve as an appropriate stabilizer for salad dressings. The incorporation of pectin in water-ice and sherbet preparations improves product acceptability by minimizing the growth of ice crystals. Galactomannans in combination with pectin serve to stabilize ice cream. Pectins are used to stabilize emulsions. Modified pectins in whey–protein emulsions (Einhornstoll *et al.*, 1996) were found to stabilize the whey protein at high enough concentrations. For any individual utilization, the most suitable pectin needs to be selected. Frozen fruit preparations are improved by incorporating pectin into the product. Pectins can be used for coating, in recipes of spray-dried instant tea and for many other products.

References

Amado, R. and Neukom, H. (1984) Isolation and partial degradation of pectic substances of potato cell walls in phosphate buffer. *Abstr.: 9th Triennial Conf. Eur. Ass. Potato Res.*, p. 103.

Anger, H. and Dongowsky, G. (1984) Uber die bestimmung der Ester gruppeverteilung in Pektin durch Fraktionierung an DAEA-cellulose. *Nahrung*, **28**, 199.

Anger, H. and Dongowsky, G. (1985) Distribution of free carboxyl groups in native pectins from fruit and vegetable. *Nahrung*, **29**(4), 397–404.

Anon. (1978) E440(a)-Pectin, E440(b)-Amidated pectin, in *Council Directive of July 1978*, laying down specific criteria of purity for emulsifiers, stabilizers, thickeners and gelling agents for use in food stuffs. *Official Journals of the European Communities*, L 223, p. 16.

Anon. (1981a) Evaluation of certain food additives. *25th report of the Joint FAO/WHO Expert Committee on Food Additives*, World Health Organization, Geneva.

Anon. (1981b) Pectin, in *Food Chemicals Codex*, 3rd edn, National Academy Press, Washington, DC, p. 215.

Anon. (1981c) Amidated pectin, pectins, in specifications for identity and purity of carrier solvents, emulsifiers and stabilizers, enzyme preparation, flavoring agents, food colors, sweetening agents and other food additives, *FAO Food and Nutrition Paper 19*, Food and Agriculture Organization of the United Nations, Rome, 10-14, pp. 152–5.

Arkhipova, A.N. and Krasnikova, L.V. (1994) Use of non-traditional additives in the manufacture of cultured milk products for therapeutic and prophylactic use. *Molochnaya Promyshlennost*, **8**, 14–15.

Ashford, M., Fell, J., Attwood, D. *et al.* (1994) Studies on pectin formulations for colonic drug delivery. *J. Controlled Release*, **30**(3), 225–32.

Attri, B.L. and Maini, S.B. (1996) Pectin from Galgal (*Citrus Pseudolimon Tan*) peel. *Bioresource Technol.*, **55**(1), 89–91.

Baker, G.L., Joseph, G.H., Kertesz, Z.I. *et al.* (1944) Revised nomenclature of the pectic substances. *Chem. Eng. News*, **22**, 105.

Baker, R.A. (1994) Potential dietary benefits of citrus pectin and fiber. *Food Technol.*, **48**(11), 133–4.

Basak, S. and Ramaswamy, H.S. (1994) Simultaneous evaluation of shear rate and time dependency of stirred yoghurt rheology as influenced by added pectin and strawberry concentrate. *J. Food Eng.*, **21**(3), 385–93.

Berth, G., Anger, H., Plashchina, I.G. *et al.* (1982) Structural study of the solutions of acidic polysaccharides. Study of some thermodynamic properties of the dilute pectin solutions with different degrees of esterification. *Carbohydr. Polym.*, **2**, 1.

Boothby, D. (1983) Pectic substances in developing and ripening plum fruit. *J. Sci. Food Agric.*, **44**, 1117.

Braconnot, H. (1825) Recherches sur un nouvel acide universellement rependu dans tous les vegetaux. *Ann. Chim. Phys., Ser.*, 2, **28**, 173.

Brake, N.C. and Fennema, O.R. (1993) Edible coatings to inhibit lipid migration in a confectionery product. *J. Food Sci.*, **58**(6), 1422–5.

Broomfield, R.W. (1988) Preserves, in *Food Industries Manual* (ed. M.D. Ranken), Blackie, Glasgow, pp. 335–55.

Buhl, S. (1990) Gelation of very low DE pectin, in *Gums and Stabilisers for the Food Industry*, vol. 5 (eds G.O. Phillips, D.J. Wedlock and P.A. Williams), IRL Press (at Oxford University Press), Oxford, pp. 233–41.

Cambell, L.A. and Palmer, G.H. (1978) Pectin, in *Topics in Dietary Fiber Research* (ed. G.A. Spiller), Plenum, New York, p. 105.

Cerda, J.J. (1993) Diet and gastrointestinal disease. *Med. Clin. N. Am.*, **77**(4), 881–7.

Christensen, O. and Towle, G.A. (1973) Pectin, in *Industrial Gums*, 2nd edn (ed. R.L. Whistler), Academic Press, New York, pp. 429–61.

Christensen, P.E. (1954) Methods of grading pectin in relation to the molecular weight (intrinsic viscosity) of pectin. *Food Res.*, **19**, 163.

Christensen, S.H. (1986) Pectins, in *Food Hydrocolloids*, vol. III (ed. M. Glicksman), CRC Press, Boca Raton, FL, pp. 206–27.

Clark, A.H., Evans, K.T. and Farrer, D.B. (1994) Shear modulus–temperature meltdown profiles of gelatin and pectin gels – a cascade theory description. *Int. J. Biol. Descript.*, **16**(3), 125–30.
Coffin, D.R. and Fishman, M.L. (1994) Physical and mechanical properties of highly plasticized pectin starch films. *J. Appl. Polym. Sci.*, **54**(9), 1311–20.
Coffin, D.R., Fishman, M.L. and Ly, T.V. (1996) Thermomechanical properties of blends of pectin and poly (vinyl alcohol). *J. Appl. Polym. Sci.*, **61**(1), 71–9.
Costell, E., Carbonell, E. and Duran, L. (1993) Rheological indexes of fruit content in jams – effect of formulation on flow plasticity of sheared strawberry and peach jams. *J. Text. Studies*, **24**(4), 375–90.
Cox, R.E. and Higby, R.H. (1944) A better way to determine the jellying power of pectins. *Food Ind.*, **16**, 441.
Dasilva, J.A.L. and Goncalves, M.P. (1994) Rheological study into the aging process of high methoxyl pectin/sucrose aqueous gels. *Carbohydr. Polym.*, **24**(4), 235–45.
Dasilva, J.A.L., Goncalves, M.P. and Rao, M.A. (1994) Influence of temperature on the dynamic and steady shear rheology of pectin dispersions. *Carbohydr. Polym.*, **23**(2), 77–87.
Dasilva, J.A.L. and Rao, M.A. (1995) Rheology of structure developments in high-methoxyl pectin/sugar systems. *Food Technol.*, **49**(10), 70, 72–3.
De Vries, J.A., Hansen, M.E., Glahn, P.E., Soderberg, J. and Pedersen, J.K. (1986) Distribution of methoxyl groups in pectins. *Carbohydr. Polym.*, **6**, 165.
De Vries, J.A., Rombouts, F.M., Voragen, A.G.J. and Pilnik W. (1982) Enzymatic degradation of apple pectins. *Carbohydr. Polym.*, **2**, 25.
De Vries, J.A., den Vijl, C.H., Voragen, A.G.J. *et al.* (1983) Structural features of the neutral sugar side chains of apple pectic substances. *Carbohydr. Polym.*, **3**, 193.
De Vries, J.A., Vorgagen, A.G.J., Rombouts, F.M. *et al.* (1984) Effect of ripening and storage on pectic substances. *Carbohydr. Polym.*, **4**, 3.
Donaghy, J.A. and Mckay, A.M. (1994) Pectin extraction from citrus peel by polygalacturonase produced on whey. *Bioresource Technol.*, **47**(1), 25–8.
Doseburg, J.J. (1965) Pectic substances in fresh and preserved fruits and vegetables. *IBVT Communication No. 25*, Institute for Research on Storage and Processing of Horticultural Produce, Wageningen, The Netherlands.
Einhornstoll, U., Glasenapp, N. and Kunzek, H. (1996) Modified pectins in whey–protein emulsions. *Nahrung Food*, **40**(2), 60–7.
Elshamei, Z. and Elzoghbi, M. (1994) Production of natural clouding agents from orange and lemon peels. *Nahrung Food*, **38**(2), 158–66.
Fernandez, M.L., Sun, D.M., Tosca, M. *et al.* (1994) Citrus pectin and cholesterol interact to regulate hepatic cholesterol homeostasis and lipoprotein metabolism in the guinea pig. *FASEB J.*, **8**(4), A153.
Food Chemicals Codex, 2nd edn (1972) National Academy of Sciences, Washington, DC, pp. 580–1.
Food Chemicals Codex, 3rd edn (1981) National Academy Press, Washington, DC, p. 215.
Fritzsch, B., Dongowski, G. and Neubert, R. (1994) Effects of structural parameters of pectin on the interactions with drugs in-vitro. *FASEB J.*, **8**(4), A188.
Goycoolea, F.M., Morris, E.R. and Gidley, M.J. (1995) Screening for synergistic interactions in dilute polysaccharide solutions. *Carbohydr. Polym.*, **28**(4), 351–8.
Guillon, F., Thaibault, J.F., Rombouts, F.M. *et al.* (1986) Structural features of the neutral sugar side chains of beet pulp pectins. *Proc. Cell Walls, Paris*, **86**, 112.
Hoefler, A.C., Sleap, J.A. and Trudso, J.E. (1994) Fat substitute. US Patent No. 5,324,531.
Huber, D.J. (1984) Strawberry fruit softening: the potential roles of polyuronides and hemicellulose. *J. Food Sci.*, **49**, 1310.
Hu-Hi and Brown, P.H. (1994) Localization of boron in cell-walls of squash and tobacco and its association with pectin – evidence for a structural role of boron in the cell-wall. *Plant Physiol.*, **105**(2), 681–9.
Hwang, J.K. and Kokini, J.L. (1995) Changes in solution properties of pectins by enzymatic hydrolysis of sidechains. *J. Korean Soc. Food Nutrit.*, **24**(3), 389–95.
Institute of Food Technologists (1959) Pectin standardisation, final report of the IFT committee. *Food Technol.*, **13**, 496.
Karpovich, N.S., Telichuk, L.K., Donchenko, L.V. *et al.* (1981) Pectin and raw material resources. *Pishch Promst.* (Moscow), **3**, 36.

Kausar, P. and Nomura, D. (1982) A new approach to pectin manufacture by copper method, Part 3. *J. Fac. Agric. Kyushu Univ.*, **26**, 111.

Kertesz, Z.I. (1951) *The Pectic Substances*, Interscience Publishers, New York.

Khalikov, S.S., Pominova, T.Y., Perevetzeva, E.I. *et al.* (1995) Preparation and physicochemical properties of polymer complexes of benzimidazolyl-2-methylcarbamate and apple pectin. *Khim. Prir. Soedin.*, **6**, 896–900.

Kirby, A.R., Gunning, A.P. and Morris, V.J. (1996) Imaging polysaccharides by atomic-force-microscopy. *Biopolymers*, **38**(3), 355–66.

Klavons, J.A., Bennett, R.D. and Vannier, S.H. (1994) Physical/chemical nature of pectin associated with commercial orange juice cloud. *J. Food Sci.*, **59**(2), 399–401.

Klurfeld, D.M., Weber, M.M. and Kritchevsky, D. (1994) Dose-response of colonic carcinogenesis to pectin and guar gum. *FASEB J.*, **8**(4), A152.

Kohn, R., Dongowsky, G. and Bock, W. (1985) Die Verteilung der freien und veresterten Carboxylgruppen in Pectinmolekul nach Einwirkung von Pektinesterasen aus *Aspergillus niger* und Orangen. *Nahrung*, **29**(1), 75–85.

Kohn, R. and Luknar, O. (1977) Intermolecular calcium ion binding on polyuronates – polygalacturonate and polyguluronate. *Collect. Czech. Chem. Commun.*, **42**, 731.

Kohn, R., Marbovich, D. and Machova, E. (1983) Deesterification mode of pectin by pectin esterase of *Aspergillus foetidus*, tomatoes and alfalfa. *Collect. Czech. Chem. Commun.*, **48**, 790.

Kratz, R. and Dengler, K. (1995) Fruit preparations for yoghurts. Pectin as a thickener – requirements posed by producers of fruit preparations and yoghurts. *Food Tech. Eur.*, **2**(2), 130–7.

Lewinska, D., Rosinski, J. and Piatkiewicz, W. (1994) A new pectin-based material for selective LDL-cholesterol removal. *Artificial Organs*, **18**(3), 217–22.

Mao, Y., Kasravi, B., Nobeak, S. *et al.* (1996) Pectin supplemented enteral diet reduces the severity of methotrexate-induced *Enterocolitis* in rats. *Scand. J. Gastroenterol.*, **31**(6), 558–67.

Markovic, O. and Kohn, R.G. (1984) Mode of pectin deesterification by *Trichoderma reesi* pectin esterase. *Experientia*, **40**, 842.

Mascaro, L.J. and Kindell, P.K. (1977) Apiogalacturonan from *Lemna minor. Arch. Biochem. Biophys.*, **183**, 139.

Matora, A.V., Korshunova, V.E., Shkodina, O.G. *et al.* (1995) The application of bacterial enzymes for extraction of pectin from pumpkin and sugar beet. *Food Hydrocolloids*, **9**(1), 43–6.

Matsuura, Y. (1984) Chemical structure of polysaccharide of cotyledons of kidney beans. *J. Agric. Chem. Soc. Japan*, **58**, 253.

May, C.D. (1992) Pectins, in *Thickening and Gelling Agents for Food* (ed. A. Imeson), Blackie A & P, Glasgow, pp. 124–52.

Melotto, E., Greve, L.C. and Labavitch, J.M. (1994) Cell-wall metabolism in ripening fruit. Biologically active pectin oligomers in ripening tomato (*Lycopersicon esculentum* Mill.) fruits. *Plant Physiol.*, **106**(2), 575–81.

Michel, F., Doublier, J.L. and Thibault, J.F. (1982) Investigation on high methoxyl pectins by potentiometry and viscometry. *Prog. Food Nutr. Sci.*, **6**, 367.

Michel, F., Thibault, J.F. and Doublier, J.L. (1981) Characterization of commercial pectins purified by cupric ions. *Sci. Aliments*, **1**, 569.

Moller, J.L. (1995) Stabilizers in special cultured products. *Maelkeritidende*, **108**(12), 318–19.

Morris, E.R., Powell, D.A., Gidley, M.J. *et al.* (1982) Conformations and interactions of pectins. Polymorphism between gel and solid states of calcium polygalacturonate. *J. Mol. Biol.*, **155**, 507.

Neukom, H., Amado, R. and Pfister, M. (1980) Neuere erkenntnisse auf dem Gebiete der pektinstoffe. *Lebensm. Wiss. Technol.*, **13**, 1.

Nwanekezi, E.C., Alawuba, O.C.G. and Mkpolulu, C.C.M. (1994) Characterization of pectic substances from selected tropical fruits. *J. Food Sci. Technol. India*, **31**(2), 159–61.

Palmer, K.J. and Hartzog, M.B. (1945) An X-ray diffraction investigation of sodium pectate. *J. Am. Chem. Soc.*, **67**, 2122.

Panchev, I.N., Kirtchev, N.A. and Kratchanov, C.G. (1994) On the production of low esterified pectins by acid maceration of pectic raw materials with ultrasound treatment. *Food Hydrocolloids*, **8**(1), 9–17.

Pathak, D.K. and Shukla, S.D. (1978) A review on sunflower pectin. *Indian Food Packer* (May–June), 49.

Pathak, D.K. and Shukla, S.D. (1981) Quantity and quality of pectin in sunflower at various stages of maturity. *J. Food Sci. Technol.*, **18**(3), 116–17.

Pedersen, J.K. (1980) Pectin, in *Handbook of Water-Soluble Gums and Resins* (ed. R.L. Davidson), pp. 15.1–15.21.

Pereyra, R.A., Schmidt, K. and Wicker, L. (1995) Stability of pectin–casein solutions. *IFT Annu. Meet. Conf. Book 1995*, p. 219.

Pienta, K.J. and Raz, A. (1994) Oral-administration of modified citrus pectin inhibits spontaneous prostate-cancer metastasis in the rat by inhibiting carbohydrate-mediated cell–cell interactions. *Clin. Res.*, **42**(3), A423.

Pilnik, W. and Zwiker, P. (1970) Pektine. *Gordian*, **70**, 202–4, 252–7, 302–5, 343–6.

Powell, D.A., Morris, E.R., Gidley, M.J. *et al.* (1982) Conformations and interactions of pectins. II. Influence of residue sequence on chain association in calcium pectate gels. *J. Mol. Biol.*, **155**, 517.

Rao, M.A. and Cooley, H.J. (1993) Dynamic rheological measurements of structure developments in high-methoxyl pectin/fructose gels. *J. Food Sci.*, **58**(4), 876–9.

Rao, M.A. and Cooley, H.J. (1994) Influence of glucose and fructose on high-methoxyl pectin gel strength and structure development. *J. Food Quality*, **17**(1), 21–31.

Redgwell, R.J. and Selvendran, R.R. (1986) Structural features of cell wall polysaccharides of onion. *Carbohydr. Res.*, **157**, 183.

Rees, D.A. (1982) Polysaccharide conformation in solutions and gels – recent results on pectins. *Carbohydr. Polym.*, **2**, 254.

Rinaudo, M. (1974) Comparison between results obtained with hydroxylated polyacids and some theoretical models, in *Polyelectrolytes* (ed. E. Selegny), Reidel, Dordrecht, p. 157.

Rolin, C. and De Vries, J.D. (1990) Pectin, in *Food Gels* (ed. P. Harris), Elsevier Applied Science, London, pp. 401–34.

Sakai, T., Sakamoto, T., Hallaert, J. *et al.* (1993) Pectin, pectinase and protopectinase – production, properties and applications. *Adv. Appl. Microbiol.*, **39**, 213–94.

Turmucin, F., Ungan, S. and Yielder, F. (1983) Pectin production from sunflower. *METU J. Pure Appl. Sci.*, **16**, 263.

Vauquelin, M. (1790) Analyse du tamarin. *Ann. Chim. (Paris)*, **5**, 92.

Vorgen, A.G.J., Schols, H.A. and Pilnik, W. (1986) Determination of the degree of methylation and acetylation of pectins by HPLC. *Food Hydrocolloids*, **1**, 65.

Walkinshaw, M.D. and Arnott, S. (1981a) Conformation and interactions of pectin. I. X-ray diffraction analyses of sodium pectate in neutral and acidified forms. *J. Mol. Biol.*, **153**, 1055.

Walkinshaw, M.D. and Arnott, S. (1981b) Conformation and interactions of pectin. II. Models for junction zones in pectinic acid and calcium-pectate gels. *J. Mol. Biol.*, **153**, 1075.

Yamaguchi, F., Shimizu, N. and Hatanaka, C. (1994) Preparation and physiological effect of low-molecular-weight pectin. *Biosci. Biotechnol. Biochem.*, **58**(4), 679–82.

6 Cellulose derivatives

6.1 Introduction

Cellulose is the most abundant naturally occurring organic substance in nature, constituting approximately one-third of the world's vegetative matter (Zecher and Van Coillie, 1992). The estimated annual amount of synthesized cellulose is $\sim 10^{11}$ tons (Walton and Blackwell, 1973). Cellulose, hemicellulose (i.e. other polysaccharides such as xylan and mannan) and lignins are the major constituents of most land plants, providing the supporting structure. Cellulose can be transformed into hundreds of products that affect every phase of daily life (Greminger and Krumel, 1980). Cellulose content in vegetative tissues varies, e.g. wood contains ~40–50% and cotton 85–97% (Ramby and Rydholm, 1956; Glicksman, 1969; Whistler, 1973).

Cellulose is a linear polymer of D-glucose monomers joined by D-β(1–4) linkages and arranged in repeating units of cellobiose, each comprising two anhydroglucoses (Fig. 6.1). Cellulose has a long molecular-chain length and the hydrogen-bonding capacity of the three hydroxyl groups is very high (Glicksman, 1986). The linear structure and uniform nature of these molecules enables them to fit together over part of their length to form crystalline regions, with resultant rigidity and strength (Ward and Seib, 1970; Whistler and Zysk, 1978).

The degree of polymerization (DP) of cellulose depends on its origin: it can be 1000–15 000 for native cellulose and 200–3200 for commercially purified cotton linters and wood pulp (Krassig, 1985; Zecher and Van Coillie, 1992); different values can be found elsewhere (Walton and Blackwell, 1973). Spatially, the polymer is arranged in long thread-like molecules. These molecules align to form fibers, some regions of which are highly ordered and have a crystalline structure owing to lateral association by hydrogen bonding (Glicksman, 1986). Each anhydroglucose contains three hydroxyl groups that are available for reaction. The polymer has a maximum of three degrees of substitution (i.e. the average number of substituted groups per anhydroglucose unit; DS) (Greminger and Krumel, 1980; Zecher and Van Coillie, 1992). Products with a wide range of functional properties can be created by controlling the degree of substitution and its type. Partial substitution is preferred and many derivatives are possible, but only some of them are of interest to industrial and food businesses (Whistler and Zysk, 1978).

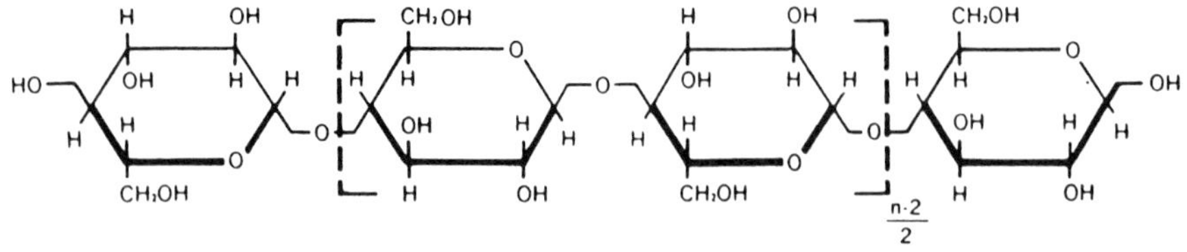

Figure 6.1 Cellulose molecule.

Cellulose is not soluble in water because of its extensive intra- and intermolecular hydrogen-bonded crystalline domains. The production of water-soluble cellulose derivatives, in contrast to those of polymers based on petrochemical resources, starts with a preformed polymer backbone of either wood or cotton cellulose, instead of a monomer (Greminger and Krumel, 1980). To make it suitable for food use, cellulose can be converted to a soluble compound via its derivatization and disruption of hydrogen bonds (Zecher and Van Coillie, 1992).

6.2 Manufacture

Cellulose ethers are produced by first preparing alkali cellulose then subjecting it to alkylation or alkoxylation. Special solvents are needed to solubilize cellulose (Greminger and Krumel, 1980; McCormick and Callais, 1987; Isogai and Atalla, 1991; Zecher and Van Coillie, 1992). The commercial derivatization reactions conducted heterogeneously at temperatures of 50–140°C, under nitrogen atmosphere, can result in high molecular weights (Michie and Neale, 1964). Access to all hydroxyls is not equal, and the distribution of substituents is, therefore, not uniform.

Swelling of the cellulose, crystalline-region disruption, reaction uniformity and catalysis of alkoxylation reactions may result from using soda to form alkali cellulose. Acetone and isopropanol (inert diluents) are used to disperse the cellulose and to moderate the reaction kinetics, facilitate recovery of the product and provide heat transfer (Zecher and Van Coillie, 1992). Two types of reaction are used to produce cellulose ethers: the Williamson etherification (reaction of alkali cellulose with organic halide) and alkoxylation, involving reaction with an epoxide (Donges, 1990). Details on these reactions can be found elsewhere (Zecher and Van Coillie, 1992). During methyl cellulose (MC) production, the use of dimethyl sulfate instead of methyl chloride results in the formation of many by-products (Heuser, 1944).

Cellulose derivatives can be prepared by reacting alkali cellulose with either methyl chloride to form MC, propylene oxide to form hydroxypropylcellulose (HPC) (a side reaction in which propylene oxide reacts with water

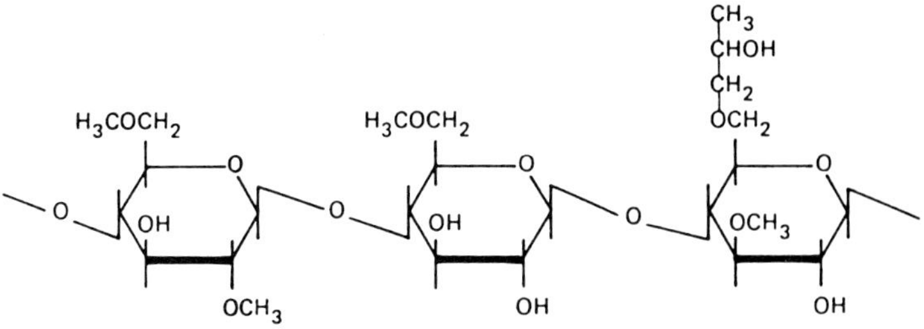

Figure 6.2 Methylhydroxypropylcellulose.

to form a mixture of propylene glycols also occurs but can be minimized by keeping the water input as low as possible), or sodium chloroacetate to form sodium carboxymethylcellulose (CMC). In this last case a side reaction, the formation of sodium glycolate, also occurs (Stelzer and Klug, 1980). Mixed derivatives such as methyl hydroxypropylcellulose (MHPC) can be formed by combining two or more of these reagents (Fig. 6.2). Of the many possible derivatives investigated and manufactured, CMC, MC, MHPC and HPC are utilized in the food industry in addition to modified forms of cellulose, which have been found to have useful functional hydrocolloidal properties and significance in several food applications. CMC is the most important cellulose-derived hydrocolloid. It is an anionic polymer that is important for thickening applications and is able to react with charged molecules within specific pH ranges (Ganz, 1977; Hercules Inc., 1978; Stelzer and Klug, 1980). HPC, a non-ionic cellulose ether, is soluble in water below 40°C and in polar organic solvents (Butler and Klug, 1980). Microcrystalline cellulose (MCC) is an acid-hydrolysed, pure α-cellulose material that has thickening and water-absorptive properties and is used in frozen and dairy-food products.

Many producers exist for all of these gums. A few examples are the Aqualon Co., Courtauls Fibers Ltd, Dow Chemical Co., Hoechst AG, Matsumoto Yushi-Seiyaku Co. Ltd, Shin Etsu Chemical Co. Ltd, Wolff Walsrode AG, Nippon Soda Co. Ltd, Akzo, Billerud AB, Carbose Corp., Daiichi Kogyo Seiyaku Co. Ltd, Daicel Chemical Industries Ltd, Fratelli Lamberti SpA and Metsa-Serla (SRI International, 1989).

6.3 Properties of methylcellulose and methylhydroxypropylcellulose

Commercial MC products have an average DS ranging from 1.4 to 2.0 (Fig. 6.3). A minimum DS of $\sim$1.4 is required for water solubility, and at 2.0–2.2 DS, solubility in organic systems is achieved (Dow Chemical Co., 1974; Greminger and Krumel, 1980; Zecher and Van Coillie, 1992). Cellulose products are of various particle sizes with a white to off-white color. The

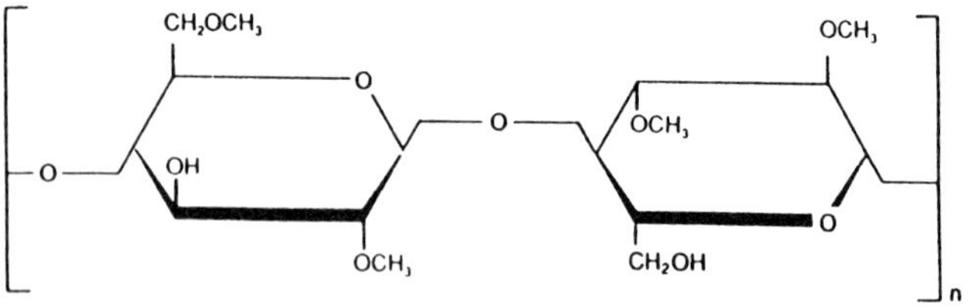

$$\left[\text{-O-} \underset{\text{OH}}{} \text{(CH}_2\text{OCH}_3\text{, OCH}_3\text{)} \text{-O-} \text{(OCH}_3\text{, CH}_2\text{OH, OCH}_3\text{)} \right]_n$$

Figure 6.3 Structure of methylcellulose (DS 2.0).

purity of the cellulose products depends on their use, either technical or in foods (higher purity). For food purposes, less than 1% sulfated ash and residual heavy metals as specified in the *European Pharmacopia* are allowed, whereas a maximum ash content of about 2.5% in cellulose products can be found. MC is metabolically inert and has a neutral taste and odor. Commercial MHPC products have an average methyl DS of 1.0–2.3 (Zecher and Van Coillie, 1992).

MC and MHPC are soluble in cold water but insoluble in hot water. When such a solution is heated, gel structures can form, at gelation temperatures ranging from 50 to 90°C. Solutions of MC can be prepared by dispersing the powder in hot water (80–90°C), adding cold water (0–5°C) or ice to the final volume and agitating until a smooth texture is achieved. MHPC products may require cooling to 20–25°C or lower. Stirring is recommended for better and quicker dissolution. MC and MHPC solutions in cold water are smooth, clear and pseudoplastic (Dow Chemical Co., 1974; Greminger and Krumel, 1980; Aqualon Co., 1989; Zecher and Van Coillie, 1992).

Solution viscosity decreases with increasing temperature to the thermal gel point, then viscosity rises sharply and the flocculation temperature is reached (Zecher and Van Coillie, 1992). The temperatures for 0.5% solutions are 50–75°C for MC and 60–90°C for MHPC. Flocculation results from the weakening of hydrogen bonding between polymer and water molecules and the strengthening of interactions between polymer chains (Zecher and Van Coillie, 1992). Gels are formed as a result of phase separation and are susceptible to shear-thinning. If the temperature is lowered, the original solution is restored. The thermal gel point is influenced by the type and degree of substitution. The flocculation temperature is influenced by concentration of salts (decreases) and alcohols (increases) (Zecher and Van Coillie, 1992).

The viscosity of MC and MHPC is stable over a very wide range of pH (2–13). Low salt concentrations have little effect on viscosity, whereas higher levels (7% sodium chloride, 15% potassium chloride or 4% sodium bicarbonate) cause salting-out of the polymer solution (2%, 7000 mPa s MC). At temperatures >140°C MC and MHPC powders darken, and at temperatures >220°C they decompose. In the dry state, they are resistant to microorganisms. If solutions are to be stored, preservatives such as sodium benzoate or potassium sorbate are recommended (Zecher and Van Coillie, 1992).

MC and MHPC films are strong and transparent, water soluble and insoluble in organic liquids, fats and oils. This phenomenon and thermogelation are an advantage in the frying of extruded food products, and when oil pickup is necessary. Dilute 0.1% MC or MHPC reduces the surface tension of water by ~40% at 20°C. MC surface activity and water retention properties are also helpful in emulsion-type sauces, whipped toppings and creams (Zecher and Van Coillie, 1992).

6.4 Hydroxypropylcellulose

Cellulose treated with aqueous sodium hydroxide followed by propylene oxide undergoes an alkoxylation reaction to yield HPC. Etherification is catalysed by the alkali. Propylene oxide reacts with water to form poly(propylene glycol) by-products, so the amount of water is minimized to improve yields. The reaction is performed at 70–100°C for 5–20 h in stirred autoclaves in the presence of organic diluents, neutralized and washed with hot water (70–90°C), then dried and ground to provide an off-white, tasteless, granular powder (Klug, 1966, 1967; Zeher and Van Coillie, 1992). Food-grade HPC is at least 99.5% pure, with a sulfated ash content of less than 0.2%. It is a physiologically inert polymer that is edible, thermoplastic and non-ionic. It is soluble in water below 40°C and in many organic solvents such as methanol, ethanol, propylene glycol and MC. Aqueous, lump-free solutions can be formed by dry blending the HPC with other powders before dispersion (Desmarais, 1973) and blending with glycerin or hot water to produce a slurry that is later agitated following its addition to cold water to dissolution (Zecher and Van Coillie, 1992).

HPC aqueous solutions are pseudoplastic, presenting little or no thixotropy and relatively high stability during shear degradation. Viscosity decreases with increasing temperature, by ~50% for every 15°C rise. HPC precipitates from solution at temperatures of 40–45°C (Desmarais, 1973; Aqualon Co., 1987). No gelation occurs during the dissolution-to-precipitation transition, in contrast to MC and MHPC. Precipitation is reversible and the polymer is redissolved below 40°C. The presence of salts or organic substances in the solution lowers the precipitation temperatures because of competition for water in the system (Klug, 1971; Aqualon Co., 1987).

The pH of a 1% HPC solution can range from 5.0 to 8.8. HPC is non-ionic; therefore, its viscosity remains stable at pH values of 2 to 11. To decrease the reduction in viscosity owing to degradation, solutions are stored at pH 6.0–8.0. The surface tension of a 0.1% HPC solution is

440 $\mu N\,cm^{-1}$, compared with 741 $\mu N\,cm^{-1}$ for water, and HPC can, therefore, function as an emulsifying and whipping additive in creams and whipped toppings. HPC solutions are compatible with most natural and synthetic hydrocolloids. A synergistic effect on viscosity is observed upon blending with CMC (Zecher and Van Coillie, 1992).

Since HPC is highly substituted with hydroxypropyl, it is resistant to microbiological attack. Nevertheless, preservatives should be added to solutions stored for prolonged periods. HPC's thermoplasticity makes it processable by all fabrication methods. It has excellent film-forming properties (flexibility, lack of tackiness, good heat sealability, barrier to oil and fat), which are useful in many food, chemical and pharmaceutical applications (Zecher and Van Coillie, 1992).

6.5 Microcrystalline cellulose

MCC only became important in the food industry in the late 1960s. The major uses of this material include pharmaceutical tableting and food stabilization. MCC is a purified native cellulose and not a chemical derivative (Thomas, 1986). During its manufacture, the dissolving α-cellulose pulp is treated with a dilute mineral acid and the cellulose microfibril is unhinged. Hydrolysis is carried out until polymerization levels off (Thomas, 1986). The paracrystalline regions that have a disordered molecular structure are weakened as the acid selectively etches this area from around the densely arranged crystalline cellulose. Subsequent shear releases the cellulose aggregates, which are the basic raw materials for commercial MCC products. Two major types of MCC are produced: the first is powdered MCC (a spray-dried, aggregated, porous, plastic and sponge-like product), and the second is a water-dispersible colloidal cellulose. The first type's primary functions lie in its ability to serve as a binder or disintegrant for tablets, as a flow aid for cheeses, as a carrier for flavors and as a fiber. The colloidal MCC's major functions are as an emulsion stabilizer, a thixotropic thickener, a controller of moisture, a foam stabilizer, an ice-cream controller, a suspension agent and a provider of cling (Thomas, 1986).

The spray-dried MCC is sold in various size ranges. These ranges (average particle sizes of 18 to 90 μm) affect the flow properties of the powder in tableting as well as its oil- and water-absorptive capacities and bulk densities. Because it is not digested, powdered MCC can be used in low-calorie foods or as a source of food fiber. The powder is white, odorless, tasteless, free-flowing and has a heavy-metal content of less than 10 ppm, less than 8 mg soluble substances per 5 g water, 0.1% residue upon ignition, and a pH range for the different types of 5.0 to 7.0.

Colloidal MCC is a food-grade submicron particle that can be used as a food stabilizer. It is a bundle of short-chain cellulose units, with a molecular weight of a few million. Different types of commercial colloidal MCC are produced and their primary use depends on their character. These different products can be dispersed by homogenization, good agitation, spoon stirring, etc., again depending on their character. MCC has no swellable or hydration-like effect, as in starch or other gums, but is dependent on shear and available water. Heat has little effect on its dispersion properties, whereas acids, salts and other ingredients play a major role. Dispersions of colloidal MCC are flocculated by small amounts of electrolytes, cationic polymers and surfactants. MCC dispersions are compatible with most non-ionic and anionic polymers. CMC, MHPC and xanthan gum are the best protective colloids for MCC dispersions. Dispersion (peptization) is very important; principal factors affecting this procedure are dependent on the type of MCC, water hardness, amount of shear available and the order of ingredient addition. MCC has to be one of the first ingredients added to the food system: it cannot be coated with fat since the agglomerate would then not swell and release the microcrystals. MCC requires considerable amounts of water and cannot compete with other gums. The fewer electrolytes in the system, the more effective and the higher the percentage of colloidal material released (Thomas, 1986).

MCC gels are highly thixotropic and have a finite yield value at low concentrations. The formation of a network by solid-particle linkage is responsible for the produced yield value and elasticity. If shear is applied, the gel shears and thins. Resting allows the gel to re-form a network. The addition of CMC reduces the thixotropic character of the gels and results in reduced yield values. Gum addition changes the rheological behavior of MCC gels. Temperature has a small effect on the viscosity of MCC dispersions.

The colloidal MCC functions after being peptized (dispersed) such that individual microcrystals are released into the aqueous phase of the food. Stabilization occurs when a sufficient number of microcrystals have been dispersed (they do not hydrate) into this aqueous phase. Dispersion of insoluble 0.2 μm cellulose particles is aided by CMC, which is used during processing to prevent hydrogen bonding and to improve taste. The CMC's rapid hydration helps link the crystallites together after rehydration to form a thixotropic gel network, which is stable to heat, shear and pH values as low as 3.5. In emulsions, the microcrystals are located in the interface around the fat globules and have emulsion-stabilization properties. The microcrystals have a structuring effect that is useful in foods such as whipped toppings, ice creams and extruded foods. In summary, colloidal MCC functions in foods as an opacifier, suspension aid, ice-crystal controller, emulsion stabilizer, thixotropic thickener, foam stabilizer, heat stabilizer and non-nutritive gelling agent.

In addition to its emulsion-stabilization effect, colloidal MCC is helpful in maintaining product consistency (especially in acidic foods) during high-temperature processing. Another advantage of MCC is its discoloration at these temperatures at low pH (which keeps food quality during heat processing), and the small effect temperature has on its viscosity. Moreover, the colloidal dispersions are either pourable suspensions or smooth thixotropic gels; therefore, via their use different textures can be achieved. If starch–MCC blends at ratios of 3–4:1 are used, a reduction of 20–25% in the amount of starch is necessary to achieve the same performance. The system masks flavor less, improves flow control, creates greater resistance to breakdown and shearing, and improves heat stability as well as that seen in low pH foods (Thomas, 1986).

MCC gel (cellulose gel) is an excellent foam stabilizer. Its addition permits fat reductions of 3–4% in the product. At cellulose gel concentrations >1% and after the addition of sucrose, thixotropic gels are formed. Thus, gelled sucrose systems change their rheological properties following MCC supplementation. MCC at a concentration of ~0.4% helps preserve the original texture of frozen desserts. Other applications involve the thickening of food systems while maintaining a favorable mouthfeel, improving clingability of starch sauces, caloric reduction and fiber addition to special foods, having a whitening effect, improving the suspension of chocolate solids in sterilized drinks, stabilizing emulsions and its use in different formed extruded foods. Cellulose gels may be used in the production of surimi (Nielsen and Pigott, 1994). The gel strength of commercial surimi was increased by the addition of phosphate blends. Further increases were obtained by complete replacement of sodium tripolyphosphate with blended phosphates. This increase in gel strength may allow a reduction in the sodium chloride used to solubilize the surimi and partial replacement of sugars, used as cryoprotectants, by protein or cellulose gels. Phosphate blends also permit non-isothermal flow of surimi, enabling the use of extrusion machinery and processing techniques to produce final products at the site of surimi production (Nielsen and Piggott, 1994).

6.6 Carboxymethylcellulose

6.6.1 General information

Sodium carboxymethylcellulose is a water-soluble, anionic, linear polymer, universally known as CMC (Klug, 1965; Ott *et al.*, 1965; Samuels, 1974). It was developed in Germany during the World War I era as a potential substitute for gelatin. During the 1930s, CMC was used to eliminate the redeposition of soil on fabric during washing and rinsing. CMC was adsorbed to the fabric by hydrogen bonding and because of its anionic nature repelled the dirt, which was also negatively charged. CMC also improved the efficiency of synthetic detergents, thereby promoting renewed interest in its production after the war. It was first produced commercially by Kalle and Co. in Wiesbaden-Biebrich in the late 1930s. Hercules developed a commercial process in 1943 and full-scale commercial production was realized in 1946.

In industries where highly purified types of CMC are desired (i.e. foods, pharmaceuticals and cosmetics), they are referred to as cellulose gum (Stelzer and Klug, 1980). Specifications for the identity and purity of these gums can be found in the *Food Chemicals Codex*, as well as in other FDA and FAO publications. CMC is sold as a white to buff-colored, tasteless, odorless, free-flowing powder. The gum is used in a more varied range of applications than any other known water-soluble polymer (Butler, 1962; Stelzer and Klug, 1980). CMC is used in detergents, drilling fluids, paper, mining, textiles, foods, coatings and cosmetics worldwide (Shelanski and Clark, 1948; Baird and Speicher, 1962; Wirick, 1968; Batdorf and Rossman, 1971). As already noted, although CMC was developed shortly after World War I as a possible replacement for some of gelatin's uses, a major boost in production began following the discovery of its efficiency in improving synthetic detergents (Stelzer and Klug, 1980). Further growth was effected after the gum's approval by the FDA as a food additive. The development of specialized types of CMC followed rapidly as a result of its film-forming ability.

CMC needs to be appropriately packaged to eliminate moisture absorbance. The water sorption properties of poly(ε-lysine) (PEL) CMC dietary complex films was determined. Three CMC samples having various degrees of substitution and molecular weights were applied to show a variety of water-sorption behaviors and physical properties (Ichikawa *et al.*, 1994). Good film swelling in water was achieved when the complex was made from PEL and CMC with a low DS and an appropriate molecular weight, and when the PEL:CMC ratio was low. This indicated that the carboxyl groups in the CMC are not readily dissociated, and that the strength of cationic and anionic interactions affects film swelling (Ichikawa *et al.*, 1994). Information on the influence of electrolytes on the electrical properties of thin cellulose acetate films can be found elsewhere (Friebe and Moritz, 1994).

The thickening ability of CMC, its moisture-binding capacity, dissolution abilities and texturization capabilities have prompted its wide use in foods such as extrudates, emulsions and frozen desserts, among many others.

6.6.2 *Chemical nature and manufacture*

Cellulose is a linear polymer of β-anhydroglucose units, each of which contains three hydroxyl groups (Stelzer and Klug, 1980). CMC is manufactured by treating cellulose with aqueous sodium hydroxide, followed by a reaction with monochloroacetic acid or sodium monochloroacetate in accordance with the Williamson etherification reaction. Esters of mono-

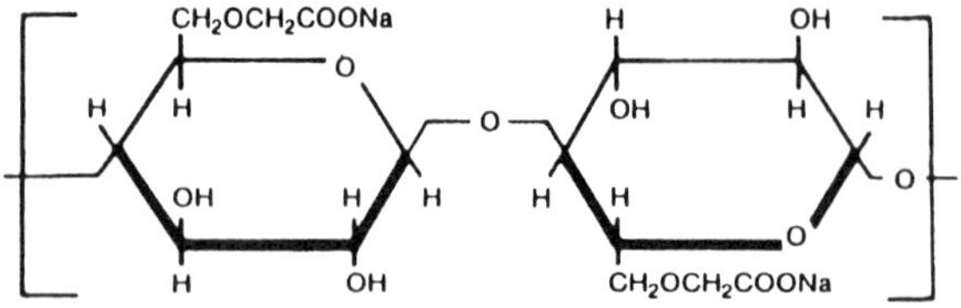

Figure 6.4 Idealized unit structure of carboxymethylcellulose (DS 1.0).

chloroacetic acid have also been used (Taguchi and Ohmiya, 1985), and alternative procedures have been reported (Klug and Tinsley, 1950). If only one of the three hydroxyl groups has been carboxymethylated, the DS is 1.0 (Stelzer and Klug, 1980) (Fig. 6.4).

Technical-grade CMC has a purity of 94–99%, and that used in foods has a minimum purity of 99.5%. Other physical properties include 8% moisture content when packed, a browning temperature of 227°C, a charring temperature of 252°C and a bulk density of 7.5 $g\,l^{-1}$. It has a biological oxygen demand (BOD) after 5 days of incubation of 11 000 ppm for high-viscosity CMC with DS 0.8, and 17 300 ppm for low-viscosity CMC with DS 0.8 (under these conditions cornstarch has a BOD of 800 000 ppm). A 2% CMC solution has a specific gravity of 1.0068 at 25°C and a refractive index of 1.3355. The typical pH of a 2% solution is 7 and the surface tension of a 1% solution at 25°C is 71 $dyn\,cm^{-1}$ (Stelzer and Klug, 1980). A DS of 0.4–1.4 is usual for commercial CMC and can be adjusted to higher values for specialized products. If the DS is less than 0.4, the resultant CMC is not water soluble (Zecher and Van Coillie, 1992). Food-grade CMC (2%) has a DS of 0.65–0.95 and a minimum viscosity of 25 mPa s. Viscosities can be controlled by oxidative degradation of the crude product with hydrogen peroxide to obtain low-viscosity CMC. The higher the DS, the higher the solubility of the polymer (Batdorf and Francis, 1963; Butler and Keim, 1965; Zecher and Van Coillie, 1992).

6.6.3 *Chemical and physical properties*

(a) Solution properties. CMC is soluble in either hot or cold water, making it quite versatile. Solubility is also possible in water-miscible organic solvents such as water–ethanol. Low-viscosity types are more tolerant to increasing ethanol concentrations than their higher viscosity counterparts. They can tolerate up to 50% ethanol or 40% acetone. This property is important in alcoholic beverages and instant bar mixes, where viscosity and clarity are desired (Keller, 1984). Sodium CMC is a carboxylic acid salt. Its dilute solution has a neutral pH and virtually all of its carboxylic acid groups are in the sodium salt form, very few in the free acid form. At pH 3.0

or lower, CMC reverts to a free-acid insoluble form. The pK of sodium CMC ranges from 4.2 to 4.4 and varies somewhat with DS (Keller, 1984). Upon dissolution, CMC passes through dispersion and hydration stages. Without good dispersion, clumps are formed with a swollen outer skin of partially hydrated gum. To eliminate this phenomenon, the gum needs to be added at a slow enough rate for each particle to separate discreetly. Other techniques include dry blending in sugar or other water-miscible non-solvent such as glycerin, sorbitol or propylene glycol, or dispersion of the gum in oil and then addition of the mixture including an emulsifier to an aqueous system. Dissolution can be accomplished by feeding the gum through a smooth-walled funnel into a waterjet eductor where it is dispersed by turbulent water flow that creates a suction effect. Such an operation permits 80–90% instant wetting and hydration of the gum. The addition of small amounts of dioctyl sodium sulfosuccinate to the gum prior to its dissolution in water inhibits its tendency to clump (Ben-Zion, 1995). CMC dissolves when the hydroxyl groups are replaced by CM groups, in the cellulose the chains separate and the unreacted hydroxyl groups become available for association with water. Complete solubility and subsequent solution properties are maintained as the gum chains are held apart by Coulomb repulsion, which allows the entrance of water and full interaction with the chain hydroxyls. Salts tend to decrease gum hydration and its corresponding viscosity (Ganz, 1973).

(b) Viscosity. Probably the most useful property of CMC is its ability to impart viscosity (Stelzer and Klug, 1980). CMC is a linear polymer and exhibits the typical rheological properties of linear polymers. Most solutions of CMC are pseudoplastic, i.e. the higher the shear rate the higher the decrease in measured viscosity. Most products having DS values below ~1.0 are also thixotropic (Stelzer and Klug, 1980). This may result from the many aggregates in the aqueous dispersion (Stelzer and Klug, 1980). In other words, viscosity depends on shear rate and time. Thixotropy is also a function of uniformity of substitution. Uniformity of substitution also increases tolerance to acidic systems and dissolved ions. The presence of salts in solution represses the disaggregation of CMC and, therefore, affects its viscosity (Brown and Henley, 1964). Higher viscosities can be obtained by dissolving CMC in a glycerin–water mixture. Another effective way of achieving high viscosities is to blend CMC with non-ionic cellulose derivatives. Such blends in a 1% solution give about twice the viscosity expected from the individual ingredients (DeButts *et al.*, 1957; Francis, 1961).

CMC flow characteristics vary from thixotropic to smooth (Batdorf and Rossman, 1973). The smooth-flowing type is preferred in food systems such as syrups or frostings where a smooth consistency is desired; thixotropic CMCs are preferred in sauces and purées, which are grainier (Keller, 1984). A finely ground CMC is preferred where viscosity in liquid and hydration need to be rapidly achieved. Coarse grinding is performed when the dispersion is to be produced from cellulose derivatives. The pH of a 1% CMC solution is typically in the range 7.0–8.5. Solution viscosity is almost uninfluenced in a pH range 5–9, whereas at pH <3, viscosity may increase and precipitation of the free acid form of CMC may occur; at pH >10 there is a slight decrease in viscosity. Therefore, cellulose gums should not be employed in highly acidic food systems (Ingram and Jerrard, 1962; Batdorf and Rossman, 1973; Stelzer and Klug, 1980). CMC solutions containing 5% acetic acid or 1.0% citric or lactic acids can be stored for months at room temperature with no significant change in viscosity (Stelzer and Klug, 1980).

CMC solutions can be heat treated, e.g. at 80°C for 30 min or 100°C for 1 min, to destroy bacteria capable of degrading the polymer. Such a treatment has almost no influence on the CMC. For long periods of storage, the addition of preservatives is recommended and the pH should be maintained at between 7 and 9, while avoiding elevated temperatures, oxygen and sunlight. The viscosity of CMC solutions decreases with increases in temperature. This phenomenon is reversible unless the heating is prolonged, in which case the damage is permanent (Zecher and Van Coillie, 1992). If salts are present in solution, CMC disaggregation is repressed and viscosity is influenced and depends on whether the metal is mono-, di- or trivalent. Monovalent ions (except for silver) have little effect, divalent ions (calcium, magnesium) lower viscosity and trivalent ions (aluminum, chromium, iron) can render CMC insoluble or cause gelation via their complexing with carboxylic groups. The order of addition of salts to the CMC solution is important in terms of the resultant viscosity. After dissolution of the gum, salts have little effect on viscosity. If the salt is dissolved before the CMC is added, it inhibits disaggregation and lowers viscosity. CMC's wide range of applications is a consequence of its compatibility with many other food ingredients. Sometimes blending CMC with other gums results in increased viscosity. CMC can interact with proteins (casein, soy proteins, etc.) (Ganz, 1974). CMC, in contrast to MC, MHPC and HPC, is not highly surface active. Because of its considerable ability to bind water, it is used in gels, pie fillings and other foods to minimize syneresis or to increase water-binding capacity (Sanderson, 1981; Aqualon Co., 1988).

CMC gelation can be favored by carefully selecting ion concentration, regulating the pH and controlling ion release via the use of appropriate chelating agents. The appropriate ions are believed to serve as cross-linking agents for polymer chains via salt formation with adjacent anionic groups (Ganz, 1966). Several different gel compositions are possible. High-DS cellulose gum gelation occurs with trivalent cations. Possible gelation of sufficiently low-DS cellulose gums is possible when they are subjected to high-shear conditions. The lowest DS type exhibits the firmest gel strength, with a smooth, greasy texture.

(c) Stability and physical data. If no bacterial contamination occurs, the cellulose gum solution remains stable, with no reduction in viscosity. Heat treatment (e.g. 1 min/100°C) is necessary to prevent microbiological contamination. Long-term storage requires the addition of an appropriate preservative such as sodium benzoate, sorbic acid or sodium propionate. In foods that are supposed to contain cellulases, the enzyme needs to be inactivated to prevent a drastic decrease in viscosity. Since the attack on cellulose gum is supposed to occur in solution, its storage as a dry powder where moisture is kept at a minimum ensures no undesirable changes. After its manufacture using solvents, the product is in an aseptic state. Since microbial attack of the gum results from microbial growth, only proper sanitary precautions can ensure its stability. Other stability factors that influence cellulose degradation are UV radiation and the presence of molecular oxygen and heavy metals, which serve as catalysts in its oxidative degradation. Therefore, chelators such as sodium citrate or hexametaphosphate are used to stabilize cellulose gum in foods in which entrained air and heavy metals are present. Three particle-size designations of Hercules cellulose gum exist: regular, coarse and fine. On the one hand, cellulose gum is a hygroscopic material, and suitable packaging is therefore necessary for producers and consumers. On the other hand, in foods requiring the retention of moisture to maintain eating quality (e.g. cakes and frostings), the higher DS-grade gums are the best water binders (Stelzer and Klug, 1980; Zecher and Van Coillie, 1992).

6.7 Food applications

Cellulose derivatives are used in many food products. The use of sodium CMC is growing in food applications, especially in developed countries where the demand for convenience has been growing rapidly since the early 1950s (Stelzer and Klug, 1980). Many, or at least one, functions (binding, thickening, stabilization, moisture retention, etc.) can be achieved by the inclusion of cellulose derivatives in foods at levels of 0.1–0.5%, generally less than 1%. Texturization is beneficial in frozen desserts, other desserts and low-calorie products; thickening in bakery batters, pet foods, gravies, low-calorie foods, soups, sauces, snack foods, instant beverages and desserts; bodying in beverages, desserts and syrups; mouthfeel in products such as beverages, frozen desserts, desserts, syrups, low-calorie products, and instant beverages; suspension in low-calorie products, beverages, fruit drinks, protein beverages, instant beverages and desserts; and binding properties in extruded foods, pet foods, snacks and speciality foods (Stelzer and Klug, 1980).

Industrial food uses include frozen desserts, soft drinks, bakery products, structured and coated products, among others. Frozen desserts include ice creams, ice milks, sherbets, water ices and frozen yogurts. In ice cream, the maximal level of CMC allowed is 0.5%. Stabilizers other than CMC are LBG, guar gum, gelatin, alginate and carrageenan. These increase viscosity by rapid hydration and postpone formation or enlargement of ice crystals. If no or only small amounts of 0.05% CMC are added to ice cream, iciness is more pronounced than in products with 0.2% CMC (Moore and Shoemaker, 1981). Viscosity increases and melting speed decreases in ice cream mixes with increasing CMC concentration. The addition of cellulose gum also enables ice cream to withstand thermal shock and improves its mouthfeel (Stelzer and Klug, 1980). Commercial-blend stabilizers based on combinations of guar gum, LBG and CMC produced lower apparent viscosities than the pure polysaccharides (Goff and Davidson, 1994).

Generally more than one gum is used as a stabilizer (Keeney, 1982; Kloow, 1985; Bassett, 1988) and to ease dispersion hydrocolloids are sometimes coated with monoglycerides. Dissolution takes place at pasteurization or homogenization temperatures following dispersion. Quick-freezing is necessary to avoid large ice crystals and the addition of stabilizer reduces the risk of ice-crystal enlargement during storage, transportation and at the consumer level; three stages at which temperature fluctuations may occur. High-viscosity CMC and LBG are a recommended blend for texture, taste and stand-up properties of ice cream (Cottrell *et al.*, 1979). The sensory perception of three sweeteners in the presence of pseudoplastic hydrocolloids was studied at 27°C, in solutions with viscosities of 230 and 500 mPa s at a shear rate of 50 s^{-1}. Comparisons were made between 10% sucrose, 9% fructose and 0.133% aspartame. Sweetness was most decreased by guar gum and least by oat gum; the identity of the thickener had more of an effect than viscosity. The sweetness of sucrose was decreased more than that of aspartame with guar gum; fructose sweetness remained equal to sucrose sweetness with guar gum, equal to aspartame sweetness with oat gum and intermediate with CMC (Malkki *et al.*, 1993).

Ice milk contains less total solids than ice cream and its content of non-fat milk solids and sweeteners is higher. To improve product smoothness and chewiness, a sucrose–corn syrup blend can be used along with a higher concentration of stabilizer relative to ice creams. CMC, guar gum and LBG are used in combination with carrageenan, the latter to prevent whey separation. A combination of CMC with other stabilizers is also recommended for stabilizing soft ices. CMC is also used to stabilize sherbet texture (sherbet contains fruits such as orange, lemon or strawberry) (Huber and Rowley, 1988; Huber *et al.*, 1989).

In fruit juice-based drinks, it is often difficult to maintain the pulp in suspension during storage. The CMC concentration that will produce good stability depends on the soluble-solids content (Zejian *et al.*, 1994). In general, the desired amount is low since the viscosity of the product is high from the beginning. CMC also reduces or prevents the formation of oily ring in bottlenecks. CMC is regularly added to mixtures of fruit products after their hydration. If other gums, such as xanthan gum and PGA are used, then low-viscosity CMC is preferable, whereas if CMC is used alone, high-viscosity types are recommended (Jackman, 1979; Deleon and Boak, 1984). CMC is recommended in dry mixes for acidic milk drinks (Sirett *et al.*, 1981) and for the preparation of effervescent drinks (Valbonesi and Cochin, 1981). CMC is also used to give body to low-calorie drinks. In the production of instant hot drinks, HPC used as a suspending agent combined with a warm liquid (e.g. alcohol), enables uniform release over long periods of use (Marmo and Rocco, 1982). Cellulosic derivatives are also used to thicken alcohols used for flavoring in other products or for cocktails.

Cellulosic derivatives are used in baked goods to overcome problems related to flour quality as well as to improve the characteristics of the final product. CMC can be used in cake mixes as a batter binder (Bayfield, 1962). CMC and other cellulose ethers increase viscosity rapidly and ease air incorporation, thereby preventing undesirable reductions in viscosity during cooking. The addition of CMC to cakes improves the homogeneity of the product, especially if raisins, pieces of crystallized fruit or chocolate chips are blended into its preliminary mixture. The moisture content retained by the cakes after such an addition is increased, resulting in a less stale product. CMC is used to adjust paste consistency if pieces or parts of raisins, grains or other ingredients are blended into the product mixture. Water retention also increases. In contrast, MC derivatives tend to lose moisture during baking and to form a gel, which leads to the production of gluten-free breads (Anon., 1985). Less gluten increases dough-mixing time, reduces the resultant loaf volume and changes the texture to an unpleasant roughness. The addition of up to 0.5% MHPC to low-gluten flour will improve these parameters. If other types of flour such as rice or potato are used for gluten-free breads, then a combination of CMC and MHPC added to the flour will give characteristics similar to those of breads made with wheat flour (Ylimaki *et al.*, 1988). High-fiber breads can be produced by including a mixture of flour, bran, CMC and guar (Foda *et al.*, 1987). CMC and MC derivatives are used in other baked goods such as low-calorie cakes, biscuits, ice-cream cones, etc. (Glicksman *et al.*, 1985). A dust-free, dried anticaking agent based on cellulose has also been described (Chappell, 1995). CMC is used in fruit pies and pastry fillings to improve water retention, and in doughnuts to reduce oil absorption during frying and to improve texture.

CMC is an anionic polymer that reacts with proteins at their isoelectric pH to form a complex. Formation is influenced by pH, molecular weight, CMC concentration and salt content. The soluble complex (resulting from the reaction between CMC and casein or other milk proteins) is stable to heat treatment and during storage. Therefore, a preparation of acidified, pasteurized dairy products and other dairy drinks (with whey or buttermilk) is possible when the gum (0.2–0.5%) is dissolved directly into the acidic milk product, or into the milk before the acid or fruit is added (Shenkenberg *et al.*, 1971). At neutral pHs, milk proteins can react with CMC to cause whey separation in low-viscosity mixes such as ice creams and ice milks. Separation can be prevented by reacting carrageenan with the caseinates instead. At a pH of ~3.2, CMC is used to precipitate whey proteins (Hansen *et al.*, 1971; Abdel-Baky *et al.*, 1981). Milk desserts favor a combination of starches and carrageenans to yield textural improvements. Combinations of MHPC and CMC can be used as a bulking agent (Hood, 1981). Whipped cream based on milk and vegetable proteins can be improved by the addition of HPC (Luzietti and Coacci, 1990).

Extruded, structured and coated products can potentially make use of cellulose derivatives as binders of small pieces of meat, fish (da Ponte *et al.*, 1987) and potatoes to produce meat sticks and many potato-based products. The potential reactivity of MC or CMC with proteins and their cohesion ability are essential in such products (Bernal and Stanley, 1989). The batter coating of such products can also contain cellulosic derivatives. This improves its adhesion to the stick, makes it stable after freeze–thaw cycles and reduces its oil absorption during deep-fat frying. Predusting of a product before cooking or reheating can help yield crispier products during microwave cooking. In coating non-structured products, CMC can be added at a maximum of 4% to the batter (Suderman *et al.*, 1981). The predust should contain at least 20% MHPC, having 27–31% methoxyl groups and 6–12% hydroxypropyl groups (D'Amico *et al.*, 1989). Chitosan, CMC, CMC–chitosan and another 12 coating materials were used to coat fruits and to eliminate growth of molds (Alzaemey *et al.*, 1993). The potential of edible-barrier technology for the extension of the shelf life of processed foods and the creation of innovative foods, as well as for health applications, is promising (Malsch and Paul, 1993; Koelsch, 1994). Several applications for cellulases or hemicellulases are also being developed for textile, food and paper-pulp processing (Beguin and Aubert, 1994). Since cellulose derivatives reduce oil absorption, they are also used as ingredients or as formers of edible barriers (Kester and Fennema, 1989), in composite foods such as ice-cream cones and ice cream, tomato paste and pizza bases. In sausage casings (Higgins and Madsen, 1986), CMC improves peelability in high-speed peeling machines. Other important uses are the thickening of emulsified and non-emulsified sauces (Vincent and Harrison, 1987) to prevent separation, and to preserve the consistency and appearance of sauces or soups that are heated before consumption, thereby utilizing the thermogelation ability of cellulose derivatives. Products that are consumed hot should be prepared cool to prevent problems with MC or MHPC hydration at concentrations of 0.25–1.0%. Other applications involve the coating of food products (meat and poultry) where water absorption properties are desired (Midkiff *et al.*, 1990) and as a water binder in semi-humid products. Cellulose derivatives can also be used in the low-calorie products (bakery, meat and dairy) most desired by many consumers. Dietary high-viscosity MHPC reduced plasma and liver cholesterol concentrations in cholesterol-fed hamsters (Gallaher *et al.*, 1993). The results suggest that materials that increase the viscosity of the intestinal contents can be effective in reducing plasma cholesterol and that only moderate increases in viscosity are necessary to achieve this effect (Gallaher *et al.*, 1993).

References

Abdel-Baky, A.A., El-Fak, A.M., Abo El-Ela. *et al.* (1981) Fortification of Domiati cheese milk with whey proteins/carboxymethylcellulose complex. *Dairy Ind. Int.*, **46**(9), 29, 31.

Alzaemey, A.B., Magan, N. and Thompson, A.K. (1993) Studies on the effect of fruit-coating polymers and organic acids on growth of *Colletotrichum-Musae* in vitro and on postharvest control of Anthracnose of bananas. *Mycol. Res.*, **97**, 1463–8.

Anon. (1985) Methylcellulose in low-gluten bread. *Food Feed Chem.*, **17**(11), 576.

Aqualon Co. (1987) *Klucel HPC, Physical and Chemical Properties.* Wilmington, DE.

Aqualon Co. (1988) *Cellulose Gum, Physical and Chemical Properties.* Wilmington, DE.

Aqualon Co. (1989) *Culminal MC, MHEC, MHPC, Physical and Chemical Properties; Benecel High Purity MC, MHEC, MHPC, Physical and Chemical Properties.* Wilmington, DE.

Baird, G.S. and Speicher, J.K. (1962) Carboxymethylcellulose, in *Water-soluble Resins*, ch. 4 (eds R.L. Davidson and M. Sittig), Reinhold, New York.

Bassett, H. (1988) Stabilization and emulsification of frozen desserts. *Dairy Field*, **171**(5), 22–5.

Batdorf, J.B. and Francis, P.S. (1963) The physical behavior of water-soluble cellulose polymers. *J. Soc. Cosmet. Chem.*, **XIV**(3), March.

Batdorf, J.B. and Rossman, J.M. (1971) Sodium carboxymethylcellulose, in *Industial Gums*, ch. 25 (ed. R.L. Whistler), Academic Press, New York.

Batdorf, J.B. and Rossman, J.M. (1973) Sodium carboxymethylcellulose, in *Industrial Gums*, 2nd edn, ch. 31 (ed. R.L. Whistler), Academic Press, New York.

Bayfield, E.G. (1962) Improving white layer cake quality by adding CMC. *Bakers Digest*, **36**(2), 50–2, 54.

Beguin, P. and Aubert, J.P. (1994) The biological degradation of cellulose. *FEMS Microbiol. Rev.* **13**(1), 25–58.

Ben-Zion, O. (1995) Physical properties of adhesed hydrocolloid systems. MSc Thesis, Hebrew University of Jerusalem.

Bernal, V.M. and Stanley, D.W. (1989) Technical note: methylcellulose as a binder for reformed beef. *Int. J. Food Sci. Tech.*, **24**, 461–4.

Brown, W. and Henley, D. (1964) The configuration of the polyelectrolyte sodium carboxymethylcellulose in aqueous sodium chloride solutions. *Makromol. Chem.*, **79**, 68–88.

Butler, R.W. (1962) *Free Acid Cellulose Ether Film.* US Patent No. 3,064,313.

Butler, R.W. and Keim, G.I. (1965) *Insolubilizing CMC with Cationic Epichlorohydrin-modified Polyamide Resin.* US Patent No. 3,224, 986.

Butler, R.W. and Klug, E.D. (1980) Hydroxypropylcellulose, in *Handbook of Water-soluble Gums and Resins*, ch. 13, (ed. R.L. Davidson), McGraw-Hill, New York.

Chappell, R.A. (1995) *Low Dust Powdered Cellulose.* US Patent No. 5,391,382.

Cottrell, J.I.L., Pass, G. and Phillips, G.O. (1979) Assessment of polysaccharides as ice cream stabilizers. *J. Sci. Food Agric.*, **30**, 1085–8.

D'Amico, L.R., Waring, S.E. and Lenchin, J.M. (1989) *Composition for Preparing Freeze–Thaw Microwaveable Pre-fried Foodstuffs.* US Patent No. 4,842,874.

da Ponte, D.J.B., Roozen, J.P. and Pilnik, W. (1987) Effects of iota carrageenan, carboxymethylcellulose and xanthan gum on the stability of formulated minced fish products. *Int. J. Food Sci. Tech.*, **22**(2), 123–33.

DeButts, E.H., Hudy, J.A. and Elliott, J.H. (1957) Rheology of sodium carboxymethylcellulose solutions. *Ind. Eng. Chem.*, **49**, 94–8.

Deleon, J.R. and Boak, M.G. (1984) *Method for Preventing the Separation in Fruit Juice-containing Products with Propylene Glycol Alginate and Sodium Carboxymethylcellulose.* US Patent No. 4,433,000.
Desmarais, A.J. (1973) Hydroxyalkyl derivatives of cellulose, in *Industrial Gums*, 2nd edn, ch. 29 (ed. R.L. Whistler), Academic Press, New York.
Donges, R. (1990) Nonionic cellulose ethers. *Br. Polym. J.*, **23**, 315–26.
Dow Chemical Co. (1974) *Handbook on Methocel Cellulose Ether Products.* Midland, MI.
Foda, Y.H., Mahmoud, R.H., Gamal, N.F. *et al.* (1987) Special bread for body-weight control. *Ann. Agric. Sci.*, **32**(1), 397–407.
Francis, P.S. (1961) Solution properties of water-soluble polymers. *J. Appl. Polym. Sci.*, **5**(15), 261–70.
Friebe, A. and Moritz, W. (1994) Influence of electrolyte on electrical-properties of thin cellulose-acetate membranes. *J. Appl. Polym. Sci.*, **51**(4), 625–34.
Gallaher, D.D., Hassel, C.A. and Lee, K.J. (1993) Relationships between viscosity of hydroxypropyl methylcellulose and plasma cholesterol in hamsters. *J. Nutrit.*, **123**(10), 1732–8.
Ganz, A.J. (1966) Cellulose gum – texture modifier. *Manuf. Confect.*, **46**(10), 23–33.
Ganz, A.J. (1973) Some effects of gums derived from cellulose on the texture of foods. *Cereal Sci. Today*, **18**(12), 398.
Ganz, A.J. (1974) How cellulose gum reacts with protein. *Food Eng.*, June, 67–9.
Ganz, A.J. (1977) Cellulose hydrocolloids, in *Food Colloids* (ed. H. Graham), AVI Press, Westport, CT, pp. 382–417.
Glicksman, M. (1969) *Gum Technology in the Food Industry*, Academic Press, New York.
Glicksman, M. (1986) *Food Hydrocolloids*, vol. III, CRC Press, Boca Raton, FL, pp. 3–121.
Glicksman, M., Frost, J.R., Silverman, J.E. *et al.* (1985) *High Quality, Reduced-calorie Cake.* US Patent No. 4,503,083.
Goff, H.D. and Davidson, V.J. (1994) Controlling the viscosity of ice cream mixes at pasteurization temperatures. *Modern Dairy*, **73**(3), 12, 14.
Greminger, G.K., Jr and Krumel, K.L. (1980) Alkyl and hydroxyalkylcellulose, in *Handbook of Water-soluble Gums and Resins*, ch. 3 (ed. R.L. Davidson), McGraw-Hill, New York, pp. 3.1–3.25.
Hansen, P.M.T., Hildalgo, J. and Gould, I.A. (1971) Reclamation of whey protein with carboxymethylcellulose. *J. Dairy Sci.*, **54**(6), 830–4.
Hercules Inc. (1978) *Cellulose Gum – Chemical and Physical Properties*, Hercules Inc., Wilmington, DE.
Heuser, E. (1944) *The Chemistry of Cellulose*, Wiley, New York, pp. 379–91.
Higgins, T.E. and Madsen, D.P.D. (1986) *Cellulosic Casing with Coating Comprising Cellulose Ether, Oil, Water-insoluble Alkylene Oxide Adduct of Fatty Acids.* US Patent No. 4,596,727.
Hood, H.P. (1981) Whipped sour cream in aerosol can. *Food Eng.*, **53**(8), 62.
Huber, C.S. and Rowley, D.M. (1988) *Soft-serve Frozen Yoghurt Mixes.* US Patent No. 4,737,374.
Huber, C.S., Rowley, D.M. and Griffiths, J.W. (1989) *Soft-frozen Water Ices.* US Patent No. 4,826,656.
Ichikawa, T., Mitsumura, Y. and Nakajima, T. (1994) Water-sorption properties of poly(epsilon-lysine)–carboxymethyl cellulose (CMC) dietary complex films. *J. Appl. Polym. Sci.*, **54**(1), 105–12.
Ingram, P. and Jerrard, H.G. (1962) *Nature*, **196**, 57.
Isogai, A. and Atalla, R.H. (1991) Amorphous cellulose stable in aqueous media: regeneration from SO_2-amine solvent system. *J. Polym. Sci.: Part A*, **29**, 113–19.
Jackman, K.R. (1979) *Citrus Fruit Juice and Drink.* US Patent No. 4,163,807.
Keeney, P.G. (1982) Development of frozen emulsions. *Food Tech.*, November, 65–70.
Keller, J. (1984) *Sodium Carboxymethylcellulose*, Special Report, NY State Agricultural Experimental Station, No. 53, pp. 9–19.
Kester, J.J. and Fennema, O. (1989) An edible film of lipids and cellulose ethers barrier properties to moisture vapor transmission and structural evaluation. *J. Food Sci.*, **54**(6), 1383–9.
Kloow, G. (1985) Viscosity characteristics of high viscosity grade carboxymethyl cellulose, in *Cellulose and its Derivatives: Chemistry, Biochemistry and Applications*, ch. 32 (eds J.F. Kennedy, G.O. Phillips, D.J. Wedlock and P.A. Williams), Halsted Press, New York.

Klug, E.D. (1965) Sodium carboxymethylcellulose, in *Encyclopedia of Polymer Science and Technology*, vol. III (eds H.F. Mark, N.G. Gaylord and N.M. Bikales), Interscience Publishers, New York, pp. 520–39.
Klug, E.D. (1966) *US Cellulose and Process*. US Patent No. 3,278,521.
Klug, E.D. (1967) *Mixed Cellulose Ethers*. US Patent No. 3,357,971.
Klug, E.D. (1971) *J. Polym Sci., Part C*, **36**, 491–508.
Klug, E.D. and Tinsley, J.S. (1950) *Preparation of Carboxyalkyl Ethers of Cellulose*. US Patent No. 2,517,577.
Koelsch, C. (1994) Edible water vapor barriers – properties and promise. *Trends Food Sci. Technol.*, **5**(3), 76–81.
Krassig, D.H. (1985) Structure of cellulose and its relation to properties of cellulose fibers, in *Cellulose and its Derivatives: Chemistry, Biochemistry and Applications*, ch. 1 (eds J.F. Kennedy, G.O. Phillips, D.J. Wedlock and P.A. Williams), Halsted Press, New York.
Luzietti, D. and Coacci, S. (1990) *A Long Life Semifinished Food Product Useful as a Vegetable Substitute for Whip Cream in Generic Pastry Articles and Ice Creams*. European Patent No. 354,356.
Malkki, Y., Heinio, R.L. and Autio, K. (1993) Influence of oat gum, guar gum and carboxymethyl cellulose on the perception of sweetners and flavor. *Food Hydrocolloids*, **6**(6), 525–32.
Malsch, G. and Paul, D. (1993) Chemical modification of cellulose membranes. *Papier*, **47**(12), 710–19.
Marmo, D. and Rocco, F.L. (1982) *Process for Producing a Flavored Heated Beverage*. US Patent No. 4,311,720.
McCormick, C.L. and Callais, P.A. (1987) Derivatives of cellulose in LiCl and *N,N*-dimethylacetamide solutions. *Polymer*, **28**, 2317–22.
Michie, R.I.C. and Neale, S.M. (1964) *J. Polym Sci., Part A*, 2063–83.
Midkiff, D.G., Twyman, N.D., Rippl, G.G. *et al.* (1990) *Absorbent Structure for Absorbing Food Product Liquids*. European Patent No. 0353334 Al.
Moore, L.J. and Shoemaker, C.F. (1981) Sensory textural properties of stabilized ice cream. *J. Food Sci.*, **46**(2), 399–402, 409.
Nielsen, G.R. and Pigott, G.M. (1994) Gel strength increased in low-grade heat-set surimi with blended phosphates. *J. Food Sci.*, **59**(2), 246–50.
Ott, E., Spurlin, H.M. and Grafflin, M.W. (1965) *Cellulose and Cellulose Derivatives*, 2nd edn, part II, Interscience, New York, pp. 937–45.
Ramby, B.G. and Rydholm, S.A. (1956) Cellulose and cellulose derivatives, in *Polymer Processes*, vol. x (ed. C.E. Schildknecht), Wiley Interscience, New York, pp. 351–429.
Samuels, R.M. (1974) Quantitative structural analysis of mechanical behavior in polycrystalline polymers. *Appl. Polym. Symp.*, **24**, 37–43.
Sanderson, G.R. (1981) Polysaccharides in foods. *Food Tech.*, July, 50–7.
Shelanski, H.A. and Clark, A.M. (1948) Physiological action of sodium carboxymethylcellulose on laboratory animals and humans. *Food Res.*, **13**(1), 29–35.
Shenkenberg, D.R., Chang, J.C. and Edmondson, L.F. (1971) Developed milk orange juice. *Food Eng.*, April, 97–8, 101.
Sirett, R.R., Eskritt, J.D. and Derlatka, E.J. (1981) US Patent No. 4,264,638.
SRI International (1989) *Chemical Economics Handbook*, Section 581.5000A (Cellulose Ethers), Menlo Park, CA.
Stelzer, G.I. and Klug, E.D. (1980) Carboxymethylcellulose, in *Handbook of Water-soluble Gums and Resins*, ch. 4 (ed. R.L. Davidson), McGraw-Hill, New York, pp. 4.1–4.28.
Suderman, D.R., Wiker, J. and Cunningham, F.E. (1981) Factors affecting adhesion of coating to poultry skin. *J. Food Sci.*, **46**(4), 1010–11.
Taguchi, A. and Ohmiya, T. (1985) *Sodium Carboxymethylcellulose Esterification of an Alkali Cellulose with an Ester of Monochloroacetic Acid and Isopropanol*. US Patent No. 2,517,577.
Thomas, W.R. (1986) Microcrystalline cellulose, in *Food Hydrocolloids*, vol. III, ch. 1 (ed. M. Glicksman), CRC Press, Boca Raton, FL, pp. 9–43.
Valbonesi, F. and Cochin, A. (1981) *Effervescent Product for Preparing Sparkling Drinks*. French Patent No. 2,478,955.
Vincent, A. and Harrison, S. (1987) Stabilising dressings and sauces. *Food Trade Rev.*, October, 527–8, 531.

Walton, A.G. and Blackwell, J. (1973) in *Biopolymers*, Academic Press, New York, pp. 464–74.
Ward, K., Jr and Seib, P.A. (1970) Cellulose, lichenan and chitin, in *The Carbohydrates – Chemistry and Biochemistry*, 2nd edn., vol. 2A (eds W. Pigman, D. Horton and A. Herp), Academic Press, New York, pp. 413–45.
Whistler, R.L. (1973) *Industrial Gums*, 2nd edn, Academic Press, New York.
Whistler, R.L. and Zysk, J.R. (1978) Carbohydrates, in *Encyclopedia of Chemical Technology*, 3rd edn, vol. 4 (eds M. Grayson and E. Eckroth), Wiley, New York, pp. 535–55.
Wirick, M.G. (1968) A study of the enzymic degradation of CMC and other cellulose ethers. *J. Polym. Sci.*, **A1**(6), 1965–74.
Ylimaki, G., Hawrysh, Z.J., Hardin, R.T. *et al.* (1988) Application of response surface methodology to the development of rice flour yeast breads: objective measurements. *J. Food Sci.*, **53**(6), 1800–5.
Zecher, D. and Van Coillie, R. (1992) Cellulose derivatives, in *Thickening and Gelling Agents for Food* (ed. A. Imeson), Blackie A & P, Glasgow, pp. 40–65.
Zejian, L., Baodong, Z. and Lijiao, C. (1994) Test on the rheological behavior of beverage stabilizer and beverage quality control. *Trans. Chinese Soc. Agric. Eng.*, **10**(2), 110–13.

7 Exudate gums

7.1 Introduction

Exudate gums have been used for centuries in a variety of fields; they have retained their importance despite the many alternative gums, with similar typical performances, that have since come into existence. The gums exude from trees and shrubs in tear-like, striated nodules or amorphous lumps, then dry in the sun, forming hard, glassy exudates of different colors, from white to pale amber for gum arabic, pale gray to dark brown for karaya gum, and white to dark brown for tragacanth. Gum production increases at high temperatures and limited moisture, and yields can be increased by making incisions in the bark or stripping it from the tree or shrub. Exudate gums have been utilized for food applications for many years, for emulsification, thickening and stabilization. Gum arabic, tragacanth and karaya gum are safe for human consumption, based on a long, safe history as well as on recent toxicological studies. Tree gum exudates are also used in non-food applications, such as pharmaceuticals, cosmetics, textiles, lithography and minor forest products (Wang and Anderson, 1994).

7.2 Gum arabic

7.2.1 Sources and producers

Gum arabic is a natural exudate of the acacia tree (the main source is *Acacia senegal* L., but minor quantities of the gum can be obtained from other species). Acacia trees can be found in the sub-Saharan zones of Africa, Australia, India and America, where they have been planted to reduce soil erosion and as a gum source. The main gum-producing countries are Senegal, Mali, Mauritania, Niger, Chad and Sudan. Gum arabic is also called 'hashab' after the local name of the tree, or 'Kordofan' after the name of the main production area in the Sudan (Imeson, 1992).

7.2.2 Manufacture and processing

The gum is exuded from the trees in large striated nodules or tears (Fig. 7.1). Collection of sun-dried gum lumps is manual, and there are two main harvests during the dry season (Thevenet, 1988). Yield per tree does not

Figure 7.1 'Tear drops' of gum arabic, the dried, gummy exudate obtained from acacia trees.

exceed ~300 g. Collection is followed by differentiation into two main grades: cleaned and hand-picked selected. Gum arabic can be sold as is (natural state), or after technological treatments such as grinding (uncleaned material), sieving or granulating. Raw gum is precleaned to remove bark, sand and fines, thereby improving its quality; foreign matter makes up less than 0.5% of food-grade powdered gum. Processing, which involves sieving, decantation, centrifugation, concentration, pasteurization and atomization followed by spray-drying, yields a product with no insoluble matter that hydrates more rapidly than its unprocessed counterpart (Williams *et al.*, 1990). Particle size and agglomeration are controlled to ensure the desired functional properties. Gum flakes can be produced by applying the solution to steam-heated rollers; this yields large flakes that are removed from the rollers with a built-in knife. These flakes have a large surface area, facilitating their hydration. The drying temperature influences the degree to which the gum's proteinaceous fraction is denatured, which in turn influences its functional behavior. Heat in general (either from spray-drying or roller-drying) causes the gum solutions to be slightly turbid or opalescent (Anderson *et al.*, 1991; Imeson, 1992).

Gum arabic contains no more than 10^3 microorganisms per gram and no pathogens (Blake *et al.*, 1988). Spray-dried preparations, because of the high temperatures involved, contain no more than ~40% of the usual count

($\sim 4 \times 10^2$ microorganisms per g). These low counts, bearing in mind the heat treatment that the food in which the gum is incorporated will be subjected to, suggest that sterilization processes are unnecessary, at least with respect to the gum. A reduction in the gum's viable bacteria is achieved by ethylene oxide treatments (no longer permitted for food use) or propylene oxide (less efficacious). Heating carried out during manufacture to reduce the microflora can lead to precipitation of the arabinogalactan–protein complex (Anderson and McDougall, 1987), which promotes stabilization and emulsification in a range of food products (Randall *et al.*, 1989).

When processed gum arabic is subjected to 10 kGy irradiation to eradicate microorganisms, no significant changes in molecular weight are detected. There is a reduction in viscosity but there is no measurable effect on emulsion stabilization (Blake *et al.*, 1988; Imeson, 1992).

Gum arabic from *A. senegal* contains ~ 3.8% ash, 0.34% nitrogen, 0.24% methoxyl, 17% uronic acid and the following sugar constituents after hydrolysis: 45% galactose, 24% arabinose, 13% rhamnose, 16% glucuronic acid and 15% 4-*O*-methyl glucuronic acid (Anderson *et al.*, 1990). The gum is a slightly acidic complex polysaccharide produced as a mixture of calcium, magnesium and potassium salts. It has a molecular mass of ~ 580 000 Da. Gums from different sources exhibit large differences in content, amino acid composition, uronic acid content and molecular weight. As the main product for food applications, gum arabic from *A. senegal* is consistent in quality. Three principal fractions have been identified by hydrophobic affinity chromatography: a low-molecular-weight arabinogalactan (AG), a very high-molecular-weight arabinogalactan–protein complex (AGP) and a low-molecular-weight glycoprotein (GI). These components represent 88%, 10% and 1% of the molecule, respectively, and they contain 20%, 50% and 30% polypeptides, respectively. The protein is located on the outside of the AGP unit. The overall conformation of the gum arabic molecule is described by the 'wattle blossom' model, in which about five bulky AG blocks, each of ~ 200 000 Da, are arranged along the GI polypeptide chain, which may contain up to 1600 amino acid residues (Connolly *et al.*, 1987; Imeson, 1992).

In order to recognize specific structural features of four exudate gums, arabic, karaya, ghatti and tragacanth, a group of monoclonal antibodies, derived for the specific recognition of developmentally regulated epitopes of plant-cell plasma membranes and cell walls, was characterized with regards to binding to exudate gums (Yates and Knox, 1994). The monoclonal antibodies were shown to recognize effectively the specific structural features of the four gums. Dialysis of the gums and competitive inhibitor ELISAs indicated that the monoclonal antibodies recognize both high- and low-molecular-weight components of each gum. The spatial relationships and possible overlap of the AG epitopes on the most antigenic gum, gum arabic, were explored in more detail via epitope mapping (Yates and Knox, 1994).

7.2.3 *Viscosity, acid stability and emulsification*

Although gum arabic is highly branched, it has a compact structure. Solutions containing less than 10% gum arabic have low viscosities and Newtonian characteristics (Williams *et al.*, 1990). Above $\sim$30%, the hydrated molecules effectively overlap and steric interactions result in much higher solution viscosities and increasing pseudoplastic behavior. This gum is unique because of the very high gum concentrations that can be used to prepare solutions. As a result, high levels of gum can be used in many food products (Imeson, 1992). Its stability in acid solutions is useful for the stabilization of citrus oil emulsions. Solution viscosities can be changed by the addition of acid or alkali, since these change the electrostatic charges on the macromolecule. A lower pH (more compact polymer volume) leads to lower viscosity, whereas a higher pH (extension of the molecule) results in maximum viscosity around pH 5.0–5.5. At still higher pH, ionic strength of the solution increases until the repulsive electrostatic charges are masked, yielding a compact conformation with lower viscosity (Anderson *et al.*, 1990; Williams *et al.*, 1990; Imeson, 1992).

The high-molecular-weight AGP complex has been shown to be adsorbed to orange oil droplets (in fact the hydrophobic amino acids are absorbed by the orange oil droplets of the emulsion), whereas the AG complex shows no affinity for the oil phase but increases the viscosity of the aqueous phase. This latter complex contains $\sim$20% protein and its emulsifying properties are correlated with nitrogen level. Only 1–2% of the gum is adsorbed at the oil–water interface and participates in emulsification, while $>$12% gum arabic is necessary to stabilize emulsions with 20% orange oil and uniformly small droplet sizes (Williams *et al.*, 1990). If not enough gum is supplied to coat all the droplets, no stable emulsion is formed, and flocculation and coalescence ensue. The higher the oil level or the smaller the droplet size, the more gum arabic is needed to provide adequate stability.

Emulsion stability is often expressed in terms of the turbidity ratio, the ratio of turbidities of a dilute emulsion estimated at two wavelengths. For vegetable oil emulsions, the turbidity ratio decreased with gum arabic concentration (Gunji *et al.*, 1992).

Gum arabic has been compared with colloidal MCC and galactomannans, in view of its ability to reduce surface and interfacial tensions and to stabilize oil-in-water emulsions via 'steric' and 'mechanical' stabilization mechanisms (Garti *et al.*, 1993). Whereas gum arabic adsorbs strongly and effectively onto the oil droplets via its proteinaceous moieties, guar gum and LBG adsorb weakly and for the most part only 'precipitate' on the oil surface, forming birefringent layers of the polymer with its hydrophobic mannose backbone facing the oil. Stabilization with MCC is thought to be achieved via adsorption of solid particles on the oil droplets (mechanical stabilization) (Garti *et al.*, 1993).

7.2.4 *Compatibility*

Gum arabic is compatible with starches and most other gums. Gum arabic and gelatin interact at low pH to give coacervates that can be used in oil encapsulation. Combinations of gum tragacanth and gum arabic at a 4:1 ratio produce a minimum viscosity that chemically, commercially and practically produces a thin, pourable emulsion with good shelf-life stability (Imeson, 1992).

7.2.5 *Nutritional aspects and food applications*

Gum from *A. senegal* that contains no bark or foreign matter is odorless, colorless and tasteless. Changes in turbidity are a consequence of previous heat treatments. In foods, gum arabic is used at levels of less than 2%, except in confectioneries, where levels can reach up to 60%. This gum is digested by the gut microflora and then absorbed; it is utilized as it passes through the body when included at levels of up to 10% of the diet (Imeson, 1992).

Another nutritional aspect involves the effect of pH and hemicellulose digestion on calcium-binding by selected gums, namely gum arabic, LBG and xanthan gum (Kolb and Kunkel, 1994). Their binding behavior was found to be pH dependent. Gum arabic and xanthan bound most of the calcium at neutral pH, whereas LBG bound most of the calcium at pH 9.5. For all three gums, binding decreased below pH 5. Gum arabic bound about 26% of the added calcium, whereas LBG bound 5–25%. There were two types of binding site for calcium with xanthan and gum arabic, and one type for LBG. Hemicellulose digestion decreased binding ability for all three gums. Gum arabic and LBG released endogenous calcium, whereas xanthan lost 35–45% of its calcium-binding capacity after a 24 h *in vitro* digestion (Kolb and Kunkel, 1994).

Several techniques, which are applicable to other gums as well, are used to solubilize gum arabic. These include blending with other dry ingredients (e.g. sugar), slurrying the gum in oil, glycerin or other non-aqueous fluid, using a high-speed stirrer to disperse the powder with an eductor and then stirring slowly, using kibbled or whole gum for ease of dispersion, and patience, because complete hydration is slow. Gum arabic is used in five main food areas: confections, beverages and emulsions, flavor encapsulation, baked goods and brewing (Imeson, 1992).

(a) Confections. Gum arabic is used to produce a wide variety of confections, from soft lozenges and pastilles to hard gums (Wolff and Manhke, 1982; Best, 1990). Since the 1970s, when climatic changes caused a pronounced and unexpected shortage of gum arabic, its replacement with different quantities of modified starches has become commonplace. Low levels of gum arabic (up to 2.0%) are included in chewy sweets based on gelatin to improve product adhesion, reduce elasticity and produce extra-

fine sugar crystallization with a smooth texture. More information on the use of gums in confectioneries and on confections based on gum arabic can be found elsewhere (Reidel, 1983, 1986; Imeson, 1992).

(b) Various other food applications. As we have already seen, gum arabic is used as a stabilizer in beverages and emulsions. Fish-oil emulsions are required as dietary supplements or in health foods; they are also emulsified with gum arabic and tragacanth. Gum arabic is also used as an encapsulating agent for flavors used in dry foods, such as soups, beverages, dessert mixes, etc. A typical formulation contains 7% oil-based flavor and 28% gum arabic and results in 20% flavor in the dried material (Thevenet, 1988). The oil droplets need to be fully coated before spray-drying to eliminate the loss or oxidation of volatile oils. Concentrated gum arabic solutions that are sprayed or brushed onto pastries or biscuits before baking yield an attractive glossy coating after the water evaporates. In icings with high sugar content, the humidity of the product is controlled in parallel to molding and rolling properties. In glazes that are applied warm to baked goods, the inclusion of gum arabic maintains adhesion between the two surfaces. In beers and lagers, the interaction between charged uronic acid residues and the proteins helps to stabilize foam, to enable it to adhere to the glass while drinking, and to eliminate cloudiness in the drink (~250 ppm of high-quality gum). In wines, low levels of gum react with inherent proteins to sediment them, followed by decantation. Since gum arabic is unique in its properties of emulsification and in the production of mouthfeel and low-solution viscosities, it is difficult to replace, especially in the area of food applications (Imeson, 1992).

The effects of sodium alginate, karaya gum and gum arabic on the foaming properties of sodium caseinate solutions have been studied (Yang *et al.*, 1993). Surface tension, specific viscosity, turbidity, foaming ability and foam stability of the caseinate solutions with added gums were determined. Optimum conditions for foaming were 0.1% gum arabic and 10 min of whipping time. For foam stability, optimal concentrations were 0.3% sodium alginate and karaya gum at pH 7.0 and 0.2% at pH 8.0. The addition of sodium alginate increased foam stability of the solution but did not increase foaming ability. Surface tensions and solution turbidities were related to foaming ability, and specific viscosities were related to foam stability (Yang *et al.*, 1993).

7.3 Tragacanth

Tragacanth is an exudate of shrubs of the *Astragalus* species, located mainly in southwest Asia. The gum is generally recognized as safe, following a long and unblemished safety record in foods and pharmaceutical products. Its main areas of production are the arid and mountainous regions of Iran and Turkey. In the past thousands of tons of tragacanth were used annually for food, technical and pharmaceutical purposes. However, as a result of its very high cost from 1982–5, and strong competition from xanthan gum, the salability of tragacanth fell dramatically. Of the annual consumption of ~ 500 tons, about 40% was targeted for food use (Robbins, 1987; Anderson, 1989; Imeson, 1992).

7.3.1 Manufacture

The best-quality gum is obtained from incisions in the shrub, rather than from natural exudations. Tapping or blazing at the stems, branches and taproots is performed in May or June, followed by collection in August or September (after ~ 6 weeks) for ribbon-grade gum, whereas flake tragacanth is tapped and collected over a shorter period, from August to November. The ideal climate for this gum's production includes abundant rains prior to tapping and arid conditions during collection. Weather changes, which include excessive rains and winds during the exudation period, yield discolored material and lower solution viscosities. Sorting is performed after collection to verify the separation into ribbon and flake grades. The best gum quality consists of high viscosity, good solution color and a low microbiological count. Blending is necessary to ensure the desired properties, which are checked by the appropriate quality control. Its main food applications include dressings, sauces, icings and confections. Good quality (ribbon grade with a low total bacterial count) is desired where solution color is important and where less or no additional processing (heat treatment, acidification, etc.) is performed, to ensure an uncontaminated food product. To reduce the counts of resistant spores from soil- and airborne contaminants, an ethylene oxide gas treatment is used on gum destined for pharmaceutical products, whereas on that destined for food uses, only the less efficient propylene oxide is permitted (Imeson, 1992).

7.3.2 Composition

Gum tragacanth is a complex, heterogeneous, acidic proteoglycan of high molecular weight. After hydrolysis, arabinose, xylose, fucose, galactose, rhamnose and galacturonic acid with traces of starch and cellulosic material are detected. Because more than 20 different species are used for gum production, there exists wide variation in composition and performance. The more viscous gum species contain high proportions of fucose, xylose, galacturonic acid and methoxyl groups and low proportions of arabinose and nitrogenous fractions. Low-viscosity products contain more arabinose and galactose but lower proportions of methoxyl and galacturonic acid. Gum tragacanth consists of water-soluble (neutral AG and tragacanthin) and water-swellable (tragacanthic acid) fractions. The ratio of these components varies from 9 : 1 to 1 : 1, explaining the viscosity differences in commercial samples. Nitrogen content may also influence viscosity, with soluble high-viscosity material containing ~0.07% nitrogen versus ~1.87% for insoluble fractions (Anderson and Grant, 1989). Hydroxyproline content in the amino acids is presumably involved in stabilizing the AG structure, as evidenced by the presence of hydroxyproline in the peptides present in highly viscous tragacanth. In gum arabic and tragacanth, the peptide sequences are likely to play a role in emulsion stabilization (Imeson, 1992).

7.3.3 Inherent properties

Tragacanth exudates are graded to provide the user with standardized functional products. At low concentrations, hydrated tragacanth produces a viscous solution, and at 2–4% it gives a thick paste. Viscosities of 3500 to 4600 mPa s have been found for 1% pseudoplastic solutions of tragacanth (Anderson, 1989). High viscosity at low shear is related to charge repulsion from the galacturonic acid residues and contributes to the emulsion stabilization and suspension abilities of the gum (Imeson, 1992). Gum tragacanth, ribbon or flake grade, is quite stable in acidic solutions, but the flakes yield lower visocity. This stability is relevant to the gum's use in salad dressings, condiments and other acidic products. Stability may be related to its backbone's resistance and to the protection afforded by the arabinofuranose side-chains (Stauffer, 1980). Low amounts of gum tragacanth in water lower the latter's surface tension. In oil–water emulsions, interfacial tension is reduced to ~ 190–230 μN cm^{-1}, depending on the kind of tragacanth added (high-viscosity tragacanth produces a smaller reduction in interfacial tension). This reduction facilitates emulsification and the viscous aqueous phase contributes to a stable emulsion. Nitrogen content (bound polypeptides) is related to emulsification properties (Stauffer, 1980; Dickinson *et al.*, 1988). The addition of gum arabic to tragacanth results in unusual viscosity reductions for reasons that are unclear; this phenomenon is exploited to achieve thin, smooth, pourable emulsions with fish and citrus oils that have a long shelf life.

7.3.4 Food applications

For food applications, dissolution without lumping is desirable. Dissolution passes through a hydration step, which can be hastened by reducing the size of the particles involved. Coarse powders are selected where good dispersion is needed and short hydration times are unimportant (as in the case of stock solutions). In such cases, the powder is introduced into the vortex of a solution, using a high-speed or high-shear mixer, to build up viscosity and achieve good hydration. Slow hydration of kibbled tragacanth fragments, 1 to 4 mm in size, can be achieved by slow continuous stirring (24 h) at ambient temperatures to achieve complete hydration. Heating can reduce dissolution times, but the prospect of degradation with excessive heat or long heating periods should be noted. Combinations of heat treatment, preservatives, pH adjustments, refrigeration and freezing are effective in maintaining solution properties throughout product preparation and shelf life (Imeson, 1992).

Tragacanth is used in many food applications. In low-pH products such as salad dressings, condiments and relishes, it serves as a stabilizer and provides a creamy oral sensation via its surface-active properties. Thickening and achieving the desired flow pattern extends product stability and prevents separation of the oil phase. These achievements can only be matched by blends of PGA, gum arabic or cellulose derivatives for emulsification, together with xanthan and guar gums for thickening. Tragacanth provides a wide spectrum of the properties necessary for condiments, dressings and sauces. Usage levels are 0.4–0.8% of the weight of the aqueous phase and depend on the oil content, the use of other thickeners and the required consistency. Product processing should be designed to minimize degradation and maintain functional properties of the gum. In confections and icings, gum tragacanth is used as an effective water-binding agent because of its high proportion of the water-swellable (insoluble) fraction. Chewy sweets are formulated with blends of tragacanth and gum arabic to yield a chewy texture, and blends of tragacanth and gelatin for a chewy and cohesive texture. By compressing the ingredients, including tragacanth (for binding purposes), fruit tablets and pastilles can be prepared with the desired consistency, mouthfeel and flavor-release properties. Tragacanth is used as a binder in highly sweetened icings, which contain fats to provide some pliability and to reduce evaporative moisture losses. Fish oil-flavored emulsions are used as dietary supplements for vitamins C and E in the Far East. Flavored oil emulsions are stabilized with 0.8–1.2% tragacanth and its blends: their shelf life is extended while the desired combination of thickening, emulsifying and mouthfeel properties is supplied. In frozen desserts, gum tragacanth (0.2–0.5%) is used to control ice-crystal growth, to reduce moisture migration and ice-crystal development during storage, and to prevent color and flavor migration during storage and consumption. In baked-good fillings, the acid stability of tragacanth is exploited to yield a creamy texture with good clarity and gloss. In a few applications, such as ready-to-spread icings, gum tragacanth cannot be successfully replaced by other gums or gum combinations (Imeson, 1992).

In nutritional studies, tragacanth, karaya gum, guar gum, xanthan gum, CMC, pectin and fiber wheat bran have served as *in vitro* models for fermentation and water-holding capacity measurements, and the results compared with those from a rat model for their ability to predict the action of dietary fiber on stool output in humans. It was concluded that the *in vitro* methods are less expensive and time-consuming than animal studies and that the log of the *in vitro* predictive index may provide a useful prescreening device for new dietary fiber sources or to detect changes in the action of dietary fibers during the manufacturing process (Edwards *et al.*, 1992).

7.4 Karaya gum

7.4.1 Origin

Karaya gum is also known as sterculia gum and is the dry exudate of *Sterculia urens* trees. Other species of *Sterculia* exist but are of less commercial interest. The trees are found in central and northern India, Senegal, Mali, Sudan and Pakistan. Annual usage is 3000–4000 tons. Most of its production (up to ~95%) is utilized for pharmaceuticals such as bulk laxatives, dental fixatives and colostomy sealing bags. No more than 5% is used in foodstuffs. This gum has been classified as GRAS in the US since 1961 and had a temporary approval for its use in foods in Europe in 1974, although it is not permitted throughout the EC (for example a ban remains

Figure 7.2 Karaya gum: the dried exudate of the *Sterculia urens* tree.

in force in Germany). Its ADI is 0–12.5 mg kg^{-1} body weight. Karaya gum is not degraded by humans: it passes through the body unmodified, acting as a bulk laxative. The gum is obtained by tapping or blazing mature *Sterculia* trees. Each tapping provides 1–5 kg, and the large, irregularly shaped ‘tears’ are collected from April to June in India (avoiding the monsoon season) and cleaned to remove bark and foreign matter before sorting (Fig. 7.2). Bark and foreign matter can be found in proportions of 0–0.5% in the hand-picked, selected white to very light tan grades, and up to 5.0–7.0% in the brown color siftings. Uses for the last grade are limited. The hygienic condition of this gum is similar to that of other exudates, and its use in sauces and dressings is safe, since the low pH of these products and the heat treatments they are regularly subjected to are sufficient to ensure safe foods (Imeson, 1992).

7.4.2 *Structure and properties*

Karaya gum is a complex, branched, partially acetylated polysaccharide, with a very high molecular mass of around 16×10^6 Da (Le Cerf *et al.*, 1990). After hydrolysis, the proportions of glucuronic acid, galacturonic acid, galactose and rhamnose vary, depending on the species, age and quality of the gum (Meer, 1980). About 40% uronic acid residues are contained within the polysaccharide, and up to ~8% acetyl groups. Because of these substituents, the gum does not fully dissolve in water, swelling instead. Via chemical deacetylation, the gum can be changed from a water-swellable to a water-soluble material. Indian karaya differs from the African variety in its higher acid value and more pronounced acetic acid odor. The former’s structure has not yet been fully characterized, but it is believed to contain a central chain of galactose, rhamnose and galacturonic acid residues with glucuronic acid side-chains. Karaya gum contains the lowest level of proteinaceous material in comparison with other exudate gums. After dispersion in water, karaya gum absorbs the water to form a viscous solution, and yield stresses of 60 and 100 μN cm^{-2} were determined for 2 and 3% gum concentrations, respectively (Mills and Kokini, 1984; Imeson, 1992). The smoothness of the gum solution is determined by its particle size, and it can be changed by prolonged stirring to achieve a smooth texture and reduced viscosity. Gum solubility can be increased by deacetylation, which gives the product a more expanded conformation. The solutions are cohesive and stringy or ropy. Heating changes the polymer conformation and solubility increases. Ropiness is accompanied by a lower acetyl content. Since heating increases solubility and the reduction in viscosity is irreversible, solution concentrations can be increased to 15%. The pH of karaya gum (of Indian origin) is 4.4–4.7. An increase in pH promotes deacetylation, and a parallel rise in viscosity and in the degree of ropiness. African karaya gum has a higher solution pH (4.7–5.2) and the solutions exhibit the high viscosity (ropy rheology) of deacetylated gum. At high concentrations (20–50%), karaya gum produces a solution that has the adhesive properties required for dental adhesives and colostomy bags. Hydrolysation of the acetyl groups in the gum is responsible for its odor and flavor. These properties and its color, ranging from light tan or gray to brown, limit its use to those applications in which there is a special need for these colors without affecting sensory properties (Imeson, 1992).

Changes in rheological properties, as measured by viscosity, of two acidic polysaccharides (karaya gum and tragacanth), and two galactomannans (guar gum and LBG) were studied under a range of irradiation doses < 10 kGy (King and Gray, 1993). Powdered samples were irradiated, and the viscosity of a 1% dispersion prepared at room temperature or by heating to 80°C for 1 h was determined over a wide ranges of shear rates. All samples showed pseudoplastic behavior that approached Newtonian with increasing irradiation dose. The viscosity of the acidic polysaccharides, karaya gum and tragacanth, following gamma irradiation at low doses (< 1 kGy) was unchanged or slightly higher relative to the non-irradiated control samples. Above 1 kGy, dispersion viscosity decreased with increasing dose. For these polysaccharides, chain hydrolysis seems to occur during irradiation at all doses, resulting in an increase in the amount of soluble polymer and hence increased viscosity at low doses, whereas at high doses viscosity decreases as a result of extensive polymer hydrolysis. Similar ESR spectra were obtained at low and high doses with a stronger signal at the higher dose (King and Gray, 1993).

A survey of the inhibitory effects of karaya gum as well as various other plant polysaccharides on the polymerase chain reaction (PCR) amplification of a 974 bp section of the gene *rbcL* (encoding ribulose bisphosphate carboxylase large subunit) in spinach revealed that most of polysaccharides tested (karaya gum, guar gum, LBG, dextran, arabinogalactan, carrageenan, inulin, mannan, pectin, starch and xylan) were not inhibitory (Demeke and Adams, 1992). In contrast, two of the acidic polysaccharides (dextran sulfate and gum ghatti) were inhibitory. The addition of Tween reversed the inhibitory effects of ghatti gum, and Tween, dimethyl sulfoxide (DMSO) and polyethylene glycol (PEG) reversed the inhibitory effects of dextran sulfate (Demeke and Adams, 1992).

7.4.3 Food applications

Karaya gum can be dissolved by blending it with other powdered ingredients and mixing in oil, alcohol or glycerin before dispersing in water. A high-shear mixer is used, with an eductor funnel for fine dispersion and coarse particle sizes for easy dispersion and slow hydration. Karaya gum is more expensive than tragacanth and differs in many practical aspects, limiting their interchangeability. The former is used mainly in dressings and sauces, dairy and bakery products, frozen desserts and for meat applications. In sauces and dressings, karaya gum is used as a thickener, suspension agent and stabilizer. In these acidic products, the odor and flavor of this gum (at concentrations of up to 1.0%) do not interfere. Other gums used in such products are tragacanth, guar gum and xanthan gum. In cheese spreads, whey exudation can be reduced with the use of karaya gum, leading to improved spreadability. Other foamed dairy products are stabilized by this gum. In frozen desserts (sherbets, popsicles), up to ~0.5% karaya gum is used to minimize moisture migration and bleeding, to control the size of the formed ice crystals and to reduce the suck-out of color and flavor (Imeson, 1992).

Karaya gum can partially replace LBG, because of its cohesive properties. Added to baked goods, karaya gum can reduce the effect of staling, thereby extending shelf life. Coatings and glazes for baked products are sometimes based on karaya gum. The gum is also used in comminuted meat products and sausages to achieve better adhesion between meat particles, to bind water during processing and storage and to achieve better consistency, as well as to provide low-calorie burgers with soluble fibers (Imeson, 1992). From a nutritional point of view, *in vitro* digestibility of fish protein was decreased by fiber constituents in the following order: pectin (9.9%), karaya gum (7.0%), sodium alginate (6.1%) and cellulose (1.5%). The order of reduction by fibrous vegetable residues ranked as follows: sea tangle (12.3%), romaine lettuce (11.1%), perillar leaf (8.9%) and green pepper (5.1%). The inhibitory effect of the dietary fibers towards fish protein digestion increased with added fiber, but the inhibition differed with the type of dietary fiber (Ryu *et al.*, 1992).

7.5 Summary

Although the tonnages of gum arabic (from *A. senegal*) and other water-soluble tree exudates traded annually have tended to decrease steadily over the past 15 years for a number of identifiable reasons, there is still some commercial demand for those few gums that meet international specifications and are available in a steady supply from year to year at competitive prices (Anderson, 1993). These include the three gum exudates approved for food use: karaya gum from India, tragacanth from West Asia and gum arabic from Africa. Gums from a wide range of other botanical sources are not approved for food use and are, therefore, restricted to technological applications (Anderson and Wang, 1994). To meet the demand for exudate gums, a few measures should be followed, e.g. the specific source tree species need to be identified and regenerated, non-permitted species need to be eradicated from areas of intensive gum production and educational programs should be initiated to help those populations still involved in gum production by traditional methods who need to become more aware of the changing requirements of the international gum trade. If scientists become more active (particularly in remote areas where the dissemination of such information is understandably slow), this will discourage the commercial utilization of gums from sources that have not been evaluated and approved (Anderson, 1993).

References

Anderson, D.M.W. (1989) Evidence for the safety of gum tragacanth (*Asiatic astragalus* spp.) and modern criteria for the evaluation of food additives. *Food Additives and Contaminants*, **6**(1), 1–12.

Anderson, D.M.W. (1993) Some factors influencing the demand for gum arabic (*Acacia senegal* (L.) Willd.) and other water-soluble tree exudates. *Forest Ecol. Management*, **58**(1–2), 1–18.

Anderson, D.M.W., Brown Douglas, D.M., Morrison, N.A. *et al.* (1990) Specifications for gum arabic (*Acacia senegal*): analytical data for samples collected between 1904 and 1989. *Food Additives and Contaminants*, **7**(3), 303–21.

Anderson, D.M.W. and Grant, D.A.D. (1989) Gum exudates from four *Astragalus* species. *Food Hydrocolloids*, **3**(3), 217–23.

Anderson, D.M.W. and McDougall, F.J. (1987) Degradative studies of gum arabic (*Acacia senegal* (L.) Willd) with special reference to the fate of the amino acids present. *Food Additives and Contaminants*, **4**(3), 247–55.

Anderson, D.M.W., Millar, J.R.A. and Wiping, W. (1991) Gum arabic (*Acacia senegal*): unambiguous identification by ^{13}C-NMR spectroscopy as an adjunct to the revised JECFA specification, and the application of ^{13}C-NMR spectra for regulatory/legislative purposes. *Food Additives and Contaminants*, **8**(4), 405–21.

Anderson, D.M.W. and Wang, W.P. (1994) The tree exudates permitted in foodstuffs as emulsifiers, stabilizers and thickeners. *Chem. Ind. Forest Products*, **14**(2), 73–83.

Best, E.T. (1990) Gums and jellies, in *Sugar Confectionery Manufacture* (ed. E.B. Jackson), Blackie, Glasgow, pp. 190–217.

Blake, S.M., Deeble, D.J., Phillips, G.O. *et al.* (1988) The effect of sterilizing doses of γ-irradiation on the molecular weight and emulsifying properties of gum arabic. *Food Hydrocolloids*, **2**(5), 407–15.

Connolly, S., Fenyo, T.C. and Vandevelde, M.C. (1987) Heterogeneity and homogeneity of an arabinogalactan–protein–*Acacia senegal* gum. *Food Hydrocolloids*, **1**(5/6), 477–80.

Demeke, T. and Adams, R.P. (1992) The effects of plant polysaccharides and buffer additives on PCR. *BioTechniques*, **12**(3), 332–3.

Dickinson, E., Murray, B.S., Stainsby, G. *et al.* (1988) Surface activity and emulsifying behaviour of some Acacia gums. *Food Hydrocolloids*, **2**(6), 477–90.

Edwards, C.A., Adiotomre, J. and Eastwood, M.A. (1992) Dietry fibre: the use of in-vitro and rat models to predict action on stool output in man. *J. Sci. Food Agric.*, **59**(2), 257–60.

Garti, N., Reichman, D., Hendrickx, H.A.C.M. *et al.* (1993) Hydrocolloids as food emulsifiers and stabilizers. *Food Struct.*, **12**(4), 411–26.

Gunji, M., Ueda, H., Ogata, M. *et al.* (1992) Studies of the oil-in-water emulsions stabilized with gum arabic by using the turbidity ratio method. *J. Pharm. Soc. Japan*, **112**(12), 906–13.

Imeson, A.P. (1992) Exudate gums, in *Thickening and Gelling Agents for Food* (ed. A.P. Imeson), Blackie A & P, Glasgow, pp. 66–97.

King, K. and Gray, R. (1993) The effect of gamma-irradiation on guar gum, locust bean gum, gum tragacanth and gum karaya. *Food Hydrocolloids*, **6**, 559–69.

Kolb, K.B. and Kunkel, M.E. (1994) Effect of pH and hemicellulose digestion on calcium binding by selected gums. *Food Chem.*, **49**(4), 379–85.

Le Cerf, D., Irinei, F. and Muller, G. (1990) Solution properties of gum exudates from *Sterculia urens* (karaya gum). *Carbohydr. Polym.*, **13**(4), 375–86.

Meer, W. (1980) Gum karaya, in *Handbook of Water-soluble Gums and Resins*, ch. 10 (ed. R.L. Davidson), McGraw-Hill, New York, pp. 10.1–10.14.

Mills, P.L. and Kokini, J.L. (1984) Comparison of steady shear and dynamic viscoelastic properties of guar and karaya gums. *J. Food Sci.*, **49**(1), 1–4, 9.

Randall, R.C., Phillips, G.O. and Williams, P.A. (1989) Effect of heat on the emulsifying properties of gum arabic, in *Food Colloids* (eds R.D. Bee, P.J. Richmond and J. Mingins), Royal Society of Chemistry, Cambridge, pp. 386–90.

Reidel, H. (1983) The use of gums in confectionery. *Confect. Prod.*, **49**(12), 612–3.

Reidel, H. (1986) Confections based on gum arabic. *Confect. Prod.*, **52**(7), 433–4, 437.

Robbins, S.R.J. (1987) *A Review of Recent Trends in Selected Markets for Water-soluble Gums*, Overseas Development Natural Resources Institute, Bulletin No. 2p.

Ryu, H.S., Park, N.E. and Lee, K.H. (1992) Effect of dietary fiber on the in-vitro digestibility of fish protein. *J. Korean Soc. Food Nutr.*, **21**(3), 255–62.

Stauffer, K.R. (1980) Gum tragacanth, in *Handbook of Water-soluble Gums and Resins*, ch. 11 (ed. R.L. Davidson), McGraw-Hill, New York, pp. 11.1–11.31.

Thevenet, F. (1988) Acacia gums, stabilisers for flavor encapsulation, in *Flavor Encapsulation*, ch. 5 (eds S.J. Risch and G.A. Reineccius), American Chemical Society, Washington DC, pp. 37–44.

Wang, W.P. and Anderson, D.M.W. (1994) Non-food applications of tree gum exudates. *Chem. Ind. Forest Prod.*, **14**(3), 67–76.

Williams, P.A., Phillips, G.O. and Randall, R.C. (1990) Structure–function relationships of gum arabic, in *Gums and Stabilisers for the Food Industry 5* (eds G.O. Phillips, D.J. Wedlock and P.A. Williams), IRL Press at the Oxford University Press, Oxford, pp. 25–36.

Wolff, M.M. and Manhke, C. (1982) Confiserie: la gomme arabique. *Rev. Fabr.* ABCD, **57**(6), 23–7.

Yang, S.T., Kim, M., Park, C. *et al.* (1993) Effects of sodium alginate, gum karaya and gum arabic on the foaming properties of sodium caseinate. *Korean J. Food Sci. Technol.*, **25**(2), 109–17.

Yates, E.A. and Knox, J.P. (1994) Investigation into the occurrence of plant cell surface epitopes in exudate gums. *Carbohydr. Polym.*, **24**(4), 281–6.

8 Seed gums

8.1 Introduction

Three galactomannans are commercially important: LBG, guar gum and tara gum. Those which are of interest but not commercially available include mesquite gum (*Prosopsis* species) and fenugreek (*Trigonella foenum-graecum*).

8.2 Locust bean gum: sources, manufacturing and legislation

LBG (*Ceratonia siliqua*) is found in Mediterranean regions. The tree is ~10–15 m in height, with roots penetrating the soil to depths of 18–27 m. It has brownish, 10–20 cm long pods containing ~10 mm long seeds (each fruit contains 10–15 seeds) that weigh ~0.2 g each. The tree thrives on rocky, semi-arid soil. It yields fruit 5 years after budding and reaches maturity at the age of 50 (Rol, 1973).

The sweet fruit (carob fruit) is consumed as it is; its sugar contents produce a liquidy concentrate that is also marketed, and the fruit can serve as animal feed. In northern Africa, carob fruit contributes to the diets of low-income people, whereas in Europe it is sometimes roasted for use as a coffee substitute. In the 1920s, carob flour was combined with wheat flour to bake special breads (Glicksman, 1969). The seeds, 10–15 per pod, contain ~38% galactomannan. They are made up of 30–33% husk, 23–25% germ, and the rest endosperm (Herald, 1986a). The seeds are covered with a dark coat that is removed from the endosperm before grinding to gum powder.

Industrial manufacture of LBG began in the 1920s, and today it is sold worldwide under a variety of trade names. Carob pods are harvested manually (picked from the ground after the tree branches have been shaken). The kernels are removed from the pod, the husks are removed, the seeds are split lengthwise and the endosperm is separated from the germ. After germ isolation, endosperms are sifted, graded, packaged and marketed as LBG or carob gum. The gum is sold in several particle sizes and can contain finely ground pieces of the dark-brown testa (seed coat), which appears as dark specks in the powder. The lower the quality of the product, the higher the content of the specks. Because of the presence of undissolved fat and proteins, the gum solution looks milky, but its appearance can be improved with an alcohol wash. If a clear product is desired (e.g. for fruit jellies), a more expensive, alcohol-precipitated polysaccharide is used. To achieve good solubility in water, the dispersion needs to be heated, or hot solutions of gum are dried after being blended with sugars to eliminate recrystallization of the polysaccharide.

LBG is considered a GRAS food ingredient by the FDA. The maximum allowable concentration in the USA is 0.80%, in cheeses (Federal Register, 1987b). Under regulation 101.4 of this code, the gum's addition to a food product must be specified on the product's label where ingredients are listed in descending order of predominance (Federal Register, 1987c).

8.3 Guar gum: sources, processing and regulatory status

Guar gum is a plant hydrocolloid extracted from two leguminous (guar) plants, *Cyamopsis tetragonolobus* and *C. psoraloides*, found in northwest India and Pakistan. The guar pods are used there for animal feed and human consumption. The guar plant is an annual (~1 m high), planted after the monsoons in June/July and harvested in December. The plants are drought resistant and can grow in semi-arid regions. The pods are green, each one containing six to nine pea-shaped seeds that are ~2–4 mm in diameter and weigh approximately 35 mg each, of which ~36% is galactomannan (Herald, 1986b). The guar plant was not introduced in the USA until the beginning of the 20th century, when it was destined for planting in the semi-arid areas of the southwest. The arid regions of Texas account for most of the guar plants in the USA (Glicksman, 1969). After 1945, when there was a shortage in the supply of LBG as a result of the Second World War, its replacement by guar gum was studied. The first of such ventures was undertaken by General Mills Inc. and Stein, Hall & Co. Inc. (Goldstein *et al.*, 1973).

The pods are harvested by hand from the plant, except in Texas where mechanical harvesters are employed. Manufacturing includes separation of the endosperm from the germ and the testa, and grinding to compose the gum. The gum is available in a wide range of sizes, which differ in their solubilization rates. Thermally degraded gums with reduced viscosities are commercially available. Steamed powders exhibit an enhanced dissolution rate and a reduction in their typical tastes. Two major grades of guar gum are marketed, food and industrial. For industrial purposes, ground endosperm is used, including small amounts of hull and germ resulting from imperfect purification. The industrial grades are manufactured with chemical additives, such as carboxymethyl, hydroxyalkyl and quaternary amine derivatives, to manipulate functional properties such as viscosity, solubility and swelling.

Guar gum is regarded as GRAS by the FDA (Federal Register, 1987a) and by other worldwide regulatory agencies. The maximum allowable concentration of guar in foods is 2% by weight, as found in vegetable products and in fats and oils. The gum's presence in foods must be listed on the label. Guar gum solutions are sometimes used as a model for non-Newtonian liquid foods (Anantheswaran and Liu, 1994; Blond, 1994).

8.4 Tara gum

Tara gum is derived from the tara bush, *Caesalpinia spinosa*, which is indigenous to Equador and Peru and is grown in Kenya. The reddish tara pods contain seeds that are about 10 mm long and weigh ~0.25 g each, of which ~18% is galactomannan (Jud and Lossl, 1986).

8.5 Galactomannan structure

Galactomannans are linear polysaccharides based on a backbone of $\beta(1-4)$-linked D-mannose residues. Single α-D-galactose residues are linked to the chain by C-1 via a glycosidic bond to C-6 of mannose. Screening gum from a single source verifies that it is not a single substance but contains molecules with different degrees of polymerization (Doublier and Launey, 1981; Lopes da Silva and Goncalves, 1990). The degree of galactose substitution varies from one botanical source to the next as well as between molecular species of one gum. In LBG (Fig. 8.1) the average galactose to mannose ratio is 1:4, compared with guar (Fig. 8.2) with an average ratio of 1:2 and tara gum with a ratio of 1:3. These differences confer different

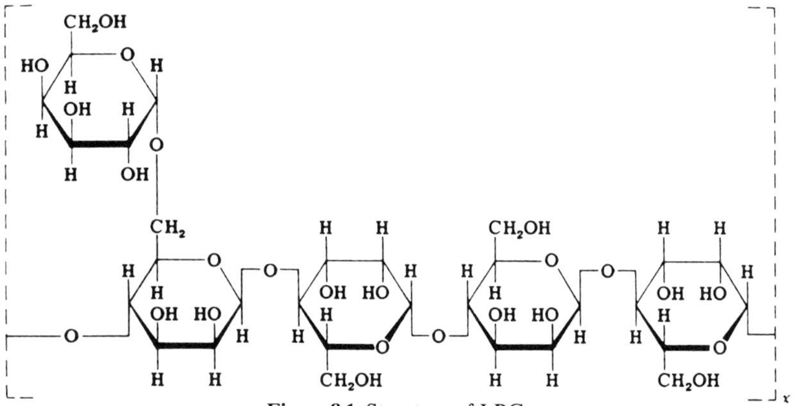

Figure 8.1 Structure of LBG.

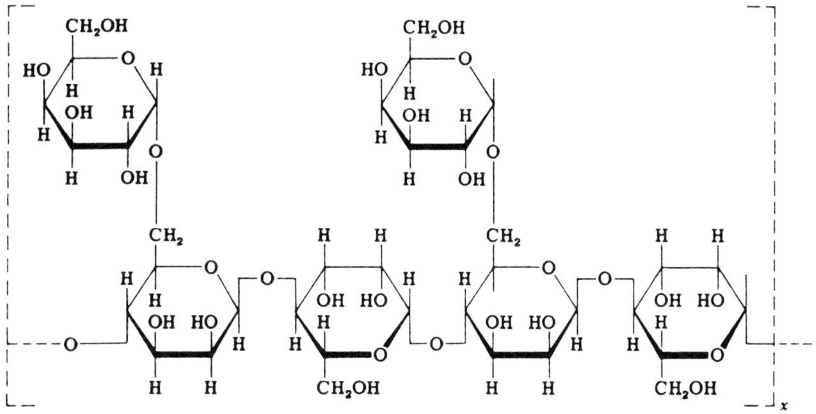

Figure 8.2 Structure of guar gum.

properties upon the molecules and this can be exploited to fractionate the gum (Morris and Ross-Murphy, 1981).

Carob gum has a 'block' structure, i.e. branching units clustered mainly in blocks of ~25 residues ('hairy' region), followed by even longer blocks of unsubstituted β(1–6)-D-mannopyranosyl units ('smooth' region). The latter are important in the formation of interchain associations (Dea, 1979). Since the separation of the testa and endosperm is limited, impurities can be found in the commercially sold gum. These include 4% pentosan, 1% cellulosic material, 1.1% ash and 6% proteinaceous material. The last is in part covalently bound to the polysaccharide (Anderson, 1986). Major amino acids are present in the proteinaceous components of LBG; in decreasing order of abundance these are glutamic acid, aspartic acid, glycine, arginine, alanine and serine (214, 96, 93, 83, 80 and 69 amino acid residues per 1000 protein-forming amino acids, respectively) (Anderson, 1986).

Unlike LBG, guar gum possesses a regular, alternating structure (Dea *et al.*, 1977; Dea, 1979). It can be schematically represented by its regular, twofold conformation. The unsubstituted D-mannopyranosyl units represent the so-called 'smooth' side, while the substituted D-galactopyranosyl units constitute the 'hairy' side. This conformation explains guar gum's functional properties.

To improve and diversify its commercial applications, guar gum was modified by introducing changes in its chain structure (Frollini *et al.*, 1995). Properties of the carboxylated polyelectrolyte obtained from guar gum were studied. The charged macromolecules formed from native guar showed all the typical characteristics of a polyelectrolyte. Viscometry results indicated that carboxylated guar had a much higher viscosity in a low salt-content medium than the native polymer, which improved its thickening properties (Frollini *et al.*, 1995).

The M_w values of LBG and guar gum are similar, 1.94×10^6 and 1.9×10^6, respectively, whereas the M_n values are $\sim 80\,000$ for LBG and $\sim 250\,000$ for guar gum (Hui and Neukom, 1964; Lopes da Silva and Goncalves, 1990). The substitution ratio is not sufficient to fully describe the molecules but gives an idea of possible interactions with other polysaccharides (Dea *et al.*, 1986; Dea, 1990). To have a clearer look at the gum, more information on the pattern of substitution is necessary. Similar to LBG, an endogenous component of the gum is its proteinaceous material, an integral part of the seed endosperm, which remains attached after milling and purification.

8.6 Gum solution properties

Commercial gums exhibiting high substitution ratios (e.g. guar gum) are better hydrated in cold water than gums with limited substitution (such as LBG), because the presence of side-chains interferes with the formation of stable crystalline regions and promotes water penetration, thereby enhancing solubility. This phenomenon is reflected in the viscosity of hot- and cold-prepared solutions. Several LBG derivatives (carboxyalkyl ethers and hydroxyethyl ether) have been synthesized to achieve better rates of swelling and hydration (Rol, 1973) via etherification, and these are utilized in non-food applications.

To achieve full viscosity in the shortest amount of time, an appropriate mixer, fine powder and elevated temperatures are adopted. Precautions should be taken not to increase the temperature to over 80°C, in order to avoid thermal degradation. The high viscosities achieved with seed gums explain their frequent use (Goldstein *et al.*, 1973). Water is used exclusively as a solvent in food applications, but LBG also dissolves well in low-molecular-weight alcohols, dimethylformamide and dimethylsulfoxide (Seaman, 1980). In practice, the viscosity of carob gum remains unchanged at pH range 3.5–11.0 (Goldstein *et al.*, 1973; Herald, 1986a).

After dissolution in water, galactomannan takes on a random coil conformation (Mitchell, 1979; Morris and Ross-Murphy, 1981). After separation by the solvent, a linear increase in viscosity with concentration is observed (Einstein, 1906). As the concentration rises, mutual entanglements as a result of random contact between polysaccharide chains occur. This causes an exponential increase in solution viscosity with concentration. In other words, the longer the molecule and the more extended the conformation in solution, the higher the chances of earlier, stronger entanglements (Fox, 1992). LBG concentrations below 0.5% do not exhibit any appreciable increase in viscosity. However, above this level, viscosity increases exponentially. Typically a 1% solution of LBG at room temperature has a viscosity of 2.4×10^3 to 3.2×10^3 cps (Dea and Morrison, 1975).

In solution, galactomannans exist as disordered mobile coils. However, galactomannans have been shown to depart from typical random-coil characteristics and exhibit intermolecular associations between unsubstituted regions of glycan chains (Goycoolea *et al.*, 1995a). Hyperentanglement was investigated by comparing solution properties in strong alkali and at neutral pH. Smaller reductions in viscosity values for guar gum are consistent with a lower content of unsubstituted sequences capable of forming intermolecular associations. In conclusion, topological entanglement in solutions of galactomannans is augmented by alkali-labile non-covalent associations. These associations give rise to departures from the general form of concentration, a dependence that is reflected by disordered polysaccharides (Goycoolea *et al.*, 1995a).

The spatial volume occupied by the molecule when it is allowed to rotate freely about its center of gravity is known as its intrinsic viscosity and can be estimated. LBG has an intrinsic viscosity of 10 dl g^{-1} (Doublier and Launey, 1981), whereas that of tara gum is 11.2 (Clark *et al.*, 1986), and high-molecular-weight (1.9×10^6) and low-molecular-weight (0.4×10^6) guar gum have intrinsic viscosities of 14 and 4.5 dl g^{-1} respectively (Robinson *et al.*, 1982). There is a marked increase in intrinsic viscosity (as expected for a linear species) as the degree of polymerization increases (Fox, 1992). The wide range of viscosities reflects the different average lengths of the mannan backbone. Of course, the intrinsic viscosity is also affected by the character of the solvent. In thermodynamically optimal solvents, polymer-chain extension is optimal, whereas in less suitable solvents, molecular extension is reduced by intramolecular binding. Therefore, when sugars, salts and even alcohols are present, solvent quality is reduced and the resultant rheological properties are influenced (Fox, 1992).

To use galactomannan effectively as a viscosity former, some basic notions need to be considered. Its concentration should lie within the entanglement domain. For galactomannan, this occurs when the space occupancy or overlap factor is slightly above unity (Doublier and Launay, 1977). Theory predicts that above this concentration, the zero-shear viscosity will rise with the third power of the concentration, such that doubling the concentration will result in an eightfold increase in viscosity (Fox, 1992). For guar, the dependency on concentration was found to be proportional to the fifth power, maybe because of the association of mannan chains in solution (Robinson *et al.*, 1982).

Galactomannans (which form an extended random coil in solution) exhibit pseudoplastic flow behavior, whereby viscosity decreases with shear (Seaman, 1980). The thickened solutions appear to thin upon mastication, giving a pleasant, light mouthfeel. Shear thinning, a result of orientation of the extended molecules under the shear gradient, means that they align themselves parallel to the direction flow. At higher gum concentrations, the overlap factor is greater than unity, and shear causes disentanglement.

When the speed of the shear increases with respect to the speed of re-entanglement the viscosity falls, but as the shear stops, it is followed by re-entanglement and recovery of its original low-shear viscosity. The reduction in viscosity, increased during the heating of galactomannan solutions of processed foods that contain these thickening agents, is a reversible process (Fox, 1992).

The effects of gamma irradiation on guar gum, LBG, tragacanth and karaya gum were studied (King and Gray, 1993). Changes in rheological properties, as measured by the viscosity, of the two galactomannans (guar gum and LBG) and the two acidic polysaccharides (tragacanth and karaya gum) were studied in a range of irradiation doses less than 10 kGy. Powdered samples were irradiated and the viscosity of a 1% dispersion prepared at room temperature or by heating to 80°C for 1 h was determined over a wide range of shear rates. All samples showed pseudoplastic behavior that approached Newtonian with increasing irradiation dose. Both galactomannans showed a decrease in viscosity with increasing gamma irradiation independent of temperature, and a hypothesis was proposed that at low gamma irradiation doses (less than 2 kGy) there is a reduction in polymer aggregation in solution, whereas at higher doses polymer hydrolysis occurs. The viscosity of the acidic polysaccharides at low doses of gamma irradiation was unchanged or slightly higher than that in unirradiated control samples. Above 1 kGy, dispersion viscosity decreased with increasing dose (King and Gray, 1993). For these polysaccharides, chain hydrolysis seemed to occur during irradiation at all doses, resulting in an increase in the amount of soluble polymer and increased viscosity at low doses. At high doses, viscosity decreased owing to extensive polymer hydrolysis.

8.7 Gelation and interactions of galactomannans

LBG solutions do not gel at any concentration. However, weak cohesive gels can be formed, even at concentrations of ~0.5%, upon freezing and thawing. The gels dissociate at temperatures of 50–55°C (Dea, 1979). At LBG concentrations of 0.75% or 1.0%, freeze–thaw gel breakdown occurs at 60–65°C, or 64–67°C, respectively. An explanation for these observations has been proposed by Dea *et al.* (1977). They stated that LBG's ability to gel in its native form results from its 'block' conformation, which allows 'smooth' regions to aggregate in order to form junction zones. The 'hairy' regions are responsible for the network's dispersibility via hydrogen bonding with water molecules. This same model explains the inability of native guar gum to gel (Dea *et al.*, 1977). The lower the concentration of β(1–6)-D-galactopyranosyl, the higher the chances of creating firm rubbery gels, which can be melted by retortion. If more than one cycle of freeze–thawing is applied to these gels, syneresis accelerates, causing them to lose up to 50%

of their water content. Weak-cohesive carob gels with a very low gum concentration ($\sim 0.2\%$) can be formed in 50% aqueous ethylene glycol (Dea *et al.*, 1977).

Because of its alternating chemical structure that sterically impedes the formation of interchain junction zones, guar gum does not produce gels under typical food-system conditions (Dea *et al.*, 1977). When ions such as borates and calcium and aluminum ions are added to the gum under alkaline conditions (pH 7.5–10.5), a range of irreversible gels of varying textures is produced. This gel formation is the result of complex formation between the cross-linking agent and the *cis*-hydroxyl groups (Seaman, 1980). Besides the addition of cross-linking agents, there are three other ways of inducing guar gum's gelation, namely reducing the water activity of the system, freeze–thaw treatments and a reduction in galactose content. Reduced water activity can be achieved by adding sucrose or other hydrophilic molecules to the system, thus creating competition for the available water between the sucrose and the gum, thereby promoting interchain binding. Freezing also promotes interchain binding of the guar gum molecules, as a result of increasing the effective hydrocolloid concentration in the residual unfrozen solution. Upon thawing, the junction zones dissociate and yield a dispersed hydrated gum. Reducing the concentration of $\alpha(1-6)$-D-galactopyranosyl units by chemical means can promote gelation of guar gum. More information can be found elsewhere (Dea, 1979; McCleary and Neukon, 1982).

Interactions of galactomannans with other hydrocolloids can be used to build firm, thermally reversible gels, as is the case when hot 1 : 1 solutions of LBG and xanthan gum are blended together and then cooled. The effect, which cannot be achieved using either hydrocolloid alone, is presumably the result of junction-zone formation between the galactose-deficient segments of the LBG and the xanthan gum (Cairns *et al.*, 1987). In support of this putative mechanism is the fact that an increase in galactose content results in decreased interaction, as is the case with tara gum (forming only a weak gel with xanthan) and guar (which produces only a synergistic rise in viscosity (Fox, 1992)). The supermolecular aspects of xanthan–LBG gels were studied using rheology and electron microscopy (Lundin and Hermannson, 1995). Results obtained with the latter showed that the xanthan–LBG network is formed from xanthan supermolecular strains, and the addition of LBG does not influence xanthan structure. Observed structural features of the gels were independent of heat treatment and LBG fraction. Structural similarities and rheological differences observed between xanthan and LBG fractions were compared with existing interaction models at the molecular level (Lundin and Hermannson, 1995). Evidence for heterotypic binding in synergistic gelation of LBG or konjac mannan with xanthan was reported (Goycoolea *et al.*, 1995c). Gels incorporating konjac mannan showed evidence of structural rearrangement after their initial formation. No such effects were seen for LBG. The same authors studied the effect of LBG and konjac mannan on the conformation and rheology of agarose and κ-carrageenan. They concluded that LBG or konjac mannan chains promote conformational ordering of carrageenan or agarose by binding to the double helix as it forms (Goycoolea *et al.*, 1995b).

Similar gel textures were achieved when LBG was blended with κ-carrageenan to produce a strong elastic gel, in contrast to the common brittle texture. Carob gum is also capable of producing cohesive gels with CMC at a total gum concentration of 0.5%. Equal amounts of both gums lead to maximal synergism (Kaletunc and Peleg, 1986).

Thermodynamic incompatibility, when mixing proteins and polysaccharides above a critical concentration, could result in separation into two phases (Tolstoguzov, 1991). This happens at concentrations at which molecular overlap begins to occur and the two molecular species are no longer totally miscible. The two phases include one rich in protein and the other containing mostly polysaccharides. Phase separation is favored by reducing the solubility of the protein and increasing the intrinsic viscosity of the galactomannan (Fox, 1992).

8.8 Stability

When galactomannans are used as thickeners in processed foods, sterilization (120°C for 10 min) under neutral conditions results in an ~10% loss in viscosity. This situation can be improved by the addition of trace elements such as sodium sulfite and propyl gallate, suggesting that the dominant mechanism is reductive-oxidative depolymerization (Mitchell *et al.*, 1992). The mechanism only applies at pH values >4.5. At low pH, hydrolysis of the glycoside bond results in an ~90% loss in viscosity when the solution is heated to 120°C for 10 min. Therefore, the buffering of processed systems, as well as a decrease in processing temperature, can reduce degradation, and hydration can be delayed by using coarsely ground galactomannan powder. Galactomannans are generally stable to shear forces unless extreme conditions are used during processing, such as those in high-pressure homogenization (Fox, 1992).

8.9 Food applications

Galactomannans are used for their inherent properties, the most widely exploited being the thickening of aqueous solutions. Other desirable properties include their synergistic interaction with other polysaccharides and their ability to reduce syneresis (Glicksman, 1969). For thickening purposes, low concentrations of 0.5–1.0% are used (versus 4.0–6.0% starch). The galactomannans produce a light texture, recover after shearing, some of them are cold-soluble and viscosity is dependent on temperature. Enzymatic attack is rare since the relevant enzymes are seldom found in foods, and following consumption they are not degraded, so little or no energy is available from their ingestion.

In many instances, starches behave differently from galactomannans; their properties are complementary and this is used to achieve special rheological properties that are influenced by other factors in the formulation such as water activity, ionic strength and pH, and by processing variables such as thermal load, shear stress and filling temperatures. The food technologist uses a blend of several hydrocolloids to achieve the desired, predetermined aim, and an awareness of gum interactions is essential for success.

Stabilization of ice cream by galactomannans is common practice compared with the use of seed gums in other milk-based products, which is limited by protein-phase separation. To an ice-cream mix that includes milk fat (or fat from another source), non-fat milk solids and added sugars, 0.5% mono/diglyceride is added (to destabilize the protein layer around the emulsified fat globules) with 0.3% high-viscosity LBG or a mixture of guar–LBG. Later on during cooling and whipping, partial churning of the fat occurs, and the free liquid fat with fat crystals forms a stabilizing layer around air pockets within the mass. The role of the galactomannans in the ice cream is to inhibit growth of larger ice crystals, by binding the fluid water to the gum and thus preventing its mass transfer, and to thicken, yielding a product with a creamier mouthfeel. Ice-crystal size in soft, flavored ices is reduced by adding ~0.1% galactomannan (Arbuckle, 1986; Barford *et al.*, 1991; Fox, 1992). Lactose or ice crystals greater than 20 μm in diameter can be detected sensorially and are perceived as sandiness. This sensorial observation indicates large temperature fluctuations during storage, and poor product formulation. As previously mentioned, guar gum combined with LBG or carrageenan prevents the formation of large lactose and ice crystals (Fox, 1992).

Galactomannans are fairly hydrophilic polysaccharides with a polymannose backbone and grafted galactose units, and a fairly rigid structure (Garti and Reichman, 1994). Most crude galactomannans contain 3–10% proteinaceous material. Surface activities of crude and purified LBG and guar gum, in correlation with their emulsification activity and stability, were studied. Crude LBG and guar gum reduced surface tension of water to ~55 $mN\,m^{-1}$ and, adsorbed to/precipitated onto oil–water interfaces, reduced their interfacial tensions. The surface activity of purified guar, where levels of proteins were reduced to a minimum of 0.8% weight, was similar to that of the crude gum. Oil-in-water emulsions of various oils with LBG or guar were prepared. Oil droplets were covered with precipitated gum layers exhibiting strong birefringency, indicating the formation of organized gum layers on the interface. The adsorption capacity and surface load were evaluated. Coalescence and flocculation were minimized by establishing the best gum: oil ratios for full droplet coverage (Garti and Reichman, 1994).

In fermented milk products, the addition of a gum mixture composed of CMC, galactomannan and gelatin is used to overcome the problem of structural loss when the fresh cheese is sheared by regular processing. In milkshakes, less than 0.1% galactomannan is used to thicken and impart a creamier mouthfeel to the product. Galactomannans are also used in milk-fortified fruit-juice drinks where the proportion of milk is low, to delay sedimentation of the casein micelles and ensure acceptance of the product.

Galactomannans, in addition to their use in gelled desserts based on carrageenans, are helpful in modifying gel texture and preventing syneresis (Herald, 1986b). A gel dessert with an identical chiffon-type gel can be achieved by mixing gelatin, LBG, soy protein, sugar and organic acids (Mancuso and Common, 1960). In mayonnaises with an oil content of less than $\sim 60\%$, rheological properties can be modified by replacing the oil with chemically modified starch and incorporating a stabilizer such as galactomannan (guar with a small addition of LBG) and xanthan to the recipe. If pieces of fish, meat or other ingredients are added to the mayonnaise, then it should be constructed such that it takes up the fluids and leaves the pieces intact. This can be achieved by increasing the inherent galactomannan content. If fresh fruits and vegetables are added to the mayonnaise, amylases may be released, causing starch hydrolysis. In such cases, the starch should be replaced with a blend of high-viscosity guar and xanthan gum, both at concentrations of not more than 0.7% (Fox, 1992).

Guar gum is used in barbecue and meat sauces, and in various salad dressings to prevent phase separation and impart a desirable mouthfeel. Galactomannans are also added to ketchups and dressings to increase viscosity and eliminate syneresis. In sterilized soups and sauces, galactomannans are used as thickeners in combination with xanthan at concentrations of ~ 0.2–0.5%. In deep-frozen foods, the addition of cold-soluble galactomannan improves freeze–thaw resistance, presumably by sterically hindering the formation of aggregates in the interstitial fluid phase. Other applications are as an ingredient in sausages (to prevent weeping), in pumped meat blends (as a suspending agent), in fish fillets (as a binder and phosphate replacer), in low-calorie jams and spreads and in baked goods (as gluten replacers). In processed cured meat products (e.g. sausages, salami, bologna), LBG is added to improve comminution of ingredients, to improve yield through the binding of free water, to extrude mixtures easily and to prevent phase separation during cooking, smoking and storage (Goldstein *et al.*, 1973; Rol, 1973; Fox, 1992). The addition of carob gum to pet foods influences the texture of the finished product. For example, a thick gravy sauce with sheen can be achieved. The same effect can be achieved with canned meat, where a rich, thick stew is desired (Seaman, 1980). Guar gum can be added at a concentration of ~0.5% of the total batch weight to canned meat products, resulting in cleaner mixing during meat cooking, easier pumping of the cooked product, cleaner technological processing and better control of the process. In stuffed meat products, guar gum offers rapid binding of free water during comminution, improved stuffing into casings, elimination of fat and free-water separation and migration during cooking, smoking and storage, and improved firmness upon cooling (Fox, 1992).

LBG can also help to achieve desired baked textures. The addition of LBG to wheat flours results in a softer, tastier product with an extended shelf life. Staling is retarded, crumbliness is reduced, and the quantity of eggs for the manufacture of biscuits, rolls and cakes is decreased (Herald, 1986a). The strength and pliability of baked tortillas can be improved by adding a blend of carob and guar gums (Gorton, 1984). In producing dry mixtures for baking, the addition of carob and guar gum at a maximum concentration of 0.15% improves both mixing and the resultant mix's characteristics (Federal Register, 1987c). When guar gum is added to dry cake, muffin, biscuit and pizza-crust mixes, it imparts shorter batter mixing times, less crumbling of the finished products (Cawley, 1964), improved ingredient mixing, reduced moisture loss during storage and the ability to freeze the finished product. Reconstituted flours, consisting of 89.5% rice, potato or tapioca starch, 9.8% vital wheat gluten, 0.5% lecithin and 0.2% ethoxylated mono/diglycerides, plus 0.1% xanthan, guar or cellulose gums, were made into batters that were evaluated for specific gravity and viscosity (Rice and Ndife, 1995). Overall, reconstituted rice flours with guar or xanthan gums produced cakes most similar to control cakes. The addition of 1% guar gum reduced the amount of absorbed oils and fats during deep-fat frying. The reduction is a consequence of protective-film formation, resulting in better-moistened products. Guar gum has been incorporated into breads, oatcakes and biscuits to yield baked products with a high soluble fiber content, which can be utilized to reduce the levels of serum cholesterol and low-density lipoproteins (Jenkins *et al.*, 1979; French and Hill, 1985). Over the years, guar gum has been used in a few miscellaneous products: as a foam stabilizer in foam-dried coffee concentrate, to produce acceptable imitation chocolate-chiffon dessert (Block, 1961) and to improve mechanical properties, water absorption and puffing characteristics of heat-extruded corn grits (Maga and Fapojuwo, 1988).

References

Anantheswaran, R.C. and Liu, L. (1994) Effect of viscosity and salt concentration on microwave heating of model non-Newtonian liquid foods in a cylindrical container. *J. Microwave Power Electromagnetic Energy*, **29**(2), 119–26.

Anderson, D.M.W. (1986) The amino acid components of some commercial gums, in *Gums and Stabilisers for the Food Industry*, vol. 3 (eds G.O. Phillips, D.J. Wedlock and P.A. Williams), Elsevier Applied Science, London, pp. 79–86.

Arbuckle, W.S. (1986) Effects of emulsifiers on protein–fat interactions in ice-cream mix during ageing. I. Quantitative analyses, in *Ice Cream*, Avi, Westport, CT, pp. 84–94.

Barfod, N.M., Krog, N., Larsen, G. *et al.* (1991) Effect of emulsifiers on protein–fat interactions in ice-cream mix during ageing. *Fat Sci. Technol.*, **93**(1), 24–35.

Block, H.W. (1961) *Chocolate Chiffon.* US Patent No. 2,983,617.

Blond, G. (1994) Mechanical properties of frozen model solutions. *J. Food Eng.*, **22**(1–4), 253–69.

Cairns, P., Miles, M.J., Morris, V.J. *et al.* (1987) X-ray fibre-diffraction studies of synergistic binary polysaccharide gels. *Carbohydr. Res.*, **100**, 411.

Cawley, R.W. (1964) The role of wheat flour pentosans in baking. II. Effect of added flour pentosans and other gums on gluten-starch loaves. *J. Sci. Food Agric.*, **15**(5), 834–9.

Clark, A.H., Dea, I.C.M. and McCleary, B.V. (1986) The effect of galactomannan fine structure on their interaction properties, in *Gums and Stabilisers for the Food Industry*, vol. 3 (eds G.O. Phillips, D.J. Wedlock and P.A. Williams), Elsevier Applied Science, London, pp. 429–40.

Dea, I.C.M. (1979) Interactions of ordered polysaccharide structures synergism, and freeze–thaw phenomena, in *Polysaccharides in Food* (eds J.M.V. Blanshard and J.R. Mitchell), Butterworths, London, pp. 229–40.

Dea, I.C.M. (1990) Structure/function relationships of glactomannans and food grade cellulosics, in *Gums and Stabilisers for the Food Industry*, vol. 5 (eds G.O. Phillips, D.J. Wedlock and P.A. Williams), IRL Press (at Oxford University Press), Oxford, pp. 373–82.

Dea, I.C.M. and Morrison, A. (1975) Chemistry and interactions of seed galactomannans. *Adv. Carbohydr. Chem. Biochem.*, **31**, 241–312.

Dea, I.C.M., Clark, A.H. and McCleary, B.V. (1986) Effect of the molecular fine structure of galactomannans on their interaction properties – the role of unsubstituted sides. *Food Hydrocolloids*, **1**, 129–40.

Dea, I.C.M., Morris, E.R., Rees, D.A. *et al.* (1977) Associations of like and unlike polysaccharides: mechanism and specificity in galactomannans, interacting bacterial polysaccharides, and related systems. *Carbohydr. Res.*, **57**(1), 249–72.

Doublier, J.L. and Launay, B. (1977) Rheological properties of guar. *Ind. Minerale*, **4**, 191.

Doublier, J.L. and Launey, B. (1981) Rheology of galactomannan solutions. *J. Text. Studies*, **12**, 151–72.

Einstein, A. (1906) Eine neue Bestimmung der Molekuldimension. *Ann. Phys.* **19**, 289–306.

Federal Register (1987a) *Guar Gum.* Code of Federal Regulations, Title 21, 184.1339, Office of the Federal Register, Washington, DC, pp. 432–3.

Federal Register (1987b) *Locust (Carob) Bean Gum.* Code of Federal Regulations, Title 21, 184.1343, Office of the Federal Register, Washington, DC, p. 433.

Federal Register (1987c) *Food; Designation of Ingredients.* Code of Federal Regulations, Title 21, 101.4a, Office of the Federal Register, Washington, DC, p. 14.

Fox, J.E. (1992) Seed gums, in *Thickening and Gelling Agents for Food*, ch. 7 (ed. A. Imeson), Blackie A & P, Glasgow, pp. 153–70.

French, S.J. and Hill, M.A. (1985) High fibre foods: a comparison of some baked products containing guar and pectin. *J. Plant Foods*, **6**(2), 101–9.

Frollini, E., Reed, W.F., Milas, M. *et al.* (1995) Polyelectrolytes from polysaccharides: selective oxidation of guar gum – a revisited reaction. *Carbohydr. Polym.*, **27**(2), 129–35.

Garti, N. and Reichman, D. (1994) Surface properties and emulsification activity of galactomannans. *Food Hydrocolloids*, **8**(2), 155–73.

Glicksman, M. (1969) *Gum Technology in the Food Industry*, Academic Press, New York, p. 130.

Goldstein, A. M., Alter, E.N. and Seaman, J.K. (1973) Guar gum, in *Industrial Gums* (eds R.L. Whistler and J.N. BeMiller), Academic Press, New York, pp. 303–21.

Gorton, L. (1984) Tortilla improvements – less fragile, more firm. *Baker's Digest*, **58**(6), 26.

Goycoolea, F.M., Morris, E.R. and Gidley, M.J. (1995a) Viscosity of galactomannans at alkaline and neutral pH: evidence of 'hyperentanglement' in solution. *Carbohydr. Polym.*, **27**(1), 69–71.

Goycoolea, F.M., Richardson, R.K., Morris, E.R. *et al.* (1995b) Effect of locust bean gum and konjac glucomannan on the conformation and rheology of agarose and κ-carrageenan. *Biopolymers*, **36**(5), 643–58.

Goycoolea, F.M., Richardson, R.K., Morris, E.R. *et al.* (1995c) Stoichiometry and conformation of xanthan in synergistic gelation with locust bean gum or konjac glucomannan. *Macromolecules*, **28**(24), 8308–20.

Herald, C.T. (1986a) Locust/carob bean gum, in *Food Hydrocolloids*, vol. 3 (ed. M. Glicksman), CRC Press, Boca Raton, FL, pp. 161–70.

Herald, C.T. (1986b) Guar gum, in *Food Hydrocolloids*, vol. 3 (ed. M. Glicksman), CRC Press, Boca Raton, FL, pp. 171–184.

Hui, P.A. and Neukom, H. (1964) Some properties of galactomannans. *Tappi*, **47**(1), 39–42.

Jenkins, D.J.A., Leeds, A.R., Slavin, B. *et al.* (1979) Dietary fiber and blood lipids: reduction of serum cholesterol type II hyperlipidaemia by guar gum. *Am. J. Clin. Nutr.*, **32**(1), 16–18.

Jud, B. and Lossl, U. (1986) Tara gum – a thickening agent with perspectives. *Int. Z. Lebensm. Tech. Verfahrenst.*, **37**(1), 28–31.

Kaletunc, G. and Peleg, M. (1986) Rheological characteristics of selected food gum mixtures in solution. *J. Text. Studies*, **17**(1), 61–70.

King, K. and Gray, R. (1993) The effect of gamma-irradiation on guar gum, locust bean gum, gum tragacanth and gum karaya. *Food Hydrocolloids*, **6**(6), 559–69.

Lopes da Silva, J.A. and Goncalves, M.P. (1990) Studies on a purification method for locust bean gum by precipitation with isopropanol. *Food Hydrocolloids*, **4**, 277–87.

Lundin, L. and Hermannson, A.M. (1995) Supermolecular aspects of xanthan–locust bean gum gels based on rheology and electron microscopy. *Carbohydr. Polym.*, **26**(2), 129–40.

Maga, J.A. and Fapojuwo, O.O. (1988) The effect of various hydrocolloids on some physical properties of extruded corn grits. *Int. J. Food Sci. Technol.*, **23**(1), 49–56.

Mancuso, J.J. and Common, L. (1960) *Chiffon.* US Patent No. 2,965,493.

McCleary, B.V. and Neukon, H. (1982) Effect of enzymic modification on the solution and interaction properties of galactomannans. *Prog. Food Nutr. Sci.*, **6**(1), 109–18.

Mitchell, J.R. (1979) Rheology of polysaccharide solutions and gels, in *Polysaccharides in Food* (eds J.M.V. Blanshard and J.R. Mitchell), Butterworths, London, pp. 51–7.

Mitchell, J.R., Hill, S.E., Jumel, K. *et al.* (1992) The use of anti-oxidants to control viscosity and gel strength loss on heating of galactomannan systems, in *Gums and Stabilisers for the Food Industry*, vol. 6 (eds G.O. Phillips, D.J. Wedlock and P.A. Williams), Oxford University Press, Oxford, pp. 303–10.

Morris, E.R. and Ross-Murphy, S.B. (1981) Chain flexibility of polysaccharides and glycoproteins from viscosity measurements. *Tech. Carbohydr. Metab.*, **B310**, 1–46.

Rice, B.O. and Ndife, M.K. (1995) Effects of the addition of hydrocolloids on the baking performance of high ratio layer microwave cakes made from reconstituted flours. *IFT Annu. Meet. 1995, Conf. Book*, p. 288.

Robinson, G.R., Ross-Murphy, S.B. and Morris, E.R. (1982) Viscosity–molecular weight relationships, intrinsic chain flexibility and dynamic solution properties of guar. *Carbohydr. Res.*, **107**, 17–32.

Rol, F. (1973) Locust bean gum, in *Industrial Gums: Polysaccharides and their Derivatives* (eds R.L. Whistler and J.N. BeMiller), Academic Press, New York, pp. 323–37.

Seaman, J.K. (1980) Locust bean gum, in *Handbook of Water-soluble Gums and Resins* (ed. R.L. Davidson), McGraw-Hill, New York, pp. 14.1–14.16.

Tolstoguzov, V.B. (1991) Functional properties of food proteins and role of protein–polysaccharide interaction. *Food Hydrocolloids*, **4**, 429–68.

9 Xanthan gum

9.1 Introduction

Xanthan gum is produced by biotechnological processes. The polymer, which is produced by the bacteria *Xanthomonas campestris*, is classified under the name B-1459 (Jeanes *et al.*, 1961). It can compete with and effectively replace other natural gums. Many other species of *Xanthomonas* have been reported to produce extracellular polysaccharides (Lilly *et al.*, 1958) and in general extracellular polysaccharides are produced by many species of microorganisms. After their production, they do not form covalent bonds with the microorganism's cell walls, being secreted instead into the culture media (Wilkenson, 1958). Xanthan gum is produced in the USA, Europe and Japan. The favored production method is fermentation because it does not depend on variable factors such as weather and a product of more consistent quality is obtained, the price of which is less sensitive to political or economic shifts. The gum is recognized as a harmless food additive for, among other purposes, thickening when its usage follows reasonable and practical manufacturing practices (Kovacs and Kang, 1977; Hart, 1988). In the early 1960s, the Kelco Company in San Diego, California began producing xanthan gum under the trade name Kelzan, and its use was approved by the FDA in 1969 (Anon., 1969; Urlacher and Dalbe, 1992).

Because of xanthan's importance, other new gums such as exopolysaccharide, produced by *Enterobacter agglomerans* grown on low-grade maple, and oat gum (extracted from oat bran) were compared with xanthan in terms of their functional properties, particle-suspension stability, emulsifying and foaming properties, to determine their potential uses (Britten and Morin, 1995; Dawkins and Nnanna, 1995).

9.2 Processing

To produce xanthan gum, pure cultures of *X. campestris* are grown using submerged aerobic fermentation in a sterilized medium composed of carbohydrates, a nitrogen source, phosphates (potassium) and trace minerals that has been preinoculated with the selected strain in a pilot-scale fermenter (Fig. 9.1). This is followed by incubation at 30°C for 3 days in an industrial-scale fermenter, then a thermal treatment to eliminate viable microorganisms. This culture is precipitated by isopropyl alcohol and the

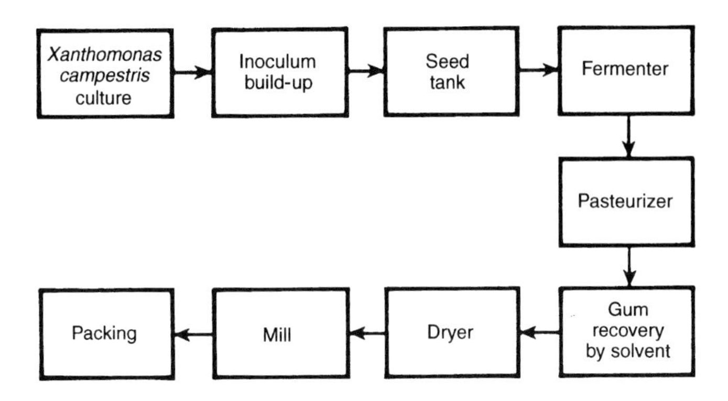

Figure 9.1 Xanthan gum production process. (Reproduced by permission of the NutraSweet Kelco Co., a unit of Monsanto Co., San Diego, CA.)

fibers are separated by centrifugation, dried, milled and sieved, then packaged (Kovacs and Kang, 1977; Sandford and Baird, 1983; Urlacher and Dalbe, 1992).

The resultant products include standard grades of xanthan with a particle size of 80 mesh, a fine-mesh grade (difficult to disperse but rapidly hydrated) with a particle size of 200 mesh, a granulated powder with improved dispersibility and a transparent xanthan powder for clear solutions at low concentrations. The gum powder includes a very low total bacterial count because of the sterilization process (Urlacher and Dalbe, 1992).

Xanthan production is influenced by the type and initial concentration of carbon and nitrogen sources, as well as by phosphate and citric concentration (de Vuyst and Vermeire, 1994). An optimal, industrially useful fermentation medium was devised for *X. campestris*. It consisted of glucose, sucrose or sirodex A as the sole carbon source, corn-steep liquor as a combined nitrogen–phosphate source and additional phosphate and citrate depending on the application the xanthan is destined for. Xanthan yields of 16.2, 15.1 and 16.5 $g\,kg^{-1}$ were obtained after 96 h fermentation on glucose (2%), sucrose (2%) and sirodex A (2.8%), respectively. The addition of citrate caused an increase in the pyruvate content of xanthan, with concomitant decreases in viscosity (de Vuyst and Vermeire, 1994). Other reports on medium formulations for xanthan production can be found elsewhere (Roseiro *et al.*, 1992).

Fermentation of *X. campestris* on a 5% glucose medium to produce xanthan in water-in-oil dispersions gave better results than control fermentation (Ju and Zhao, 1993). Whereas control fermentation gave a maximum xanthan concentration of 26 $g\,l^{-1}$ after 60 h, feed-batch operation with a water-in-oil dispersion (slow feed) of 50% glucose resulted in better bulk mixing and oxygen transfer, yielding a still-rising xanthan concentration of 65 $g\,l^{-1}$ after 202 h (Ju and Zhao, 1993).

Diffusion studies of nutrients such as ammonium and oxygen in xanthan gum solutions are needed for a better comprehension of mass-transfer limitations during culture (Brito *et al.*, 1995). Therefore, diffusion of ammonium nitrate in xanthan solutions under conditions similar to those occurring during typical fermentation was examined. Ammonium diffusivity in xanthan solutions drastically decreased as xanthan concentration increased. The ratio between the diffusivity in polymer and the diffusivity in water was almost independent of pH and only weakly a function of ionic strength. It was concluded that this diffusional behavior reflects the typical case of the diffusion of a small molecule in a non-Newtonian macromolecular solution (Brito *et al.*, 1995).

9.3 Chemical structure

Xanthan is a microbial polysaccharide composed of a 1–4-linked β-D-glucose backbone (like cellulose), with side-chains containing two mannoses and one glucuronic acid (Fig. 9.2). The pyruvic acid residues carried on half of the terminal mannose units represent ~ 60% of the molecule and give the gum many of its unique properties, e.g. its extraordinary resistance to hydrolysis and its uniform physical and chemical properties (Melton *et al.*,

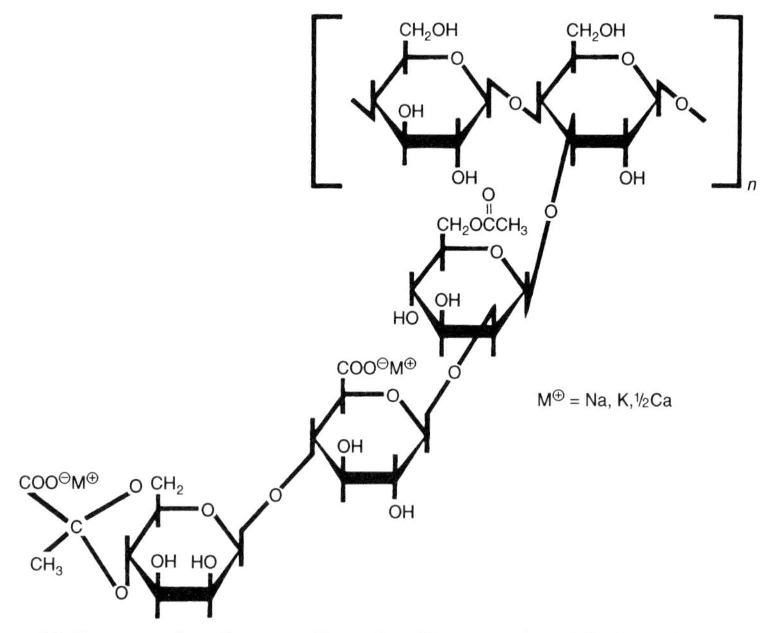

Figure 9.2 Structure of xanthan gum. (Reproduced by permission of the NutraSweet Kelco Co., a unit of Monsanto Co., San Diego, CA.)

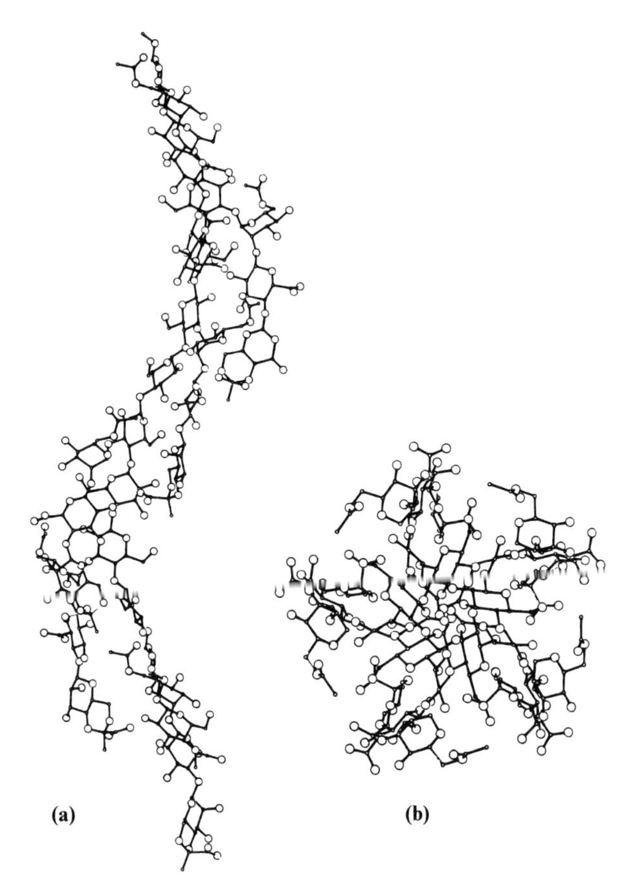

Figure 9.3 Helical conformation of xanthan gum viewed perpendicular (a) and parallel (b) to the helix axis. (Reproduced by permission of the Nutra-Sweet Kelco Co., a unit of Monsanto Co., San Diego, CA.)

1976; Jansson *et al.*, 1977; Kovacs and Kang, 1977). Commercial xanthans have a substitution degree of 30–40% for pyruvate and 60–70% for acetate, although substantial variations in these proportions exist (Smith *et al.*, 1981). The covalent structure of xanthan gum is related to its function as a thickening, suspending or gelling agent (Symes, 1980). The major amino acids in its proteinaceous component are alanine, glutamic acid, aspartic acid and glycine (Anderson, 1986).

Xanthan has a molecular weight of $\sim 2.5 \times 10^6$ with low polydispersity. Hydration in water is complete owing to its side-chains (Urlacher and Dalbe, 1992). Hydration rate ranges of xanthan gum were determined from its viscosity–time curves and used to study the effect of salt on gum-hydration rate (Kar Mun *et al.*, 1994). Xanthan's molecular conformation, as determined by X-ray diffraction studies, consists of a helix (Fig. 9.3) with

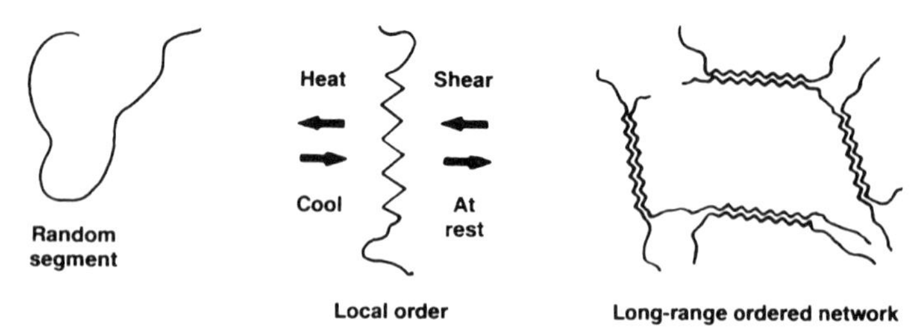

Figure 9.4 Conformational ordering in xanthan gum polysaccharide. (Reproduced by permission of the NutraSweet Kelco Co., a unit of Monsanto Co., San Diego, CA.)

a pitch of 4.7 nm and stabilization via hydrogen bonds (Moorhouse, 1992). It was proposed that xanthan macromolecules in solution should be considered as single rigid helices, while not rejecting the existence of a double or triple helix. Under the influence of temperature, xanthan passes from a rigid ordered state to a more flexible disordered state (Fig. 9.4).

Xanthan can be in one of two ordered conformations: native and renatured. The latter is more highly viscous at the same concentration, even though its molecular weight is the same as the former. The transition from native to denatured state is irreversible, whereas the transition from denatured to renatured state is reversible. The transition temperature depends on the gum concentration, ionic strength and pyruvic and acetic acid contents of the xanthan. The transition can be monitored by optical rotation, calorimetry, circular dichroism or viscometry. The thermal transition at low concentrations, up to 0.3% in distilled water, generally occurs at close to 40°C, but the transition temperature can increase to > 90°C when small amounts of salt are included (Urlacher and Dalbe, 1992).

9.4 Xanthan gum solutions

Commercial xanthan gum powder is a yellowish material that is soluble in cold or hot water. The gum is easily dissolved in 8% solutions of sulfuric, nitric and acetic acids, 10% hydrochloric acid and 25% phosphoric acid (Carnie, 1964). These solutions remain stable at room temperature for several months. Up to 50% solvents, such as ethanol and propylene glycol, can be added to aqueous xanthan solutions and still be tolerated. Hot glycerol (65°C) can also be used to solubilize the gum. Xanthan's dissolution in water yields a highly viscous, opaque solution, which exhibits pseudoplastic properties resulting from its conformation (rod-like) in solution and its high molecular weight (Whitcomb and Macosko, 1978).

The viscosity of xanthan gum is a function of its concentration in the dispersion (Pettitt, 1982). Viscosity versus xanthan gum concentration is comparable to other natural gums, such as tragacanth, guar gum and alginate (Anon., 1960). The pseudoplastic character of xanthan gum solutions (Pettitt, 1982) presents an advantage, since a reduction in viscosity values follows an increase in shear rate, making it easier to treat the food during processing. The pseudoplasticity is important in suitable food sensory qualities (Kelco Co., 1975). Some equations relating apparent viscosity, concentration, shear rate and temperature have been reported (Speers and Tung, 1986). The molecular structure of xanthan can be degraded by high shear rates, resulting in a decrease (loss) of functional properties.

Xanthan solutions exhibit higher viscosities than other gums at the same low concentration. Shear-thinning of xanthan solutions is more pronounced than with other gums owing to its semi-rigid conformation (compared with the random coil conformation found in the other gums). Formation of a weak network in solution results in high yield-point values. The interactions are shear reversible and, therefore, not permanent. At a 1% concentration in a 1% KCl solution, the yield values for xanthan gum, guar gum, HMC, LBG, CMC and sodium alginate are 11 300, 4000, 830, 360, 410 and 210 mPa, respectively. Xanthan exhibits a yield value of 500 mPa at a concentration of 0.3%, whereas the other previously mentioned gums at this concentration have no significant yield values. This quality is responsible for xanthan's ability to stabilize emulsions and dispersions (Launay *et al.*, 1986).

9.4.1 Stability in different media, under different technological treatments

Because of its secondary structure (side-chains wrapped around the cellulose backbone), xanthan is quite stable against degradation by acids or bases, heat treatments (high temperatures), freeze–thaw cycles, enzymes and long mixing (Urlacher and Dalbe, 1992). Xanthan solutions are stable over a wide range of pH, becoming affected only at pH values > 11 and < 2.5 (Pettitt, 1982). At $\sim$ pH 9 or above, xanthan gum gradually deacetylates. Its stability depends on gum concentration: the higher the concentration the more stable the solution. Gum solutions are degraded by high levels of strong oxidizing agents such as persulfates, hypochlorite and hydrogen peroxide (Pettitt, 1982). Xanthan gums are compatible with high concentrations of various salts (McNeely and Kang, 1973). Xanthan has also been reported to exhibit iron-binding properties and antioxidative activity upon autoxidation of soybean oil in emulsion (Shimada *et al.*, 1994).

In the range 10–90°C, xanthan viscosity is almost unaffected in the presence of salts. After sterilization (30 min at 120°C) of food products containing different gums, only 10% of the viscosity is lost in products containing xanthan which is less than the loss seen with products containing other hydrocolloids, such as guar gum, alginate and CMC. Enzymes such as amylase, pectinase and cellulase, found inherently in the food or added later, do not degrade xanthan. In microwaveable foods, the addition of xanthan eliminates moisture separation. Moreover at a concentration of

0.4%, the viscosity of xanthan is not affected by electrolytes, whereas at 1% there is a significant increase in viscosity. With sugar concentrations as high as 60%, xanthan hydration is unmodified. Interactions and precipitation can occur in acidic or heat-processed dairy–protein systems (Urlacher and Dalbe, 1992).

Stress–relaxation behavior of frozen 20% sucrose solutions in the presence of 0.5% xanthan gum, gelatin or guar gum was investigated using a thermomechanical analyser (Sahagian and Goff, 1995). The calculated relaxation time and asymptotic modulus at temperatures within and above the glassy state were significantly affected by freezing rate and stabilizer identity. Xanthan and gelatin exerted the greatest effect. It is postulated that, above the critical concentration, stabilizer action may derive from a modification of the kinetics of the unfrozen phase via changes in free-volume distribution specifically altering the rheological and viscoelastic response of systems at subzero temperatures (Sahagian and Goff, 1995).

9.4.2 Solution preparation

The preparation of xanthan solutions depends on their dispersibility and solubility. Dispersion occurs when the gum particles are introduced into liquid and separate from one another. Solubility is the process whereby particles swell and become viscous. Since the two processes are negatively related (easy dispersion means slow hydration), a suitable compromise needs to be found. Hydration depends on dispersion effectiveness, and the size of the gum particles relative to other ingredients in the mix. Lumping can be avoided by using a high-shear mixer and introducing the gum powder (previously mixed with other dry components of the mix if possible) through the top of the vortex. In commercial preparations, dispersion funnels or cyclone chambers are used to achieve rapid dispersion (Urlacher and Dalbe, 1992).

9.5 Xanthan interactions

Interactions between xanthan gum and other polysaccharides can have synergistic effects, such as enhanced viscosity or gelation with galactomannans and glucomannans. Synergistic interactions with all galactomannans, e.g. guar gum, LBG, tara gum and cassia gum, can occur.

Most important are the guar– and LBG–xanthan interactions (Fig. 9.5). LBG reacts more strongly, because its mannose to galactose ratio is 4:1 compared with 2:1 in guar gum. The interaction is suggested to occur between xanthan molecules and the 'smooth' regions of galactomannans, explaining the strong interaction with LBG, which is less branched than guar and has a more favorable galactose distribution (Thomas and Murray,

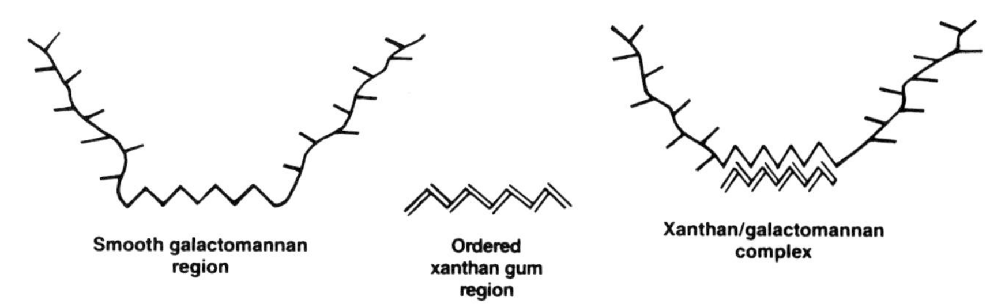

Figure 9.5 Schematic representation of the interaction between xanthan gum and galactomannans. (Reproduced by permission of the NutraSweet Kelco Co., a unit of Monsanto Co., San Diego, CA.)

1928; Schuppner, 1971; Jordan and Lester, 1973). Xanthan gum also exhibits a definite reactivity with tara gum (Glicksman and Farks, 1973, 1974). The literature contains some ideas on the role of pyruvate and acetate within the xanthan in the synergism (Shatwell *et al.*, 1991; Tako, 1991).

The xanthan–guar interaction leads to an increase in the solution's viscosity and elastic modulus. Such systems can be described by a measurement of the in-phase modulus G' (called elastic or storage modulus), and the out-of-phase modulus G'' (called viscous or loss modulus). Tan $\delta = G''/G'$ also provides information on the elasticity of such systems (a low value for tan δ indicates an elastic network). Determinations of elastic modulus and apparent viscosity are important for the stabilization of emulsions or suspensions (e.g. salad dressings require a high elastic modulus for stabilization, whereas in sauces and soups a high apparent viscosity is required while a high degree of elasticity is undesirable). Maximal synergism in terms of viscosity (at several shear rates, 0.2 up to $21\,s^{-1}$) can be achieved for blends with a total gum content of 1.0%, of which xanthan makes up no more than 20%. The synergism decreases in the presence of salts and is also influenced by gum concentration: the higher the latter, the greater the synergism. Xanthan–guar interactions are also important in cases in which blend ratios have lower elastic modulus values than xanthan gum alone, except with blends containing up to 30% guar gum, which possess properties similar to those of pure xanthan and can, therefore, serve as valuable stabilizers. Although the guar–xanthan blends improve guar's heat stability, xanthan gum alone remains more thermally stable. Xanthan–guar blends have a wide variety of applications owing to their good sensory qualities and lower cost compared with xanthan alone (this is because only a small proportion of xanthan is needed to change guar's rheological properties); although xanthan gum alone provides good thermal stability and efficient stabilizing properties, the blend is widely utilized in foods.

With xanthan–LBG interactions, at gum concentrations of $>0.2\%$, strong, cohesive thermoreversible gels exhibiting very little syneresis can be produced. The optimum gum ratio is between 40:60 and 60:40. Since the gel is elastic, cohesive and non-brittle, in terms of its sensory evaluation its product is undesirable. However, these properties can be improved by a suitable addition of starch or protein. Such gels are preferentially used in pet-food products (Urlacher and Dalbe, 1992).

Rheological and ultracentrifugation studies were conducted on heated and unheated mixtures of xanthan with whole LBG, and temperature fractions of the latter possessing different mannose:galactose ratios (Mannion *et al.*, 1992). Results suggest that xanthan and galactomannan interact via two distinct mechanisms. One mechanism takes place at room temperature, gives weak elastic gels and has little dependence upon the galactose content of the galactomannan, whereas the second mechanism requires significant heating of the polysaccharide mixture, gives stronger gels and is highly dependent upon galactomannan composition (Mannion *et al.*, 1992).

Mixed gels of xanthan gum and galactomannans (guar and fractionated LBG) were compared by means of oscillatory shear measurements (Fernandes, 1995). Whatever the proportion of LBG, maximal synergism was observed when the ratio of xanthan to galactomannan was 1:1. However, the magnitude of this maximum varied with LBG content in the mixture. Thermal behavior of 1:1 xanthan-fractionated LBG fractions was characterized by cooling–heating cycles between 10 and 80°C. Slight thermal hysteresis was systematically described for the several xanthan–LBG fraction systems. The mannose to galactose (M:G) ratio of LBG fractions determined the sol–gel transition (gelation and melting temperatures) of xanthan–galactomannan mixed systems. From the galactomannan fine structure and thermorheological profiles, it was evident that the higher the M:G ratio of galactomannan samples, the higher the synergistic interaction and transition temperature of the xanthan–LBG systems. The author also described a linear relationship between maximum synergism and the galactose content of the galactomannan samples (Fernandes, 1995).

The sedimentation rate in aqueous mixtures of xanthan with enzyme-modified galactomannans (the galactose content being decreased by the enzymes) was studied (Luyten *et al.*, 1994). Results were related to rheological properties of the systems at large deformations. The shear modulus increased while strain at yielding and sedimentation velocity decreased with decreasing galactose content of galactomannan. Properties of the xanthan–enzyme-modified galactomannan system were compared with those of xanthan–LBG systems. LBG was suggested to be a better stabilizer than the enzyme-modified galactomannans studied (Luyten *et al.*, 1994).

Xanthan–konjac mannan (glucomannan) interactions are strong above concentrations of 0.2% (Dalbe, 1992). The optimum ratio is between 40:60 and 30:70 (xanthan:konjac mannan). The presence of salt decreases the interaction. At temperatures $< 50°C$, the system's G' is greater than its G'', indicating well-defined elastic properties. At $\sim 55°C$, the gel structure collapses and there is a sharp decrease in the value of G'. At temperatures

> 55°C, the behavior is typical of a liquid which has G'' significantly higher than G'. At ~1.0% total gum concentration, maximum gel strength is achieved, with no further increase in this property with increasing gum concentration. The mechanical properties of xanthan–LBG are also similar to those of xanthan–konjac mannan. Small amounts of xanthan are sufficient to stabilize starch solutions during storage. Xanthan at 0.1–0.2% prevents starch solution retrogradation and increases its stability, especially at low pH (around 3). These small amounts of xanthan also improve starch stability during freeze–thaw cycles.

Reactions of xanthan gum with specific dextrins and incompatibility between xanthan and gum arabic at pH values below 5 have been reported. Reactivity is also apparent with casein below its isoelectric point. Interactions with proteins and other hydrocolloids are reported elsewhere (Sanderson, 1982).

A brief discussion of some relevant concepts and terminology describing the influence of protein–polysaccharide (p/p) interactions in the stabilization of emulsions, with particular reference to cases of weak reversible p/p interaction and permanent p/p covalent linkage, can be found elsewhere (Dickinson and Walstra, 1993). Mixtures of sodium caseinate and xanthan have been used to study weak interactions. The use of soluble p/p complexes to stabilize emulsions requires two biopolymers to be linked together to form a stable conjugate combining surface-active properties of the hydrophobic protein component with potential steric-stabilizing properties of the hydrophilic polysaccharide (Dickinson and Walstra, 1993).

A knowledge of the behavior of starch pastes during frozen storage is necessary to understand more complex systems such as sauces, dressings and desserts. Effects of storage at subzero temperature on properties of 10% corn pastes, prepared in the presence or absence of 0.3% xanthan gum, were studied (Ferrero *et al.*, 1994). Pastes were frozen at different rates (0.3–270 cm h^{-1}) and stored at −5°C (i.e. close to glass transition temperature), −10 and −20°C. Syneresis, rheological properties, ice recrystallization and amylopectin retrogradation were analysed. Samples stored at −5°C or higher deteriorated owing to amylose and amylopectin retrogradation and ice recrystallization. Starch retrogradation was not detected in samples stored at −10 or −20°C, although ice recrystallization was observed. The addition of xanthan gum improved the quality of frozen pastes by preventing the formation of a spongy structure associated with amylose retrogradation and by maintaining the smooth texture characteristic of unfrozen samples. However, the presence of xanthan gum did not prevent ice recrystallization or amylopectin retrogradation (Ferrero *et al.*, 1994).

Xanthan can serve to stabilize aloe vera gel. Aloe vera gel (used in health foods) interacts with algal polysaccharides and xanthan gum. This is demonstrated by increased apparent viscosities, apparent yield points and hysteresis in some cases; these phenomena do not occur with guar gum.

These properties did not deteriorate during storage, demonstrating that xanthan's influence and structure stabilization are long term (Yaron *et al.*, 1992).

9.6 Food applications

Following its approval as a food additive, xanthan gum found many uses in the food industry, since at low concentrations it provides storage stability, water-binding capacity and esthetic appeal (Urlacher and Dalbe, 1992). Dilute solutions of xanthan and other gums (e.g. guar gum) also serve as model systems for studies of natural and synthetic sweeteners (Pastor *et al.*, 1994).

Xanthan is used as a stabilizer for dressings (oil–water emulsions). Ideally the product has high yield value (permitting the suspension of spices, herbs and vegetables and enabling the dressing to cling to the salad as well as to have body) and strong pseudoplasticity. These requirements make xanthan favorable for such an application. Product stability is not influenced by low pH (~ 3.5 in some dressings) and high salts (15% or less), or thermal treatment. The uniform viscosity of xanthan from 5 to 70°C also helps yield a uniform texture and good stability. The amounts of xanthan required depend on oil content, with lower oil content requiring a higher amount of xanthan in the product. In fact the same flow properties can be achieved with different oil levels by adjusting the level of xanthan.

Xanthan gum can be used to control the rheological properties of mayonnaise (Ma and Barbosa Canovas, 1995a,b). Flow and viscoelastic properties of mayonnaise at different oil and xanthan gum concentrations (75–85% and 0.5–1.5%, respectively) were investigated using a plate–plate rheometer in rotational and oscillatory mode. Viscoelastic properties of mayonnaise were characterized using small-amplitude oscillatory shear, showing weak gel-like properties. Gel strength depended on oil and xanthan gum concentrations. The magnitude of elastic modulus and complex viscosity increased with increased oil or xanthan gum concentrations (Ma and Barbosa Canovas, 1995a,b).

Xanthan gum improves the mouthfeel of citrus and fruit-flavored beverages. In beverages that include flavor emulsions, the addition of xanthan at concentrations of up to 0.5% can help stabilization and mouthfeel (Schuppner, 1968).

The use of xanthan in sauces, gravies, relishes, canned soups and dairy products has been previously described (Section 9.5). Its heat stability and excellent stabilizing and suspension properties are important in canned foods. In whipped creams and mousses, the high yield values of xanthan help stabilize air cells, and whipping is made easier by the gum's pseudoplasticity.

In instant mixes (drinks, soups and low-calorie and aerated desserts, instant milkshakes, sauces and breakfast drinks), xanthan is used as a thickener and a suspension and body agent at a concentration of about 0.1–0.2%; in such beverages, CMC– or guar–xanthan blends are used (Katz, 1971; Edlin, 1972; Farkas, 1974).

Xanthan's pseudoplastic behavior is important in bakery products during dough preparation, i.e. pumping, kneading and molding. The xanthan prevents lump formation during kneading and improves dough homogeneity. It also increases product volume and makes the distribution and size of the air cells in baked goods more uniform. In syrups, xanthan is used as a particle suspender whereas in toppings and fillings xanthan improves texture, controls syneresis and improves freeze–thaw stability (Kovacs and Kang, 1977). In frozen desserts and confections (ice cream and sherbets), the addition of ~0.2% xanthan gum results in products with better mouthfeel and heat-shock resistance (Yaki and Okamura, 1969; Moorthy and Balachandran, 1992). Ice cream is frequently stabilized by a combination of xanthan, LBG and guar at a final concentration of ~0.1%. Xanthan is also used to coat frozen confections (Jenkenson and Williams, 1973). Control of ice-cream viscosity can be achieved with xanthan as well as with commercial blend stabilizers based on combinations of guar gum, LBG and CMC (Goff and Davidson, 1994). Milk reactivity, defined as gum's ability to produce significantly higher viscosity in milk than in water, was determined for carrageenan, guar and xanthan gums in dried skim milk or whey protein concentrate (reconstituted to 11% solids). Xanthan gum exhibited milk reactivity with whey protein concentrate at 0.05% (Schmidt and Smith, 1992). The Kelco company has developed a food stabilizer comprising xanthan and guar gum synergy. The stabilizer binds and thickens at low concentration, controls syneresis, disperses and hydrates easily in cold or hot milk or water; it is stable over a wide range of pH and provides good heat-shock and freeze–thaw protection. It can be used in processed and low-fat cheeses and in instant low-fat mixes for desserts and puddings (Kelco Co., 1992). Xanthan gum has been used in frozen desserts to protect against heat shock and to control the development of a coarse, icy texture. However, its functionality is limited in lower-pH milk products because long, narrow fibers appear, resulting from a complex formed between xanthan gum and whey proteins. This reaction has been exploited to develop a form of fat replacer for use in some foods (Milani and Bradley, 1992). Stabilization of chocolate milk, yoghurt-based beverages and low-fat spreads using galactomannans, pectins, alginates, carrageenans and xanthan gum has been reported (Tilly, 1991). The optimum concentration of stabilizers such as gelatin, guar gum and xanthan gum for the improvement of the quality of soy-whey yoghurt was evaluated by sensory panel and a study of viscosity and consistency. The optimum stabilizer combination ($g\,l^{-1}$) was gelatin 1.27, guar gum 2.32 and xanthan gum 0.19. This combination gave a better texture than that obtained with gelatin alone (Rossi *et al.*, 1990).

Other applications include the stabilization of cottage-cheese creamy emulsions with a blend of xanthan–LBG (0.10–0.12%) (Kovacs and Titlow, 1976). The same combination at ~0.5% has been reported as an aid in the manufacture of starch-jelly candies (Cornan *et al.*, 1956), reducing their setting time. The addition of xanthan to meat batters, at concentrations up to 1.0, contributes to their textural properties (Whiting, 1984; Foegeding and Ramsey, 1987). Other uses of xanthan include the production of a pourable aerated dairy dessert (Tilly, 1995), as part of compositions enabling proteins to remain heat stable (Colarow *et al.*, 1994) and in ready-to-serve frozen desserts for soft-serve dispensing (Malone and Sage, 1994).

References

Anderson, D.M.W. (1986) The amino acid components of some commercial gums, in *Gums and Stabilisers for the Food Industry*, vol. 3 (eds. G.O. Phillips, D.J. Wedlock, P.A. Williams), Elsevier Applied Science, London, pp. 79–86.

Anon. (1960) New polysaccharide gums produced by microbial synthesis. *Manuf. Chem.*, **31**(5), 206–8.

Anon. (1969) Xanthan gum. *Fed. Reg.*, **34**(53), 5376–7.

Brito, E., Torres, L. and Galindo, E. (1995) Diffusion behavior of ammonium in xanthan gum solutions. *Biotechnol. Prog.*, **11**(2), 221–3.

Britten, M. and Morin, A. (1995) Functional characterization of the exopolysaccharide from *Enterobacter agglomerans* grown on low-grade maple sap. *Lebensm. Wiss. Tech.*, **28**(3), 264–71.

Carnie, G.A.R. (1964) Evaluation of a new polysaccharide gum. *Aust. J. Pharm.*, **45**(19), 580–3.

Colarow, L., Dalan, E. and Kusy, A. (1994) *Composition Enabling Proteins to Remain Heat-stable and Product thus Obtained.* European Patent Application, EP 0 628 256 A1.

Cornan, J., Tsuchiya, H.M. and Stringer, C.S. (1956) *Enzymatic Preparation of High Levulose Food Additives.* US Patent No. 2,742,365.

Dalbe, B. (1992) Interactions between xanthan gum and konjac mannan, in *Gums and Stabilisers for the Food Industry*, vol. 6 (eds G.O. Phillips, D.J. Wedlock and P.A. Williams), Oxford University Press, Oxford, pp. 201–8.

Dawkins, N.L. and Nnanna, I.A. (1995) Studies on oat gum: composition, molecular weight estimation and rheological properties. *Food Hydrocolloids*, **9**(1), 1–7.

de Vuyst, L. and Vermeire, A. (1994) Use of industrial medium components for xanthan production by *Xanthomonas campestris*. *Appl. Microbiol. Biotechnol.*, **42**(2/3), 187–91.

Dickinson, E. and Walstra, P. (1993) Protein–polysaccharide interactions in food colloids, in *Food Colloids and Polymers, Stability and Mechanical Properties* (ed. E. Dickinson), Royal Society of Chemistry, Cambridge, UK, pp. 77–93.

Edlin, R.L. (1972) *Method of Producing a Dehydrated Food Product.* US Patent No. 3,694,236.

Farkas, E. (1974) *Method for Preparing a Low-calorie Aerated Structure.* US Patent No. 3,821,428.

Fernandes, P.B. (1995) Influence of galactomannan on the structure and thermal behavior of xanthan/galactomannan mixtures. *J. Food Eng.*, **2**, 269–83.

Ferrero, C., Martino, M.N. and Zaritzky, N.E. (1994) Corn starch–xanthan gum interaction and its effect on the stability during storage of frozen gelatinized suspensions. *Starch*, **46**(8), 300–8.

Foegeding, E.A. and Ramsey, S.R. (1987) Rheological and water-holding properties of gelled meat batters containing iota-carrageenan, kappa-carrageenan, or xanthan gum. *J. Food Sci.*, **52**(3), 549–53.

Glicksman, M. and Farks, H.E. (1973) *Pudding Compositions.* US Patent No. 3,721,571.

Glicksman, M. and Farks, H.E. (1974) *Gum Gelling System: Xanthan–Tara Dessert Gel.* US Patent No. 3,784,712.
Goff, H.D. and Davidson, V.J. (1994) Controlling the viscosity of ice cream mixes at pasteurization temperatures. *Modern Dairy*, **73**(3), 12, 14.
Hart, B. (1988) Fixing formulations. *Food Proc.* (UK), **57**(1), 15–16, 21.
Jansson, P.E., Kenne, L. and Lindberg, E.B. (1977) *Carbohydr. Res.*, **45**, 275–82.
Jeanes, A.R., Pittsley, J.E. and Samtis, F.R. (1961) Polysaccharide B-1459: a new hydrocolloid polyelectrolyte produced from glucose by bacterial fermentation. *J. Appl. Polym. Sci.*, **5**, 519–26.
Jenkenson, T.J. and Williams, T.P. (1973) *Gel-coated Frozen Confection.* US Patent No. 3,752,678.
Jordan, W.A. and Lester, W.H. (1973) *Blends of* Xanthomonas *and Guar Gum.* US Patent No. 3,765,950.
Ju, L.K. and Zhao, S. (1993) Xanthan fermentations in water/oil dispersions. *Biotechnol. Tech.*, **7**(7), 535–40.
Kar Mun, T., Mitchell, J.R., Hill, S.E. *et al.* (1994) Measurement of hydration of polysaccharides. *Food Hydrocolloids*, **8**(3/4), 243–9.
Katz, M.H. (1971) *Edible Mix Composition for Producing an Aerated Product.* US Patent No. 3,582,357.
Kelco Co. (1975) Xanthan gum/Keltral/Kelsan, in *Natural Biopolysaccharides for Scientific Water Control*, 2nd edn, Kelco Co., San Diego, CA.
Kelco Co. (1992) Gum synergy K.O. Competition. *Dairy Foods*, **93**, 8, 78.
Kovacs, P. and Titlow, B.D. (1976) Stabilizing cottage cheese emulsions with a xanthan gum blend. *Am. Dairy Rev.*, **38**(4), 34J.
Kovacs, P. and Kang, K.S. (1977) Xanthan gum, in *Food Colloids* (ed. H.P. Graham), AVI Publishing, Westport, CT, pp. 500–21.
Launay, B., Doublier, J.L. and Cuvelier, G. (1986) in *Functional Properties of Food Macromolecules* (eds J.R. Mitchell and D.A. Ledward), Elsevier Applied Science, London, pp. 1–78.
Lilly, V.G., Wilson, H.A. and Leach, J.G. (1958) Bacterial polysaccharides. II. Laboratory-scale production of polysaccharides by species of *Xanthomonas. Appl. Microbiol.*, **6**, 105–8.
Luyten, H., Kloek, W. and Van Vliet, T. (1994) Yielding behavior of mixtures of xanthan and enzyme-modified galactomannans. *Food Hydrocolloids*, **8**(5), 431–40.
Ma, L. and Barbosa Canovas, G.V. (1995a) Rheological characterization of mayonnaise. I. Slippage at different oil and xanthan gum concentrations. *J. Food Eng.*, **25**(3), 397–408.
Ma, L. and Barbosa Canovas, G.V. (1995b) Rheological characterization of mayonnaise. II. Flow and viscoelastic properties at different oil and xanthan gum concentrations. *J. Food Eng.*, **25**(3), 409–25.
Malone, M.J. and Sage, J.G. (1994) *Ready to Serve Frozen Dessert for Soft-serve Dispensing.* US Patent No. 5,328,710.
Mannion, R.O., Melia, C.D., Launay, B. *et al.* (1992) Xanthan/locust bean gum interactions at room temperature. *Carbohydr. Polym.*, **19**(2), 91–7.
McNeely, W.H. and Kang, K.S. (1973) Xanthan and some other biosynthetic gums, in *Industrial Gums* (eds R.L. Whistler and N.J. BeMiller), Academic Press, New York, pp. 473–521.
Melton, L.D., Mindt, L., Rees, D.A. *et al.* (1976) Covalent structure of the extracellular polysaccharide from *Xanthomonas campestris*: evidence from partial hydrolysis studies. *Carbohydr. Res.*, **46**, 245.
Milani, F.X. and Bradley, R.L., Jr (1992) Modification of induced complex formation between xanthan gum and whey proteins at reduced pH. *J. Dairy Sci.*, **75**(1), 121.
Moorhouse, R. (1992) in *Thickening and Gelling Agents for Food*, ch. 9 (ed. A. Imeson), Blackie A & P, Glasgow, pp. 202–26.
Moorthy, P.R.S. and Balachandran, R. (1992) Certain non-conventional stabilizers – their effect on whipping ability of ice cream mix. *Cheiron*, **21**(5–6), 180–2.
Pastor, M.V., Costell, E. and Duran, L. (1994) Effect of viscosity on the detection, recognition and differential thresholds of sucrose and aspartame. *Revista Espanola de Ciencia y Technologia de Alimentos*, **34**(1), 91–101.
Pettitt, D.J. (1982) Xanthan gum, in *Food Hydrocolloids*, vol. 1 (ed. M. Glicksman), CRC Press, Boca Raton, FL, pp. 127–49.

Roseiro, J.C., Esgalhado, M.E., Amaral Collaco, M.T. *et al.* (1992) Medium development for xanthan production. *Process Biochem.*, **27**(3), 167–75.
Rossi, E.A., Faria, J.B., Borsato, D. *et al.* (1990) Optimization of a stabilizer system for a soya-whey yoghurt. *Alimentos e Nutricao*, **2**, 83–92.
Sahagian, M.E. and Goff, H.D. (1995) Influence of stabilizers and freezing rate on the stress relaxation behavior of freeze-concentrated sucrose solutions at different temperatures. *Food Hydrocolloids*, **9**(3), 181–8.
Sanderson, G.R. (1982) The interaction of xanthan gum in food systems. *Prog. Food Nutr. Sci.*, **6**, 77–87.
Sandford, P.A. and Baird, J. (1983) Industrial utilization of polysaccharides, in *The Polysaccharides*, vol. 2 (ed. G.O. Aspinall), Academic Press, New York, pp. 411–90.
Schmidt, K.A. and Smith, D.E. (1992) Milk reactivity of gum and milk protein solutions. *J. Dairy Sci.*, **75**(12), 3290–5.
Schuppner, H.R. (1968) *Non-alcoholic Beverage.* US Patent No. 3,413,125.
Schuppner, H.R. (1971) *Heat-reversible Gel.* US Patent No. 3,507,664.
Shatwell, K.P., Sutherland, I.W., Ross-Murphy, S.B. *et al.* (1991) *Carbohydr. Polym.*, **14**, 29–51.
Shimada, K., Muta, H., Nakamura, Y. *et al.* (1994) Iron-binding property and antioxidative activity of xanthan on the autoxidation of soybean oil in emulsion. *J. Agric. Food Chem.* **42**(8), 1607–11.
Smith, I.H., Symes, K.C., Lawson, C.J. *et al.* (1981) Influence of the pyruvate content of xanthan on macromolecular association in solution. *Int. J. Biol. Macromol.*, **3**, 129.
Speers, R.A. and Tung, M.A. (1986) Concentration and temperature dependence of flow behavior of xanthan gum dispersions. *J. Food Sci.*, **51**(1), 96–8, 103.
Symes, K.C. (1980) The relationship between the covalent structure of the *Xanthomonas* polysaccharide (xanthan) and its function as a thickening, suspending and gelling agent. *Food Chem.*, **6**, 63–76.
Tako, M. (1991) Synergistic interaction between a deacylated xanthin and galactomannan. *Carbohydr. Polym.*, **16**, 239–52.
Thomas, A.W. and Murray, H.A. (1928) Gum arabic. *J. Phys. Chem.*, **32**(6), 676–97.
Tilly, G. (1991) Stabilization of dairy products by hydrocolloids, in *Food Ingredients Europe: Conference Proceedings*, pp. 105–21.
Tilly, G. (1995) *Stabilizer Composition enabling the Production of a Pourable Aerated Dairy Dessert.* European Patent Application, EP 0 649 599 A1.
Urlacher, B. and Dalbe, B. (1992) Xanthan gum, in *Thickening and Gelling Agents for Food* (ed. A. Imeson), Blackie A & P, Glasgow, pp. 206–26.
Whitcomb, P.J. and Macosko, C.W. (1978) Rheology of xanthan gum solutions *J. Rheol.*, **22**(5), 493–505.
Whiting, R.C. (1984) Addition of phosphates, proteins and gums to reduced-salt frankfurter batters. *J. Food Sci.*, **49**(5), 1355–7.
Wilkenson, J.F. (1958) The extracellular polysaccharides of bacteria. *Bacteriol. Rev.*, **22**(1), 46–113.
Yaki, K. and Okamura, T. (1969) *Manufacturing Method for Ice Cream and Sherbet.* Japanese Patent No. 44-10149.
Yaron, A., Cohen, E. and Arad, S.M. (1992) Stabilization of aloe vera gel by interaction with sulfated polysaccharides from red microalgae and with xanthan gum. *J. Agric. Food Chem.*, **40**(8), 1316–20.

10 Agricultural uses of hydrocolloids

10.1 Agricultural chemicals

10.1.1 Controlled release

The release of agricultural chemicals (including pesticides) can be controlled by the use of polymers. These polymers control the rate of delivery, the mobility and the length of time for which the component is considered effective. The greatest advantage of controlled-release formulations is that less chemical is used during a given time interval, thereby lowering its impact on non-target species and limiting leaching, volatilization and degradation (Kydonieus, 1980).

Controlled-release systems are divided into two categories. In the first, the active agent is dissolved, dispersed or encapsulated by coating in a polymer matrix. In the second, the polymers contain the active agent as part of their macromolecular backbone or pendent side-chain. In the latter case, release results from biological or chemical cleavage of the bond between the bioactive agent and the polymer. Many natural and synthetic polymers and synthetic elastomers are used for controlled-release purposes. Natural polymers include CMC, gelatin, gum arabic, starch, arabinogalactan and many others. Synthetic elastomers include polybutadiene, polyisoprene, neoprene, polysiloxane, styrene–butadiene rubber and silicone rubber. Many synthetic polymers exist including PVA, polystyrene, polyacrylamide and polyvinyl chloride (PVC) (Kydonieus, 1980; Mark *et al.*, 1985a,b).

Controlled release can be useful for fungicides, germicides, growth regulators, herbicides, insect diets, insecticides, etc. (Orzolek and Kaplan, 1987). To adjust a polymer to a specific system, the following parameters need to be taken into consideration: cost, seasonal conditions, desired release rate, duration and ease of formulation and application. Other issues include the nature of the polymer, its thermostability or thermoplasticity, its melting and glass transition temperatures, its compatibility with bioactive agents, the stability of the combination, processing conditions and desired shape and size of the final product (Paul, 1976).

Physical systems that incorporate agricultural chemicals include microencapsulation, physical blends, dispersion in plastics or rubbers, laminates, hollow fibers and membranes. Models for the release of active materials have been developed, and processing techniques are discussed elsewhere (Kydonieus, 1980; Harris *et al.*, 1975).

10.1.2 Herbicides, insecticides, nematicides and pheromones

Controlled-release systems can include natural polymers, such as starch, cellulose, chitin, alginic acid and lignin, among many others. These polymers are abundant, relatively inexpensive and biodegradable. Their advantage lies in their potential to participate in derivatization and formulation; their disadvantage is their insolubility in solvents suitable for encapsulation, dispersion, reaction and formulation. In some instances, this can limit the amount of bioactive agent per unit weight of polymer. This problem has been solved in a few cases by *in situ* encapsulation, which involves the cross-linking of gelatinized starch containing the pesticide by either calcium chloride or boric acid, or by xanthation followed by oxidation. These processes result in the entrapment of a pesticide within a granular particle. Many herbicides have been formulated via this type of technique: vernolate, butylate, cycloate, Surpass, pebulate, metribuzin, trifluralin, chloramben, picloram, dicamba and thiocarbonates, the nematicide DBCP and the insecticides Dimilin, Diazinon, parathion, malathion, Captan and Zineb (Harris *et al.*, 1975; Allan *et al.*, 1980; Kydonieus, 1980; Mark *et al.*, 1985a,b).

Water-soluble gums and resins are used in agricultural products such as herbicides. Carrageenans are used as suspension and sticking agents in pesticide and herbicide formulations. PEGs and their fatty-acid esters can be used in many agricultural formulations as emulsifiers, spreader-stickers or carriers in insecticide and herbicides sprays. Polyvinylpyrrolidone (PVP), xanthan gum and combinations of xanthan gum and LBG are used to stabilize emulsions in herbicides, fungicides and similar products (Davidson, 1982). Alkyl- and hydroxyalkylcellulose are used in agriculture as dispersing agents for wetting powders, as dust stickers, seed stickers, spray-drift controllers, spray stickers and weed killers. An old Japanese patent made use of gum arabic as a binder for insecticides (Matsudaira, 1953). Tragacanth gum finds industrial applications in the production of insecticide emulsions. The gum increases the aqueous-phase viscosity and lowers the interfacial tension between oil and water. The same attributes are put to use in insecticide sprays.

Lignin, sawdust and bark, composed to a large extent of lignin and cellulose, have been combined with pesticides such as 2,4-D, 2,4-DB, 2,4,5-TMCPA, MCPB, TCA, dalapon, metribuzin, picloram, silvex and mecoprop. Many synthetic polymers have been used for controlled-release purposes. Commercial controlled-release products include Brocon CR, Tox-Hid, Ocusert and Precise, among many others (Mark *et al.*, 1985a,b). A special polymer application is the controlled release of pheromones from hollow fibers or plastic baits. Other potential uses of this idea are the control of pink bollworms, fruit flies, moths, corn-root worms, cucumber-beetle larvae, cabbage maggots, white grubs and corn borers. Research is being conducted to develop or improve such preparations. Information on the development of controlled-release technology in repellents and plant-growth regulation is available (Mark *et al.*, 1985a,b).

10.1.3 Fertilizers

A major application of the controlled-release technology in agriculture is with fertilizers. PVP is a polyamide that possesses unusual complexing and colloidal properties and is physiologically inert. Because of its non-phytotoxicity, it can be used in agricultural sprays, fertilizing compositions and wettable formulations for dusting and seeding (Blecher *et al.*, 1980). PVP is used with mercurial herbicides for the eradication of crab grass and has been shown to have some effect as a soil conditioner (Blecher *et al.*, 1980). Xanthan gum solutions exhibit yield value and rheological properties that enable stabilization of solid, liquid and gaseous dispersions. It is, therefore, used as a suspending agent for herbicides, pesticides, fertilizers and fungicides. Xanthan gum undergoes very little drift, enabling the spraying of a precisely defined area. Its cling properties are also important, because they increase the contact time between the pesticide and the crop (Cottrell *et al.*, 1980).

10.2 Uses for water-soluble, swellable or dispersible polymers

10.2.1 Seed coatings

The coating of seeds is an important application of polymers in which the seed surface is coated with a hydrophilic polymer. After planting, the polymer absorbs water and the rate as well as the chances for germination increase. Because coating composition and manufacture can be controlled they can be designed to delay germination, inhibit rot, control pests, fertilize or bind the seed to the soil. Seed coatings have been composed of agar, various water-soluble cellulose ethers and hydrolysed starch-γ-polyacrylonitrile copolymers (e.g. HSPAN). Many seeds have been coated, among them soybean, cotton (DeVay *et al.*, 1991), corn, sorghum, sugar beet and a number of vegetables. As a result, cotton and soybean yields increased 20–30%. The poorer the growing conditions, the more pronounced the advantage of the coated seeds versus controls (Mark *et al.*, 1985a,b).

Coatings may include entrapped microorganisms. An optimized process for the production of crop inoculants was developed with an *Azospirillum lipoferum* strain. This process involves the entrapment of living cells in alginate beads and dehydration. The influence of several parameters, i.e. alginate concentration, the addition of adjuvants at different stages, culture-broth dilution, water activity and dehydration method, on bacterial survival was studied. The highest survival rate was obtained with the addition of skim milk and controlled air-dehydration of the alginate beads. A powdered inoculant was obtained, containing more than 10 billion cells g^{-1}, which is easy to store and handle and which can be used in the field as microgranules or as a seed coating. Moreover, its biodegradability ensures that there is no environmental pollution (Fages, 1989).

10.2.2 Equipment for seed coating and sowing

Procedures and a list of the equipment needed for seed coating can be found in the literature (Garrett *et al.*, 1989, 1991, 1994). A technique for mechanically encapsulating individual hybrid tomato seeds in drops of sodium alginate gel is a good example. Tomato seeds climb spiral paths in a vibrating bowl. Air jets and seed sensors send one seed at a time to nozzles where alginate, pumped in discrete drops from an annular nozzle, forms a meniscus which catches the seed. Air inflates the meniscus slightly as a pulse of gum material traps the seed. Capsules, falling into a hardening bath, are inspected by digital camera. An air jet blows capsules without seeds into a reject receptacle; 'good' capsules fall into a bath of calcium chloride and are hardened. This system is capable of producing 100 μl gel capsules at a maximum rate of one capsule every 1.2 s per nozzle. Capsules of 80 μl could not be produced, and a production rate of 500 000 capsules daily was not possible (Garrett *et al.*, 1991).

Another example of a seed coating technique consists of a wire loop, dipped into a solution of sodium alginate, which produces a meniscus of gum. A seed, placed on top of the meniscus, is driven inside the drop, which is then detached from the loop during impact. The detached gel drop containing the seed is dropped into a solution of calcium chloride where a skin of insoluble calcium alginate forms around it. The diameters of the wire and loop as well as the gel concentration affect the volume of the gel in the formed capsules. Wire loops ranging in diameter from 3.2 to 11.1 mm were fabricated from high-strength wire with a diameter of 0.56 to 1.04 mm. Gels at concentrations ranging from 1.25 to 2.50% by weight were used. Capsules ranging in volume from 2.5 to 240 μl were formed from the gel at temperatures ranging from 5 to 35°C. A device with 16 loops, complete with a vacuum-powered seed-singling unit, was capable of producing at least 160 capsules min^{-1} (Garrett *et al.*, 1994).

Coated seeds are sown in the field with different machines. An evaluation of these machines can be found in the literature. Two machines designed for precision sowing of sugar beet have been described and assessed. Two types of individually gel-coated seeds have been used and the results were compared with the distribution of untreated seeds. The Harmonie machine sowed with higher accuracy when coated seed was used (98–99%), but the mechanism proved unsuitable for untreated seeds. The Aeromat machine gave better results with the untreated seeds than the Harmonie machine (Jilek and Velda, 1991).

10.2.3 Different film coatings for seeds

Hydrophilic gels have been used as a water-holding medium for seeds and seedlings, but chemical and physical reactions occur with some of them. Physical properties of commercially produced and naturally occurring gels were therefore evaluated, i.e. pourability, pH, surface tension, water activity and density. Double-cropped soybeans planted with a gel overlay exhibited faster emergence times than seeds planted dry. The optimum pH for gel overlays was found to be 7–9. Surface tension, water activity and density were consistent for all gels. Water applied at a rate of 2 ml per soybean seed improved emergence relative to dry seeds in a dry soil (Castaldi and Krutz, 1989).

Many examples of seed coatings can be found in the recent literature (Orr and Underwood, 1987). Seeds of eight rice cultivars were uncoated or coated with $50\,g\,l^{-1}$ of an aqueous emulsion of sodium alginate and then dipped in $10\,g\,l^{-1}$ calcium oxide. Coated seeds exhibited a higher percentage of germination and better seedling growth than the uncoated seeds when germinated under moisture stress caused by the application of a 20% PEG 8000 solution (Dadlani *et al.*, 1992). Under rain-fed upland conditions with 164 mm of rainfall for the 15 days following sowing, field emergence of the coated seeds was faster and seedling growth more vigorous than in uncoated seeds. Moisture absorption by the coated seeds was 121% higher than that by their uncoated counterparts within 2 h of placement over a moist blotter, but there was no difference in the patterns of moisture absorption by the coated and uncoated seeds in a moist vapor-saturated atmosphere (Dadlani *et al.*, 1992). An accelerated aging test (100% RH and 43°C), resulted in higher levels of viability and vigor in coated seeds than in their uncoated counterparts, which may be attributable to the presence of calcium oxide and antibiotic compounds in the coating emulsion (Dadlani *et al.*, 1992).

Another example deals with seeds that had been surface-sterilized with 3% sodium hypochlorite before resident bacteria and fungi were isolated. Among the 26 bacteria, four isolates of *Bacillus subtilis* were antagonistic to *Rhizoctonia zeae*, *Fusarium moniliforme*, *Macrophomina phaseolina* and *Diplodia zeae* on maize-seed agar. The degree of antagonism varied among isolates. Coating the seed with some of the isolates increased germination and promoted seedling growth, as indicated by measurements of shoot and root length (Adetuyi and Olowoyo, 1993).

Pea seeds were treated to reduce root rot. To prepare the desired coating, *Trichoderma harzianum* isolate ThzID1 was grown in liquid culture, formulated with alginate and PEG 8000, and milled into fine granules (average diameter, 500 μm). Granules contained chlamydospores, conidia and hyphal fragments. Viability of the encapsulated fungus remained high for at least 6 months when stored at 5°C (> 90% of the granules produced hyphal growth when incubated on agar, but viability was significantly reduced when granules were stored at 22°C). Application of the granular formulation to pea seeds reduced root rot caused by *Aphanomyces euteiches* in growth-chamber experiments and also increased plant top weights compared with those from uncoated seeds. The results provide a potentially improved formulation methodology for coating seeds with biocontrol organisms and methods for evaluating the compatibility of fungal and bacterial biocontrol agents applied to seeds (Dandurand and Knudsen, 1993).

In another manuscript, two methods of applying a biological control agent (*T. harzianum* T 581 from soil) directly to asparagus seeds were evaluated. Seed coating was achieved using talcum and sodium alginate as dispersant and thickening agent for the conidial suspension, respectively, with or without the addition of a food-based compound (chitin). *Trichoderma* was then encapsulated by dropping the coated seeds into a 0.1 M calcium gluconate solution. Seed treatment with *T. harzianum* was concluded to enhance asparagus seed germination (Nipoti *et al.*, 1990).

Encapsulation is also useful in the preparation of artificial seeds: somatic embryos grown on MS medium containing 3.0 mg l^{-1} 2,4-D and 0.5 mg l^{-1} kinetin were encapsulated in calcium alginate beads. Germination frequencies of 96% and 45% were achieved *in vitro* and in the soil, respectively. The *in vivo* plantlet conversion frequency increased to 56% following an additional coating of mineral oil on the alginate beads. Germinated artificial seeds could be raised into plantlets (Mukunthakumar and Mathur, 1992).

Although most of the reports on seed coatings deal with alginate coatings, reports on other gels can also be found. In field trials, sorghum seeds were left dry, soaked in water for 4 h and dried, or pregerminated. They were then soaked for 18 h until radicle extrusion and sown in a 1.4% Laponite gel suspension into a dry seedbed of self-mulching Mywybilla clay soil with different levels of irrigation. Emergence increased with irrigation and was higher with dry seeds than with primed or pregerminated seeds. At intermediate soil contents (0.2 g g^{-1}), gel-treated seeds exhibited enhanced emergence relative to dry sowing (Ferraris, 1989). The authors found that at 9 ml 1.2% Laponite gel per seed, the gel application overcame the limitations of a dry soil, and seedling-emergence rate was improved. Priming or pregermination of sorghum seeds provided no significant advantage in either field or laboratory studies (Ferraris, 1989).

10.2.4 Gel planting

HSPAN is unique in its water-absorbing capacity: it holds over 800 times its weight of distilled water. Other water-soluble polymers have found other applications in planting and growth development (Banyai, 1987). Gel coatings have been used to eliminate the drying of seedling root zones.

Advantages of such techniques are prevention of root drying, reduction in wilting and transplant shock, and improvement in plant survival via the incorporation of fungicides, pesticides or nutrients in the hydrogel slurries (Mark *et al.*, 1985a,b).

10.2.5 Soil conditioning

Hydrophilic polymers such as polyacrylamide, PVA, CMC and HSPAN are also used as soil conditioners in a technique called hydromulching (Mark *et al.*, 1985a,c). The water-holding capacity of these polymers enables them to be sprayed or their slurries to be blown with nutrients and other mulching materials. Other polymers can be used to hold the mulch in place (Al-Afifi *et al.*, 1991). Thatch is sometimes formed, protecting the seeds and soil against erosion. Polyethylene oxide (PEO) resins serve as binders for soil stabilization (see further on). By making use of their thermoplastic effect, they can also be used as seed-type and water-soluble packaging ingredients.

10.2.6 Polymeric adjuvants

Many polymers are used in pesticide and fertilizer formulations. They serve as sticking, wetting or compatibility agents. They are used to improve formulation spreadability, as antitranspirants, as activators or to prolong activity. Unique uses such as protection of citrus trees from frost damage by copolymers of methyl acrylate or methyl methacrylate with *N*-vinylpyrrolidinone have been reported. These copolymers inhibit ice nucleation and suppress the freezing of aqueous solutions. Other uses of water-soluble or water-compatible polymers (e.g. polyoxyalkylenes, PVA, partially hydrolysed polyacrylates) are for defoaming, and as coagulation or drift-control agents (Mark *et al.*, 1985a,b).

10.2.7 Special agricultural applications of gums and resins

Carrageenans have the ability to gel, thicken and suspend. They are capable, therefore, of emulsion stabilization, syneresis control, bodying, binding and dispersion. In agriculture, carrageenans are used to stabilize suspensions and their continuous supply is dependent on the success of the aquaculture seaweed industry. As previously mentioned, MC polymers can be used as seed-stickers. The viscosity of $\sim 15\,\text{cP}$ produced by MC solutions helps adhere fungicides to the seed. Protective films are made with MC of the same viscosity. Different concentrations are used to achieve wetting action. Agricultural dust can include 63 to 125 g MC or MHPC to stick the dust. The incorporation of agar has been reported to increase and prolong the activity of nicotine plant sprays. Agar is also used regularly to produce culture media and its use in orchid culture is standard. PEO gel can absorb ~100 times its weight in water. The absorbed water is also readily desorbed by drying these hydrogels. They can, therefore, be used as soil amendments. Mixing with soil at a ratio of (2×10^{-4}):1 (wt% of the soil) reduces evaporation from the soil, and enables less frequent watering in addition to the full development of plants grown in these treated soils (Herrett and King, 1967).

10.3 Coatings for fruits and vegetables

10.3.1 Introduction

It is variously estimated that 25–80% of freshly harvested fruits and vegetables are lost through spoilage (Willis *et al.*, 1981). Fruits and vegetables continue their metabolic activity after harvest and will either ripen (if climacteric) or senesce (if non-climacteric) rather rapidly unless special procedures are adopted to slow down these processes. One method to extend postharvest shelf life is to coat fruits and vegetables with edible materials: semi-permeable barriers to gases and water vapor. Edible coatings can enhance or even replace many of the techniques used to preserve fresh commodities, such as controlled atmosphere (CA) or modified atmosphere (MA) (Barkai-Golan, 1990; Smith *et al.*, 1987), and can provide partial barriers to moisture and gas exchange (carbon dioxide and oxygen) (Baldwin, 1994). Coatings also improve mechanical handling properties (Mellenthin *et al.*, 1982), help maintain structural integrity, help retain volatile flavor compounds (Nisperos-Carriedo *et al.*, 1990) and can serve as carriers of additives such as antioxidants and antimicrobial agents (Kester and Fennema, 1988).

10.3.2 Coatings of the past

Wax coatings were developed to mimic the natural coating of fruits and vegetables. The coating of fresh citrus fruits with wax to extend their shelf life can be traced back to 12th to 13th century China (Hardenburg, 1967). However, these coatings inhibited respiratory gas exchange to such an extent that fermentation was induced. Gelatin films were proposed in the 19th century to preserve meats and other foods. In the 1930s, hot-melted paraffin waxes were commercially available for coating citrus fruits and vegetables, and in the 1950s carnauba wax oil emulsions were developed for this purpose (Kaplan, 1986). Wax coatings control decay when fungicides are incorporated within them (Brown, 1984; Miller *et al.*, 1988; Radnia and Eckert, 1988). Other incorporated ingredients include chilling-injury protectants, coloring agents, antisenescence substances, growth regulators and biological control agents (McQuilken *et al.*, 1992; Baldwin, 1994). Successful lipid coatings (made of acetylated monoglycerides, waxes and surfactants) for blocking the transport of moisture, reducing surface abrasion through handling and controlling soft scald formation in apples, coreflush in MacIntosh apples and spots on Jonathan apples have been reported (Fisher and Britton, 1940; Hardenburg, 1967; Paredes-Lopez *et al.*, 1974; Lawrence and Lyengar, 1983; Warth, 1986; Kester and Fennema, 1988). Many reports discuss the effect of wax, oil and shellac coatings on internal atmosphere, weight loss and ethanol content in apples and citrus fruit (Magness and Diehl, 1924; Hitz and Haut, 1938; Trout *et al.*, 1953; Eaks and Ludi, 1960; Cohen *et al.*, 1990; Nisperos-Carriedo *et al.*, 1990). Wax-type coatings have been reported to prolong the shelf life of banana (Lawson, 1960; Ben-Yehoshua, 1966; Blake, 1966; Dalal *et al.*, 1970), mango, pineapple, papaya, guava and avocado (Bose and Basu, 1954; Mathur and Srivastava, 1955; Dalal *et al.*, 1971). Similar wax coatings for vegetables (e.g. carrots, cucumbers, pumpkins, eggplants) and melons have also been reported (e.g. Hartman and Isenberg, 1956; Nelson, 1942; Wu and Salunkhe, 1972).

Polysaccharide-based coatings were developed for fruit and nut products. Low-methoxy-pectin (LMP) coatings have been used for nuts and dried dates (Swenson *et al.*, 1953). A powdered formulation of hydroxypropyl starch was used for prunes (Jokay *et al.*, 1967). Amylose starch and plasticizer were used to coat raisins and dates (Moore and Robinson, 1968). A laminated coating composed of an initial layer of an amylose ester of fatty acid and a layer of soy or zein protein was used to coat freeze-dried peas, carrots and apple slices (Cole, 1965). Pieces of freshly cut fruits and vegetables were coated with CMC dust and starch to prevent juice loss and to preserve texture (Mason, 1969). Water loss and browning of cut apple slices were inhibited by coatings of chitosan and lauric acid (Pennisi, 1992). Information about synthetic polymers and derivatives of natural polymers used in coatings can be found in reports and patents from the late 1960s and early 1970s (Rosenfield, 1968; Mulder, 1969; Shillington and Liggett, 1970; Delong and Shepherd, 1972; Danials, 1973; FDA, 1991).

10.3.3 Coatings used today

Many of the coatings used today are similar to those used in the past. Hardenburg (1967) lists the commercial uses of wax preparations, e.g. polyethylene wax is used to make emulsion coatings for citrus (Kaplan, 1986). Commercial wax coating companies, products and applications are listed elsewhere (Krochta *et al.*, 1994). A few water-soluble composite coatings composed of CMC with fatty acid ester emulsifiers are used for fruits such as pears and bananas (Ukai *et al.*, 1975; Banks, 1984a,b; Smith and Stow, 1984; Drake *et al.*, 1987; Mitsubishi-Kasei, 1989). In the latter, delayed ripening and changes in internal gas levels are observed (Banks,

1983, 1984a,b, 1985). This type of commercial coating was first called Tal Pro-long and later, Pro-long. It increases resistance to some types of fungal rot in apples, pears and plums (Lowings and Cutts, 1982) but is not effective at decreasing respiration rate and water loss in tomato and sweet pepper (Nisperos-Carriedo and Baldwin, 1988; Lowings and Cutts, 1982).

Another coating with similar composition – Semperfresh – contains a higher proportion of short-chain unsaturated fatty acid esters in its formulation (Drake *et al.*, 1987). These coatings retard color development and retain acids and firmness compared with controls when tested on apples (Smith and Stow, 1984). Semperfresh also extends the storage life of citrus (Curtis, 1988); Valencia oranges coated with Tal Pro-long have a better

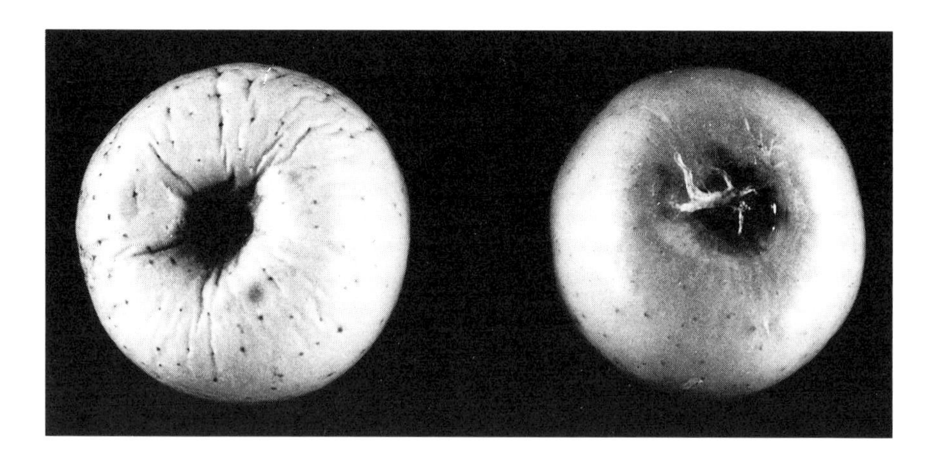

Figure 10.1 Gellan-coated (right) versus uncoated (left) apples. Note the wrinkles on the uncoated apple. (Courtesy of V. Hershko.)

flavor and lower ethanol levels than controls (Nisperos-Carriedo *et al.*, 1990). By adding waxes to Semperfresh, the coating gives a better shine and the fruit has higher turgidity, less decay and good flavor. However, Semperfresh is not effective at retarding water loss in melons (Edwards and Blennerhassett, 1990).

The carrageenan-based coatings Soageena and Soafil (edible polysaccharides) were developed by Mitsubishi International Corp. (IFT, 1991), but aside from their designation for fresh produce, no information is available (Baldwin, 1994). Other carrageenan coatings have been used to retard moisture loss from coated foods (Torres *et al.*, 1985). Protein, lipid and composite films have been studied intensively (e.g. Kumins, 1965; Schonherr, 1982; Deasy, 1984; Kamper and Fennema, 1984; Roth and Loncin, 1985; Guilbert, 1986; Hagenmaier and Shaw, 1990). Alginate-based coatings have been used to coat meats (Allen *et al.*, 1963; Earle, 1968; Williams *et al.*, 1978; Wanstedt *et al.*, 1981; Glicksman, 1983; King, 1983). Information on MC and MHPC can also be found elsewhere (Vojdani and Torres, 1990; Rico Pena and Torres, 1991). Other sodium alginate and gellan-based coatings have been investigated (Nussinovitch and Kampf, 1992, 1993; Nussinovitch, 1994; Nussinovitch *et al.*, 1994). As a barrier to moisture loss (Fig. 10.1) these hydrocolloid coatings postponed the drying of mushroom tissue, thereby preventing changes in its texture during short periods of storage. Coated mushrooms were found to have a better appearance and gloss. Garlic coatings enjoyed similar advantages. The incorporation of ingredients which can be found naturally in garlic skin, or are chemically similar to these, into the gum solution before coating improves adhesion of the film

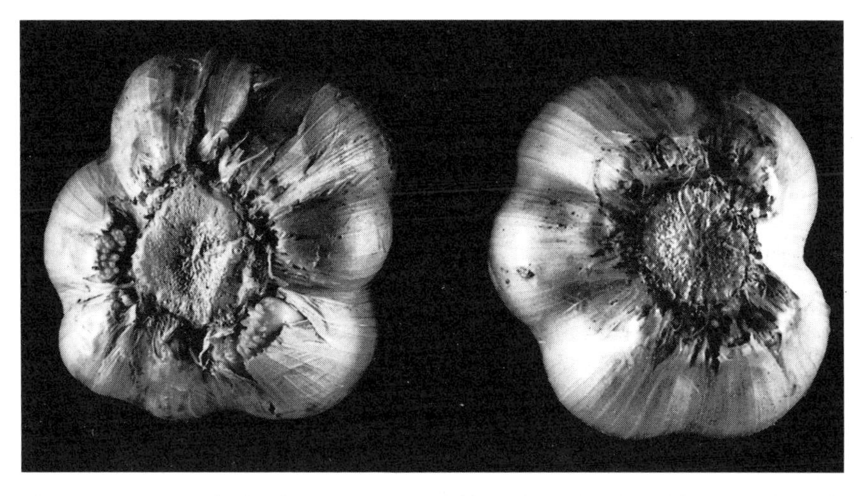

Figure 10.2 Uncoated (right) versus coated (left) garlic bulb. The alginate-based coating eliminates clove opening, extends shelf life and decreases metabolism of fresh and dried garlic. (Courtesy of V. Hershko.)

Figure 10.3 Uncoated garlic (left) in a shed (worst conditions) which has undergone browning and has been attacked by insects (two holes where the roots were). The gellan-based coating served as a mechanical barrier against insects (right). (Courtesy of V. Hershko).

to the surface of the coated commodity (Hershko *et al.*, 1996; Nussinovitch and Hershko, 1997) (Figs 10.2 and 10.3). Adhesion strengths were ~25% higher than those recorded for a film made of gum and cross-linking agent alone. Distances between the film and the vegetable were measured by image processing, and they could sometimes be reduced by varying the film composition.

10.3.4 Coating effects on fruit and vegetable physiology

If the coating is designed to affect permeability to oxygen and carbon dioxide, the fruit or vegetable becomes an individual package with a modified atmosphere. Respiration of the commodity causes a depletion of oxygen and a build-up of carbon dioxide. Care must be taken in desigining the coating: if oxygen levels drop too low anaerobic reactions will proceed, resulting in off-flavors and abnormal ripening (Kader, 1986). This can be monitored by measuring ethanol and acetaldehyde concentrations in the tissue, the final and next-to-final products of anaerobic respiration, respectively. Environmental levels of oxygen below 8% decrease ethylene production, and carbon dioxide levels above 5% delay or prevent many responses to ethylene in the fruit tissue, including ripening (Kader, 1986). Therefore, if the coating can create these moderate modified atmosphere conditions, a climacteric fruit will exhibit decreased respiration, lower ethylene production, slower ripening and an extended shelf life.

The commercial uses of coatings are rather extensive with apples, and may involve shellac, carnauba or resin-based waxes (Hitz and Haut, 1942). In a comparative study of these waxes on Delicious and Golden Delicious apples, no quality differences were observed. The waxed apples exhibited improved color, higher internal levels of carbon dioxide and ethylene, and reduced weight loss compared with their unwaxed counterparts, but no effect was observed on firmness (Drake and Nelson, 1990). More recent work on apple coatings has involved coatings that contain a polysaccharide base. These compounds, because of their hydrophilicity, provide only minimal moisture barriers (Kester and Fennema, 1988). The carbon dioxide and oxygen permeabilities of the polysaccharide-based coatings result in retardation of ripening in many climacteric fruits, thereby increasing shelf life without creating anaerobic conditions (Banks, 1984a,b; Smith and Stow, 1984; Drake *et al.*, 1987). The idea of these coatings is to mimic controlled or modified atmosphere (plastic- or shrink-wrapped) storage, which is labor-intensive and expensive, with an environmentally friendly biodegradable edible coating.

The other gas that coatings can affect is water vapor. Fruits and vegetables are very vulnerable to water loss through respiration once they are harvested. They are, therefore, usually stored in highly humid environments (90–98% RH) to minimize water loss, subsequent weight loss and shrivelling (Woods, 1990). The development of the first commercial coatings was primarily to reduce moisture loss by including hydrophobic components such as waxes, oils or resins, and adding sheen by including shellac or carnauba wax (Baldwin, 1994).

Reducing transpiration is particularly important in non-climacteric fruits and vegetables, which do not continue ripening after harvest. Water stress in a tissue accelerates senescence processes and shortens the shelf life of these commodities. Pepper is stored commercially at 8°C because lower temperatures cause chilling injury. At 8°C, optimum storage life is 2 to 3 weeks, because of water loss. After 3% of the fruit weight is lost, the fruit becomes flaccid and its quality decreases (Ben-Yehoshua *et al.*, 1983). Seal-packaging bell peppers has been found to create a saturated atmosphere around the fruit and extend its postharvest life threefold (Ben-Yehoshua *et al.*, 1983). If instead of film seal-packaging a coating were to be used, shrivelling, respiration and decay would be reduced, and there would be better retention of ascorbic acid (vitamin C) in the waxed fruit (Habeebunnisa *et al.*, 1963).

Another potential benefit of edible coatings for fruit physiology is a reduction in their sensitivity to low-temperature storage (Ben-Shalom *et al.*, 1993). Subtropical fruits have a limited storage life because storage at temperatures low enough to arrest ripening or senescence causes them to develop chilling injury. Chilling injury symptoms can be greatly alleviated if the fruit is held in a water-saturated atmosphere (Lurie *et al.*, 1986). Coatings that reduce transpiration can be used to create this atmosphere around individual fruits. Coating grapefruit with a polyethylene wax emulsion reduced incidence of rind pitting when the fruit was stored at 0 to 10°C (Davis and Harding, 1960). Coatings have been shown to impart cold protection in other citrus fruits and thereby prevent pitting, which is a symptom of chilling injury (Aljuburi and Huff, 1984; McDonald, 1986).

10.3.5 Film-application techniques

Films can be applied by dipping, spraying or casting, to name only a few methods. Fruits and vegetables are dipped in one or several baths (if cross-linking of the hydrocolloid is required), followed by draining and drying. This method has been used to apply wax coatings to agricultural produce. Although films can be produced by spraying, this method is more suitable to applying films to planar surfaces (Krochta *et al.*, 1994).

The casting technique is used in many laboratories to produce films of nearly constant thickness. The gel or hydrocolloid solution is poured into a 'sandwich' prepared from glass plates, assembled with screw clamps and aligned using the casting stand's alignment slot and an alignment card. After the gels are cast, they are left to equilibrate under high humidity at room temperature. The films are then dried with warm air until a predetermined amount of moisture remains within them. Specimens are then prepared with a dumb-bell-shaped cutter for tensile tests.

Brushes, falling-film enrobing technique, panning or rollers can also be used to apply films to the surfaces of fruits and vegetables (Guilbert, 1986).

10.3.6 Testing methods of coatings

The gas permeability of packaging films can be measured by several methods (Anon., 1975; Karel, 1975; Stern *et al.*, 1964). Many devices for measuring film permeability to oxygen have been invented (Quast and Karel, 1972; Landrock and Proctor, 1952). Water vapor transmission rates through dried coatings can be determined by ASTM E393-83 (standard test methods for water vapor transmission of materials) (ASTM, 1983).

Peel testing, i.e. the force necessary to peel the coating, is used to estimate the degree of adhesion of the film to the fruit or vegetable. The coating is peeled at 90° from the substrate, and the adhesion strength is estimated by the force per unit width necessary to peel the coating.

To obtain 'good' coatings, it is important to study surface wetting and adhesion properties of coated commodities. Roughness of fruits and vegetables and of film surfaces can be studied using a roughness tester, and using electron and atomic-force microscopes (the last for finer mapping of surface roughness) (Fig. 10.4). Parameters that can be studied include R_t, the distance between the highest peak and the deepest valley of the roughness profile within an evaluation length of tested surface, and R_a, the

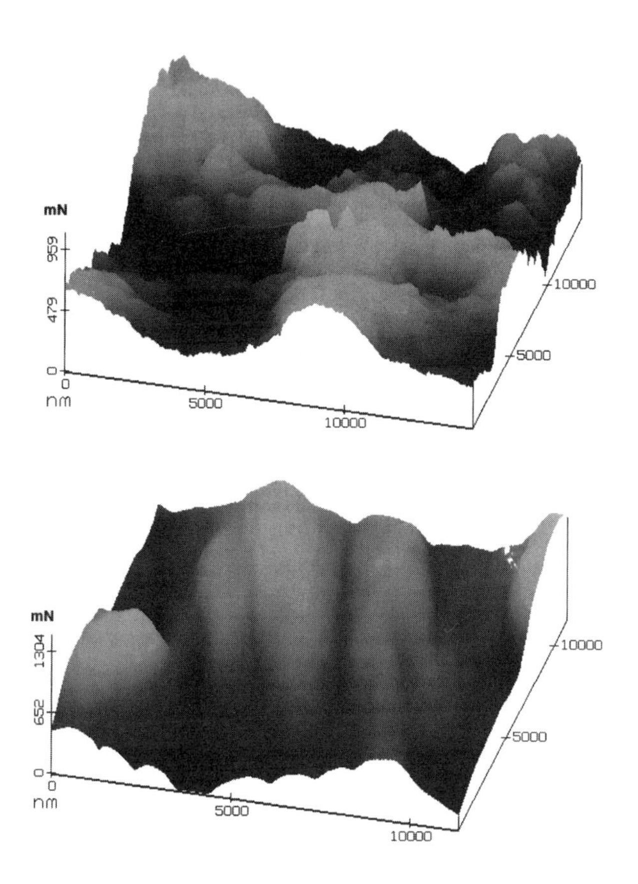

Figure 10.4 Atomic force microscopy: two typical three-dimensional maps of garlic surface. Information on commodity roughness can be very helpful in the development of new coatings and their adherence to the surface of the commodity. (From Hershko *et al.*, 1996; reproduced by permission of the IFT.)

arithmetic mean of the absolute values of the roughness profile deviation from the center line within the evaluation length.

The surface tension of gelling and inducing solutions, and their contact angles on fruit and vegetable surfaces, can be studied with surface-tension instruments (maximum adhesion requires a contact angle of 0°). It is

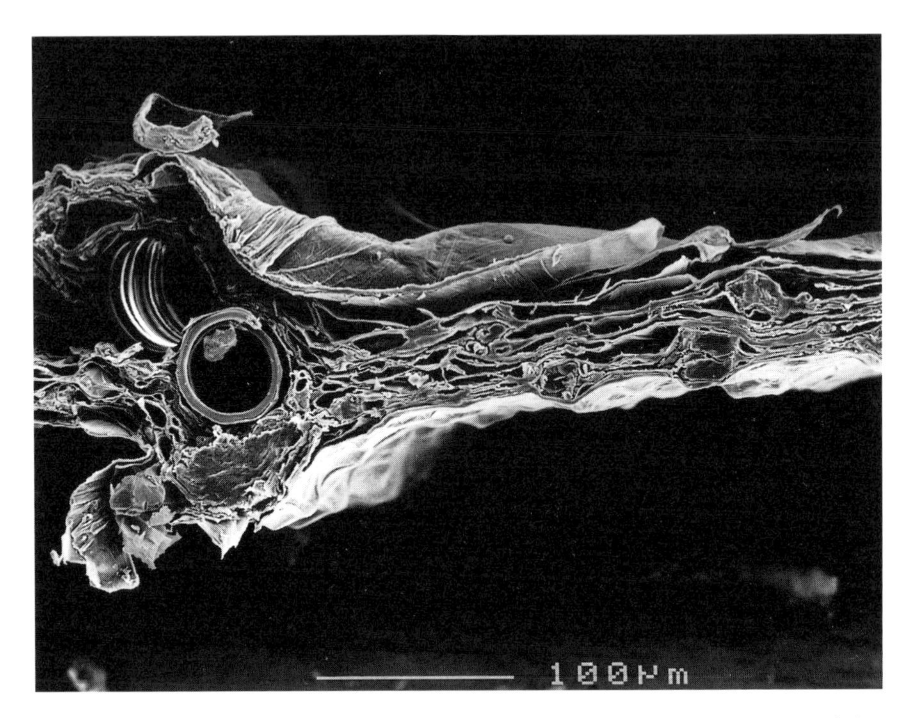

Figure 10.5 Coated garlic in cross-section. The transparent coating glued to the top of the garlic epidermis is strong and transparent. (Adapted from Nussinovitch and Hershko, 1996.)

important to note that 'If the coating does not spread spontaneously over the substrate surface, so that there is intermolecular [as in Fig. 10.5] contact between the substrate surface and the coating, there cannot be interactions and hence no contribution to adhesion' (Wicks *et al.*, 1994).

References

Adetuyi, F.C. and Olowoyo, O.O. (1993) Antagonistic microflora from dried maize seed and their effect on seed germination. *Ind. J. Mycol. Plant Pathol.*, **23**(2), 157–61.

Al-Afifi, M.A., Al-Masoum, A., Shawky, M.E. *et al.* (1991) Preliminary studies on testing soil conditioners for vegetable use in UAE. *Bull. Fac. Agric., Univ. Cairo*, **42**(3), 1061–77.

Aljuburi, H.J. and Huff, A. (1984) Reduction in chilling injury in stored grapefruit (*Citrus paradisi Macf.*) by vegetable oils. *Sci. Hort.*, **24**, 53–8.

Allan, G.G., Beer, J.W., Cousin, M.J. *et al.* (1980) in *Controlled Release Technologies: Methods, Theory and Applications*, vol. II, ch 2 (ed. A.F. Kydonieus), CRC Press, Boca Raton, FL, p. 7.

Allen, L., Nelson, A.I., Steinberg, M.P. *et al.* (1963) Edible corn-carbohydrate food coatings. Evaluation on fresh meat products. *Food Technol.*, **17**, 442.

Anon. (1975) *Instruction for the Oxtran-100 Oxygen Permeability Tester.* Mocon Instruments, Inc., Minneapolis, MN.

ASTM (1983) Standard test method for water vapor transmission rate of sheet materials using a rapid technique for dynamic measurements. E398-83, *ASTM Book of Standards*, **4.06**: 712–16.

Baldwin, E.A. (1994) Edible coatings for fresh fruits and vegetables: past, present and future, in *Edible Coatings and Films to Improve Food Quality* (eds J. Krochta, E. Baldwin and M. Nisperos-Carriedo), Technomic, Basel, Switzerland.
Banks, N.H. (1983) Evaluation of methods for determining internal gases in banana fruit. *J. Exp. Bot.*, **44**, 871–9.
Banks, N.H. (1984a) Some effects of Tal Pro-long coating on ripening bananas. *J. Exp. Bot.*, **35**, 127–37.
Banks, N.H. (1984b) Studies of the banana fruit surface in relation to the effects of TAL Pro-long coating on gaseous exchange. *Sci. Hort.*, **24**, 279–86.
Banks, N.H. (1985) Responses of banana fruit to Pro-long coating at different times relative to the initiation of ripening. *Sci. Hort.*, **26**, 149–57.
Banyai, B.E. (1987) N-Gel (TM) polymers for agricultural fluid drilling, in *Acta Hortic. Wageningen: Int. Soc. Hort. Sci.*, June (1987), pp. 111–20.
Barkai-Golan, R. (1990) Postharvest disease suppression by atmospheric modifications, in *Food Preservation by Modified Atmospheres* (eds M. Calderon and R. Barkai-Golan), CRC Press, Boca Raton, FL, pp. 238–65.
Ben Shalom, N., Hanzon, J., Klein, J.D. *et al.* (1993) Postharvest heat treatment inhibits cell wall degradation in apples during storage. *Phytochemistry*, **34**, 955–8.
Ben-Yehoshua, S. (1966) Some effects of plastic skin coating on banana fruit. *Trop. Agric. Trin.*, **43**, 219–32.
Ben-Yehoshua, S., Shapiro, B., Even-Chen, Z. *et al.* (1983) Mode of action of plastic film in extending life of lemon and bell pepper fruits by alleviation of water stress. *Plant Physiol.*, **73**, 87–93.
Blake, J.R. (1966) Some effects of paraffin wax emulsions on bananas. *Queensland. J. Agric. Animal Sci.*, **23**, 49–56.
Blecher, L., Lorenz, D.H., Lowd, H.L. *et al.* (1980) Polyvinylpyrrolidone, in *Handbook of Water-soluble Gums and Resins*, ch. 21 (ed. R.L. Davidson), McGraw-Hill, New York, pp. 21.1–21.17.
Bose, A.N. and Basu, G. (1954) Studies on the use of coating for extension of storage life of fresh Fajli mango. *Food Res.*, **19**, 424–8.
Brown, E. (1984) Efficacy of citrus postharvest fungicides applied in water or resin solution water wax. *Plant Dis.*, **68**, 415–18.
Castaldi, A. and Krutz, G.W. (1989) Interactions of gel with soybean seeds. *Appl. Eng. Agric.*, **5**(4), 496–500.
Cohen, E., Shalom, Y. and Rosenberger, I. (1990) Postharvest ethanol buildup and off-flavor in 'Murcott' Tangerine fruits. *J. Am. Soc. Hort. Sci.*, **115**, 775–8.
Cole, M.S. (1965) *Method for Coating Dehydrated Food.* US Patent No. 3,479,191.
Cottrell, I.W., Kang, K.S. and Kovacs, P. (1980) Xanthan gum, in *Handbook of Water-soluble Gums and Resins*, ch. 24 (ed. R.L. Davidson), McGraw-Hill, New York, pp. 24.1–24.30.
Curtis, G.J. (1988) Some experiments with edible coatings on the long-term storage of citrus fruits, in *Proc. 6th Int. Citrus Congr. 3*, pp. 1514–20.
Dadlani, M., Shenoy, V.V. and Seshu, D.V. (1992) Seed coating to improve stand establishment in rice. *Seed Sci. Technol.*, **20**(2), 307–13.
Dalal, V.B., Eipeson, W.E. and Singh, N.S. (1971) Wax emulsion for fresh fruits and vegetables to extend their storage life. *Ind. Food Packer*, **25**, 9–15.
Dalal, V.B., Thomas, P., Nagaraja, N. *et al.* (1970) Effect of wax coating on bananas of varying maturity. *Ind. Food Packer*, **24**, 36–9.
Dandurand, L.M. and Knudsen, G.R. (1993) Influence of *Pseudomonas fluorescens* on hyphal growth and biocontrol activity of *Trichoderma harzianum* in the spermosphere and rhizosphere of pea. *Phytopathology*, **83**(3), 265–70.
Danials, R. (1973) *Edible Coatings and Soluble Packaging.* Noyes Data Corp., Park Ridge, NJ, p. 360.
Davidson, R.L. (1982) *Handbook of Water-soluble Gums and Resins*, chs 3, 5, 7, 9, 12, 18, 19, 22, 24, McGraw-Hill, New York.
Davis, P.L. and Harding, P.C. (1960) The reduction of rind breakdown of Marsh grapefruit by polyethylene emulsion treatments. *J. Am. Soc. Hort. Sci.*, **75**, 271–4.
Deasy, P.B. (1984) *Microencapsulation and Related Drug Processes*, Marcel Dekker, New York.
Delong, C.F. and Shepherd, T.H. (1972) *Product Coating.* US Patent No. 3,669,691.

DeVay, J.E., Wakeman, R.J., Paplomates, E.J. *et al.* (1991) Effectiveness of alginate gel and Opadry seed coatings for use with biological controls of cotton seedling diseases, in *Proc. Beltwide Cotton Conf.* Memphis, Tenn., National Cotton Council of America, vol. 1, pp. 160–2.
Drake, S.R., Fellman, J.K. and Nelson, J.W. (1987) Postharvest use of sucrose polyesters for extending the shelf life of stored Golden Delicious apples. *J. Food Sci.*, **52**, 1283–5.
Drake, S.R. and Nelson, J.W. (1990) Storage quality of waxed and nonwaxed Delicious and Golden Delicious apples. *J. Food Qual.*, **13**, 331–41.
Eaks, I.L. and Ludi, W.A. (1960) Effects of temperature, washing and waxing on the composition of the internal atmosphere of orange fruits. *Proc. Am. Hort. Soc.*, **76**, 220–8.
Earle, R.D. (1968) *Method for Preserving Food by Coating.* US Patent No. 3,395,024.
Edwards, M.E. and Blennerhassett., R.W. (1990) The use of postharvest treatments to extend storage life and to control postharvest wastage of Honey Dew melons (*Cucumis melo L.* var. inodorus Naud.) in cool storage. *Aust. J. Exp. Agric.*, **30**, 693–7.
Fages, J. (1989) An optimized process for manufacturing an *Azospirillum* inoculant for crops. *Appl. Microbiol. Biotechnol.*, **32**(4), 473–8.
FDA (1991) Food and drugs, in *Edible Coatings and Films to Improve Food Quality*, ch. 21 (eds J. Krochta, E. Baldwin and M. Nisperos-Carriedo), Technomic, Basel, Switzerland, p. 59.
Ferraris, R. (1989) Gel seeding of sorghum into Mywybilla clay, in *Proc. Aust. Sorghum Workshop*, Toowoomba, Queensland, 28 Feb.–1 March (eds M.A. Foale, B.W. Hare and R.G. Henzell), Occasional Publications, Australian Institute of Agricultural Science No. 43, Australian Institute of Agricultural Science, Brisbane, Australia.
Fisher, D.V. and Britton, J.E. (1940) Apple waxing experiments. *Sci. Agric.*, **21**, 70–9.
Garrett, R.E., Mehlschau, J.J., Smith, N.E. *et al.* (1991) Gel encapsulation of tomato seeds. *Appl. Eng. Agric.*, **7**(1), 25–31.
Garrett, R.E., Shafii, S. and Upadhyaya, S.K. (1994) Encapsulation of seeds in gel by impact (presented as ASAE paper No. 92-1073). *Appl. Eng. Agric.*, **10**(2), 183–7.
Garrett, R.E., Smith, N.E. and Mehlschau, J.J. (1989) *Apparatus and Method for Encapsulating Seeds and the Like.* US Patent No. 4,806,357.
Glicksman, M. (1983) Red seaweed extracts, in *Food Hydrocolloids*, vol. 2 (ed. M. Glicksman), CRC Press, Boca Raton, FL, p. 73.
Guilbert, S. (1986) Technology and application of edible protective films, in *Food Packaging and Preservation Theory and Practice* (ed. M. Mathlouthi), Elsevier Applied Science, London, p. 371.
Habeebunnisa, M., Pushpa, C. and Srivastava, J.C. (1963) Studies on the effect of protective coating on the refrigerated and common storage of bell peppers (*Capsium frutescense*). *Food Sci.*, **12**, 192–6.
Hagenmaier, R.D. and Shaw, P.E. (1990) Moisture permeability of edible films made with fatty acid and hydroxypropyl methylcellulose. *J. Agric. Food Chem.*, **38**(9), 1799.
Hardenburg, R.E. (1967) Wax and related coatings for horticultural products. A bibliography. *Agr. Res. Bull.* No. 51-15. US Department of Agriculture, Washington, DC.
Harris, F.W., Feld, W.A. and Bowen, B.K. (1975) *Proc. Int. Controlled Release Pesticide Symp.*, Wright State University, Dayton, OH.
Hartman, J. and Isenberg, F.M. (1956) Waxing vegetables. *New York Agric. Exten. Ser. Bull.* No. 965, pp. 3–14.
Herrett, R.A. and King, P.A. (to Union Carbide Corp.) (1967) *Plant Growth Medium.* US Patent No. 3,336,129.
Hershko, V., Klein, E. and Nussinovitch, A. (1996) Relationships between edible coating and garlic skin. *J. Food Sci.*, **61**(4), 769–77.
Hitz, C.W. and Haut, I.C. (1938) Effect of certain waxing treatments at time of harvest upon the subsequent storage quality of 'Grimes Golden' and 'Golden Delicious' apples. *Proc. Am. Soc. Hort. Sci.*, **36**, 440–7.
Hitz, C.W. and Haut, I.C. (1942) Effects of waxing and prestorage treatments upon prolonging the edible and storage qualities of apples. *Univ. MD Agric. Exp. Sta. Tech. Bull.*, No A14.
IFT (1991) New from Mitsubishi *Annu. Meet. Food Expl. Prog. Exhibit Directory*, Dallas Convention Center, June 1–5, 1991, Chicago, IL.
Jilek, J. and Velda, K. (1991) Accuracy of the sowing mechanism of Harmonie and Aeromat sowing machines. *Mechnizace Zemedellstvi*, **41**(4), 161–4.

Jokay, L., Nelson, G.E. and Powell, E.C. (1967) Amylaceous coatings for foods. *Food Technol.*, **21**, 1064–6.

Kader, A.A. (1986) Biochemical and physiological basis for effects of controlled and modified atmospheres on fruits and vegetables. *Food Technol.*, **40**, 99–104.

Kamper, S.L. and Fennema, O. (1984) Water vapor permeability of an edible, fatty acid, bilayer film. *J. Food Sci.*, **49**, 1482.

Kaplan, H.J. (1986) Washing, waxing and color adding, in *Fresh Citrus Fruits* (eds W.F. Wardowski, S. Nagy and W. Grierson), AVI Publishing, Westport, CT, p. 379.

Karel, M. (1975) Protective packaging of foods, in *Principles of Food Science* (ed. O. Fennema), Marcel Dekker, New York, pp. 399–464.

Kester, J.J. and Fennema, O.R. (1988) Edible films and coatings: a review. *Food. Technol.*, **42**, 47–59.

King, A.H. (1983) Brown seaweed extracts, in *Food Hydrocolloids*, vol. 2 (ed. M. Glicksman), CRC Press, Boca Raton, FL, p. 115.

Krochta, J.M., Baldwin, E.A. and Nisperos-Carriedo, M. (1994) *Edible Coatings and Films to Improve Food Quality*, Tachnomic, Lancaster, Basel.

Kumins, C.A. (1965) Transport through polymer films. *J. Polym. Sci.* (Part C), **10**, 1.

Kydonieus, A.F. (ed.) (1980) *Controlled Release Technologies: Methods, Theory and Applications*, vols I and II, CRC Press, Boca Raton, FL.

Landrock, A.H. and Proctor, B.E. (1952) Gas permeability of films. *Mod. Packag.*, **25**(10), 131–5, 199–201.

Lawrence, J.F. and Lyengar, J.R. (1983) Determination of paraffin wax and mineral oil on fresh fruits and vegetables by high temperature gas chromatography. *J. Food Safety*, **5**, 119–24.

Lawson, J.A. (1960) Banana packing and waxing. *West Aust. Dept. Agric. J.* (Ser. 4), **1**, 11–5.

Lowings, P.H. and Cutts, D.G. (1982) The preservation of fresh fruits and vegetables, in *Proc. Inst. Food Sci. Tech. Annu. Symp.*, July 1981, Nottingham, UK, p. 52.

Lurie, S., Shapiro, B. and Ben-Yehoshua, S. (1986) Effects of water stress and degree of ripeness on rate of senescence of harvested bell pepper fruit. *J. Am. Soc. Hort. Sci.*, **111**, 880–5.

Magness, J.R. and Diehl, H.C. (1924) Physiological studies on apples in storage. *J. Agric. Res.*, **27**, 1–38.

Mark, H.F., Bikales, N.M., Overberger, C.G. *et al.* (1985a) *Encyclopedia of Polymer Science and Engineering*, vol. 1. Wiley-Interscience, New York, pp. 611–21.

Mark, H.F., Bikales, N.M., Overberger, C.G. *et al.* (1985b) *Encyclopedia of Polymer Science and Engineering*, vol. 2. Wiley-Interscience, New York, pp. 179–82.

Mark, H.F., Othmer, D.F., Overberger, C.G. *et al.* (1985c) *Kirk-Othmar Encyclopedia of Chemical Technology*, 3rd edn, vol. 20. Wiley-Interscience, New York, pp. 219–20.

Mason, D.F. (1969) *Fruit Preservation.* US Patent No. 3,472,662.

Mathur, P.B. and Srivastava, H.C. (1955) Effect of skin coatings on the storage behavior of mangoes. *Food Res.*, **20**, 559–66.

Matsudaira, T. (1953) Japanese Patent No. 2997 (*Chem. Abstr.*, **48**, 7840 (1954)).

McDonald, R.E. (1986) Effects of vegetable oils, CO_2, and film wrapping on chilling injury and decay of lemons. *Hort. Sci.*, **21**, 476–7.

McQuilken, M.P., Whipps, J.M. and Cooke, R.C. (1992) Use of oospore formulations of *Pythium oligandrum* for biological control of Pythium damping-off in cress. *J. Phytopathol.*, **135**(2), 125–34.

Mellenthin, W.M., Chen, P.M. and Borgic, D.M. (1982) In line application of porous wax coating materials to reduce friction discoloration of 'Bartlett' and 'd'Anjou' pears. *Hort. Sci.*, **17**, 215–17.

Miller, W.R., Chun, D., Risse, L.A. *et al.* (1988) Influence of selected fungicide treatments to control the development of decay in waxed or film wrapped Florida grapefruit, in *Proc. 6th Int. Cong.*, **3**, 1471–7.

Mitsubishi-Kasei (1989) *Ryoto Sugar Ester*, Technical Information, Mitsubishi-Kasei, pp. 1–20.

Moore, C.O. and Robinson, J.W. (1968) *Method for Coating Fruit.* US Patent No. 3,368,909.

Mukunthakumar, S. and Mathur, J. (1992) Artificial seed production in the male bamboo *Dendrocalamus strictus* L. *Plant Sci. Limerick*, **87**(1), 109–13.

Mulder, T.H. (1969) US Patent No. 3,451,826.

Nelson, R.C. (1942) A method of studying the movement of respiratory gases through waxy coatings. *Plant Physiol.*, **17**, 509–14.

Nipoti, P., Manzali, D., Gennari, S. *et al.* (1990) Activity of *Trichoderma harzianum* Rifai on the germination of asparagus seeds. I. Seed treatments. *Acta Hort.*, **271**, 403–7.
Nisperos-Carriedo, M.O. and Baldwin, E.A. (1988) Effect of two types of edible films on tomato fruit ripening, in *Proc. Fl. State Hort. Soc.*, **101**, 217–20.
Nisperos-Carriedo, M.O., Shaw, P.E. and Baldwin, E.A. (1990) Changes in volatile flavor components of pineapple orange juice as influenced by the application of lipid and composite film. *J. Agric. Food Chem.*, **38**, 1382–7.
Nussinovitch, A. (1994) Extending the shelf life of mushrooms by hydrocolloid coating. *Hassadeh*, **74**(10), 1131–2.
Nussinovitch, A. and Hershko, V. (1996) Gellan and alginate vegetable coatings. *Carbohydr. Polym.*, **30**, 185–92.
Nussinovitch, A., Hersko, V. and Rabinowitch, H.D. (1994) *Protective Coatings for Food and Agricultural Products.* Israel Patent Application No. 109,328.
Nussinovitch, A. and Kampf, N. (1992) Shelf-life extension and texture of alginate coated mushrooms. Presented at the *IFTEC*, 15–18 Nov., the Hague, the Netherlands, p. 118.
Nussinovitch, A. and Kampf, N. (1993) Shelf-life extension and conserved texture of alginate coated mushrooms (*Agaricus Bispouus*). *J. Food Sci. Technol.*, **26**, 469–75.
Orr, J.P. and Underwood, T. (1987) Influence of gel-coated seeds on germination time and tolerance of seedling tomatoes to postemergence herbicides. *Res. Prog. Rep. West Soc. Weed Sci. Western Society of Weed Science*, p. 130.
Orzolek, M.D. and Kaplan, R.C. (1987) Effect of the addition of growth regulators in gel on growth and yield of tomatoes, in *Acta Hort. Wageningen: Int. Soc. Hort. Sci.*, June (1987), pp. 135–40.
Paredes-Lopez, O., Camargo-Rubio, E. and Gallardo-Navarro, Y. (1974) Use of coatings of candelilla wax for the preservation of limes. *J. Sci. Food Agric.*, **25**, 1207–10.
Paul, D.R. (1976) Controlled-release polymeric formulations, in *Am. Chem. Soc. Symp. Ser.*, No. 33, American Chemical Society, Washington DC, p. 2.
Pennisi, E. (1992) Sealed in edible film. *Sci. News Washington*, Jan. 4, **141**(1), 12–13.
Quast, D.G. and Karel, M. (1972) Technique for determining oxygen concentration within packages. *J. Food Sci.*, **37**, 490–1.
Radnia, P.M. and Eckert, J.W. (1988) Evaluation of imazalil efficacy in relation to fungicide formulation and wax formulation, in *Proc. 6th Int. Cit. Cong.*, **3**, 1427–34.
Rico Pena, D.C. and Torres, J.A. (1991) Sorbic acid and potassium sorbate permeability of an edible methylcellulose palmitic acid film: Water activity and pH effects. *J. Food Sci.*, **56**(2), 497.
Rosenfield, D. (1968) *Storage Life Improvement.* US Patent No. 3,410,696.
Roth, W.B. and Loncin, M. (1985) Fundamentals of diffusion of water and rate of approach of equilibrium water activity, in *Properties of Water in Foods in Relation to Quality and Stability* (eds D. Simatos and J.L. Multon), Martinus Nijhoff Publishing, Dordrecht, The Netherlands, p. 331.
Schonherr, J. (1982) Resistance of plant surfaces to water loss: transport properties of cutin, suberin and associated lipids, in *Physiological Plant Ecology. Water Relations and Carbon Assimilations* (eds O.L. Lange, P.S. Nobel, C.B. Osmond and H. Ziegler), Springer-Verlag, New York, p. 153.
Shillington, W.L. and Liggett, I.J. (1970) *Coating Process.* US Patent No. 3,533,810.
Smith, S., Geeson, J., Browne, K.M. *et al.* (1987) Modified atmosphere retail packaging of Discovery apples. *J. Sci. Food Agric.*, **40**, 165–78.
Smith, S.M. and Stow, J.R. (1984) The potential of a sucrose ester coating material for improving the storage and shelf-life qualities of Cox's Orange Pippin apples. *Ann. Appl. Biol.*, **104**, 383–91.
Stern, S.A., Sinclair, T.P. and Gareis, T.P. (1964) An improved permeability apparatus of the variable-volume type. *Mod. Plastics*, **10**, 50–3.
Swenson, H.A., Miers, J.C., Schultz, T.H. *et al.* (1953) Pectinate and pectate coatings. Application to nut and fruit products. *Food Technol.*, **7**, 232–5.
Torres, J.A., Motoki, M. and Karel, M. (1985) Microbial stabilization of intermediate moisture food surfaces. Control of surface preservative concentration. *J. Food Proc. Pres.*, **9**, 75.
Trout, S.A., Hall, E.G. and Sykes, S.M. (1953) Effect of skin coatings on the behavior of apples in storage. *Aust. J. Agric. Res.*, **4**, 57–81.

Ukai, N.T., Tsutsumi, T. and Marakami, K. (1975) *Preservation of Agricultural Products*. US Patent No. 3,997,674.
Vojdani, F. and Torres, J.A. (1990) Potassium sorbate permeability of methylcellulose and hydroxypropyl methylcellulose coatings: effect of fatty acids. *J. Food Sci.*, **55**(3), 841.
Wanstedt, K.G., Seideman, S.C., Donnelly, L.S. *et al.* (1981) Sensory attributes of precooked calcium alginate-coated pork patties. *J. Food Prot.*, **44**, 732.
Warth, A.H. (1986) *The Chemistry and Technology of Waxes*, Reinhold, New York, pp. 37–192.
Wicks, Z.W., Jones, F.N. and Peter-Pappas, S. (1994) *Organic Coatings, Science and Technology*, Wiley, New York.
Williams, S.K., Oblinger, J.L. and West, R.L. (1978) Evaluation of calcium alginate film for use on beef cuts. *J. Food Sci.*, **43**, 292.
Willis, R.H., Lee, T.H., Graham, D. *et al.* (1981) *Postharvest, an Introduction to the Physiology and Handling of Fruit and Vegetables*. AVI Publishing, Westport, CT. pp. 1–2.
Woods, J.L. (1990) Moisture loss from fruits and vegetables. *Postharv. News Info.*, **1**, 195–9.
Wu, M.T. and Salunkhe, D.K. (1972) Control of chlorophyll and solanine synthesis and sprouting of potato tubers by hot paraffin wax. *J. Food Sci.*, **37**, 629–30.

11 Ceramics

11.1 Introduction

The word 'ceramic' is most often used to describe dishes, pottery and figurines. But wall tiles, building bricks, high-voltage insulators and glass products are also ceramics. Ceramics are widely used in a host of other specialized applications: nuclear reactors, space vehicles, electronic modules, computers, pumps, valves, metal-processing furnaces and ladles, optical equipment, supports (Walter *et al.*, 1989; Saito *et al.*, 1994), filtration equipment (Eriksson *et al.*, 1993; Lehtovaara and Mojtahedi, 1993; Burrell *et al.*, 1994), lasers and protective coatings (Allen *et al.*, 1993; Dialdo *et al.*, 1995). The world would be a different place without ceramics. Such critical areas of technology as communications, construction, transportation, power generation and transmission, sanitation, space exploration and medicine (Eisner, 1995) owe their development in part to ceramics technology (Jones and Berard, 1972). Ceramics are products made from natural or synthetic minerals. Most of the important ceramics comprise complex oxides and silicates, although a number of useful carbide, nitride and boride ceramics are also produced (Jones and Berard, 1972). Mica is a common mineral in nature and is utilized as electrical insulation: an example of a naturally occurring ceramic. The great majority of ceramics, however, are synthetic materials produced by the careful blending of raw materials followed by heat treatment to produce new mineral forms (Jones and Berard, 1972). Blending and firing raw materials allow the ceramic engineer to produce useful products including bricks and cements, plasters, abrasives, heat-resistant materials, tableware (Matte *et al.*, 1994; Baczynskyj and Yess, 1995), glass products of all kinds, single crystals and a host of materials having unique electrical and magnetic properties for use in the electronics industry (Jones and Berard, 1972).

11.2 Origins of Egyptian faience: the first 'high-tech' ceramics

Egyptian and Near Eastern faience are considered to be the first 'high-tech' ceramics (Vandiver and Kingery, 1986) (Fig. 11.1). The use of fired-clay containers began between 7000 and 6000 BC (Watson, 1965). Non-clay ceramic ware (produced by stone-working techniques) was manufactured

Figure 11.1 At the southern Tomb of Saqqara, faience tiles were the first large-scale faience project, exhibiting convex outer surfaces. (Adapted from *High-technology Ceramics, Past, Present and Future*, Vol. 3, ed. W.D. Kingery, 1986.)

from about 4000 BC. Their compositions were based on mixtures of powdered quartz or sand containing a lime impurity, with potassium and sodium salts and a copper coloring. The technology was developed independently of traditional pottery.

Beads made of Egyptian faience (Mohs, hardness of 6–7) have been excavated from grave deposits (*c.* 4000–3100 BC), together with beads of glazed soft steatite (stone) and semi-precious stones such as turquoise, carnelian, quartz and lapis lazuli (Vandiver, 1983). Some of the faience glazes were formed by firing (800–1000°C) (Beck, 1986) an effloresced salt layer on the surface of the bead in an oxidizing atmosphere. Other methods were also practiced (Fig. 11.2). The same technology was used to glaze beads of flaked quartz and rock crystal. The faience material also served as the foundation for the wrapping and fitting of gold-foil beads. In cross-section, a faience body exhibits angular quartz particles coated with and held together by lenses of a soda-lime copper silicate glass. A few layers can be detected: the glaze (green to blue), the white quartz layer and the interior, made up of impure quartz (Kingery and Vandiver, 1986a).

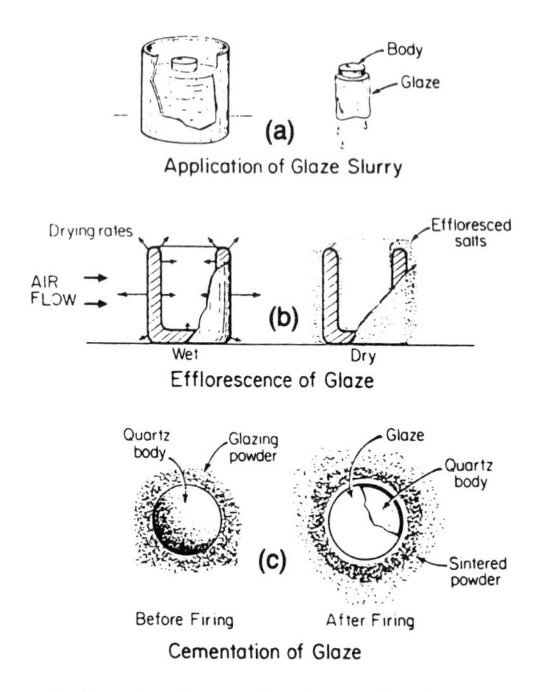

Figure 11.2 Three methods used to form a glaze in Egyptian faience. (Adapted from *High-technology Ceramics, Past, Present and Future*, vol. 3, ed. W.D. Kingery, 1986.)

Glazing by firing was also used to produce tiles, figures, game pieces and vessels. Changes in raw materials (sand with iron impurities), processes (rapid methods of molding), compositions (changes in the proportions of lime, alkali and copper, and the use of calcium oxide) resulted in the development of three different methods of glaze application. The first used prefritted ground glass added to water to form a slurry, which was applied by painting and dipping. The second was to incorporate soluble salts in the mixing water or the body's composition. As the water moved to the surface and evaporated, the salts deposited as an effloresced layer. The third consisted of embedding and firing the quartz body in a special glazing powder. The glaze constituents migrated to and reacted with the quartz surface (Kingery and Vandiver, 1986b).

The development of faience in the Near East, particularly in Syria, Iraq and Iran (Mesopotamian heartland) is similar to that in Egypt, except that the preservation of the objects was not as good. Faiences found in the Indus valley sites appeared later. The Egyptian techniques for designing faience to replace the highly prized lapis lazuli and turquoise represent the first successful 'high-tech' ceramics.

11.3 Roman concrete, Chinese celadon and European porcelain

Roman concrete had a huge impact on architecture and was used for Roman imperial buildings. The concrete was held together by hydraulic setting, which also served as an attractive ceramic medium for the buildings (Langton and Roy, 1984; Lechtman and Hobbs, 1986). The rheological plasticity of the cement in the paste state is responsible for its properties. This ceramic was composed of pozzuolanic aluminosilicate sands and limestone. It was castable, cohered at ambient temperatures and yielded integral monolithic structures that could be incrementally extended. Further information on chemical analyses of the cementitious binder phase, on reaction sequences for slaked-lime mortar and on the hydraulic setting reaction in pozzuolanic cement can be found elsewhere (Langton and Roy, 1983; Lechtman and Hobbs, 1986).

Chinese celadon is an important branch of Chinese ceramics. It can be traced back for about 1800 years. Green glazed porcelain, the predecessor of celadon, evolved from glazed pottery and proto-porcelain developed in the Shang and Zhou dynasties during the Spring, Autumn and Warring States periods. Porcelain could be found in China as early as the Eastern Han dynasty (AD 20 to 220). Characteristic features of different Chinese celadons, their chemical compositions and physical properties, the chemical composition of the glazes, the microstructure of body and glaze, and spectral reflectance and chromaticity of the glazes can be found in the literature (Guozhen *et al.*, 1985; Fukang, 1985).

Soft-paste porcelain was first successfully produced in Europe at Saint Cloud in about 1695 and the first hard porcelain was made at Meissen in 1708 (Kingery, 1986). These innovations began a revolution in ceramics manufacture, science and engineering, which included the industrialization of the ceramics industry in Europe. Whereas previously porcelain had been imported in large quantities from China, Japan and Persia, at the end of the 16th century, high-quality porcelain and fine earthenware were being exported over much of the world. European porcelain was manufactured in Florence from 1575–87, and named 'Medici porcelain' after its patronage. This porcelain was composed of 78% SiO_2, 9.5% Al_2O_3, 0.5% Fe_2O_3, 0.7% MgO, 2.5% CaO, 3.4% Na_2O, 5.4% K_2O, 0.3% PbO, 0.2% TiO_2 and 0.1% P_2O_5. The ware was fired at $\sim$1100°C (Kingery and Vandiver, 1984, 1986a).

French soft-paste manufacturing is similar to that used for the Medici porcelain. The changes that led to its success were the use of high lime content and a very plastic clay with low aluminum content. The first successful European hard-porcelain production was achieved in Meissen in 1708 by a research team under the direction of Johann-Friedrich Bottger (Kingery and Smith, 1985). The porcelain was composed of 61.5% SiO_2,

30.1% Al_2O_3, 1.1% MgO, 6.3% CaO, 0.1% Na_2O, 0.3% K_2O and 0.6% other. The glaze was composed of 71.6% SiO_2, 19.2% Al_2O_3, 1.1% MgO, 5.7% CaO, 0.9% Na_2O, 0.6% K_2O and 0.9% other. The ware was fired at $\sim$1350°C, a level never before achieved in Europe. Its success required the simultaneous development of both compositions and firing methods: ceramics science and furnace engineering. More new compositions had been developed by the end of the 18th century in Europe and the quality and reasonable cost of these products made them a major export product (d'Albis, 1983).

11.4 Spark plug insulators, refractory materials and television tubes

About 110 years ago, the internal combustion engine was invented. In 1860, J. J. Lenoir of France was the first man to use a ceramic spark plug to ignite the fuel in this engine. His design was similar to that of the current spark plugs. Early low-compression engines used mica, steatite or porcelain-ceramic materials. Since then, many other materials have been developed and used, because increasing engine needs have changed the compression ratios. New ceramic materials were developed based on high alumina compositions to build the basic ceramic insulator body, and clay was no longer used as a primary agent of workability. New tunnel kilns also permitted higher firing temperatures and improved insulator quality. The developments in these areas were later utilized for ceramic sensors and electronic substrates (Jeffrey, 1932; Givens, 1976; Owens *et al.*, 1977).

Spark plug insulators and intricate shapes are produced by injection molding. For the successful application of this method, synthetic resins and special plasticizers can be obtained in a range of molecular weights which allows control of the melting range. On drying, the low-molecular-weight materials evaporate first, providing a pore system to faciliate the removal of the higher-molecular-weight materials as the temperature is increased. If at least one resin has a slightly thermosetting character it will help retain the dimensional stability of the part during drying (Jones and Berard, 1972).

The introduction of silicon carbide (SiC) as a refractory material was important in the development of ceramics. Developments began with the first oxide-bonded products and continued to the dense, high-strength, high-performance materials of today. Silicon carbide was a new material whose characteristics differed considerably from those of the oxides used previously. It decomposed above 2400°C, retaining strength to higher temperatures than oxides; it exhibited high thermal conductivity, low thermal expansion, a high modulus of elasticity, lower electrical resistivity, better thermal-shock resistance, corrosion resistance and high hardness values (Butler, 1957; Lloyd and Howard, 1968; Yajima *et al.*, 1977).

In 1929, RCA and other companies were engaged in the development of electronic television. This brought about a need to increase the capabilities of the luminescent materials used in television cathode-ray tubes. An important product of RCA's television-luminescence research was the zinc–beryllium silicate phosphor system, which became the major component of light-emitting coatings in luminescent lamps. This study also evidenced the production of white light through the mechanical mixtures of certain blue-emitting and yellow-emitting phosphors. Thus 'black-and-white' television was developed. Television-tube technology represented an interaction of art, ceramics science and human cooperation. More information on these topics can be found in Smith (1976) and Leverenz (1940).

11.5 Nuclear fuel

In 1954, uranium dioxide (UO_2) was selected as the fuel for the first commercial power-producing nuclear reactor. It combines a high concentration of fissionable uranium atoms with good resistance to corrosion by hot water and with the ability to withstand having up to 5% of its metallic atoms fissioned without severe degradation. Used as a ceramic fuel, this material must have a few desired properties, adjusted to its operating conditions. Uranium dioxide clad in Zircaloy (a group of zirconium alloys developed by the Westinghouse Electric Corp., Pittsburgh, PA) in pellet form can withstand high levels of burnup without catastrophic failure, retaining the greater part of the fission gas generated. Moreover, its thermal conductivity is not adversely affected by the exposure. This paved the way for the use of this fuel element design in pressurized water reactors. In a short time, improvements were made in the powder preparations and production methods of density fuels. Research is now aimed at preparing powders with improved reproducibility (Belle, 1961; Rohn *et al.*, 1985). As a ceramic material, uranium dioxide is fascinating, but its long-term future is less clear. For the present water-cooled reactors, changes are aimed at producing a fuel that has a longer life. If reactor types change, then uranium dioxide might be replaced with uranium carbide (in a graphite matrix) or by other fuel types (Burke *et al.*, 1957).

11.6 Silicon nitride, Sialons and related ceramics

The nitriding of compacted silicon-powder shapes to achieve reaction-bonded components and other related methods has had limited success. The goal of silicon nitride technology is the ceramic gas turbine: American programs began this research in 1971, and the British have been working on it since the mid-1950s. The ceramic turbine, running at 1370°C, is probably the most difficult application to have been proposed for ceramics; other applications are also being pursued. Previous experiments involving substitution of materials led to the conclusion that the takeover of current automotive engines by ceramic gas turbine should not be expected before the year 2050.

Silicon nitride has been recognized as the ceramic of the 1990s. However, there is still no commercial process for producing complex silicon nitride shapes with the combination of strength, oxidation resistance and creep resistance required for the gas turbine, together with the necessary reliability and reproducibility (Hardie and Jack, 1957; Parr *et al.*, 1959; Godfrey, 1974). Modifications of silicon nitride and the development of Sialons are approaches to solving the problem. Sialon is made up of silicon and aluminum, the two most abundant metallic elements, and oxygen and nitrogen, which make up the atmosphere: therefore, there is no shortage of raw materials for Sialon manufacture. Although Sialons include a large array of materials, β'-Sialon is the only one that has been explored in depth for its technological potential. Nevertheless, other Sialons are equally promising and offer microstructural control and tailoring. The most successful commercial application of Sialons has been as a cutting tool for machining metals. Excellent thermal-shock resistance, high-temperature strength and electrical insulation combine to make Sialon eminently suitable to welding applications.

The reader interested in Sialons and silicon nitrides, their compositions, variety, densification and commercialization, is referred to Ziegler and Wotting (1985) and Trigg and Jack (1984).

11.7 Multi-layer ceramics

Laminated multi-layer semiconductors and laminated capacitors were introduced by RCA. The value of laminated-ceramics production grew from next to nothing in 1960 to over one billion dollars in 1983. This new technology shaped the present ceramics and electronics industries (Fig. 11.3). Laminated multilayer ceramic circuits changed the way large computers are designed and constructed. Studies to improve binder and dispersant systems and powder preparations, in addition to green-sheet-lamination processes to obtain novel system properties, are being conducted at many research institutes (Schwartz and Wilcox, 1967; Cadenhead and DeCoursey, 1985; Kahn, 1985).

11.8 Role of hydrocolloids in the ceramics industry

Ceramic pieces are more difficult to form than metal pieces because of the inherent brittleness and high melting temperatures of ceramics. The most

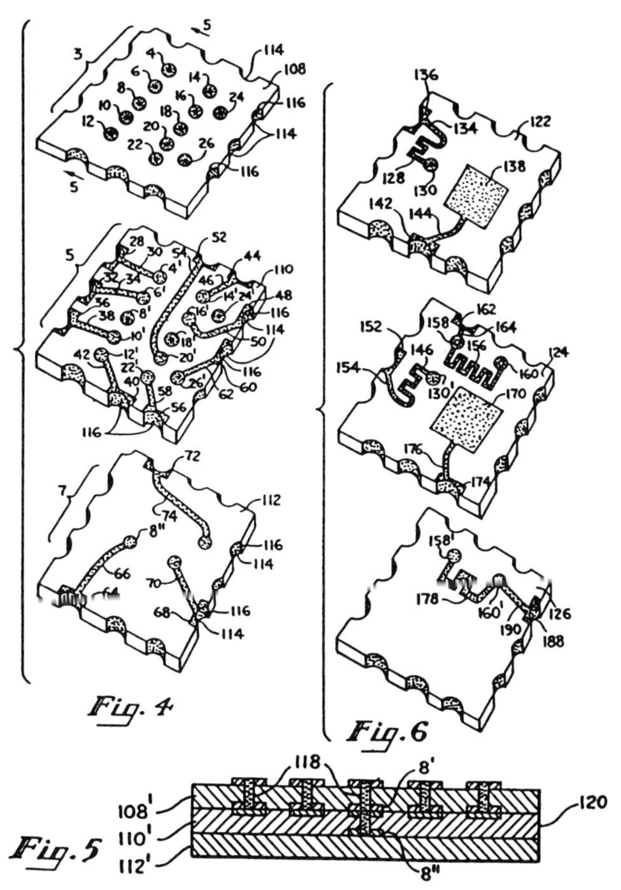

Figure 11.3 Patent drawings showing multi-layer ceramic via construction of internal capacitors and resistors. (Adapted from *High-technology Ceramics, Past, Present and Future*, vol. 3, ed. W.D. Kingery, 1986.)

common ceramic raw materials making up the bulk of those processed by the ceramics industry are clays, fluxing minerals, silica, refractory oxides and several hundred others. The tonnages of these materials used are considerably less than for the major materials already mentioned, but their dollar value is certainly not insignificant. They include graphite, gypsum, sillimanite minerals, silicon carbide, zirconia and zircon. Gums (hydrocolloids) are included in ceramic applications as stabilizers, thickeners, adhesives, binders and lubricants.

11.8.1 Binders

Non-clay ceramic materials, although not rendered plastic by the simple addition of water, can be brought into a plastic condition by the addition of certain organic materials. These materials may be classified as binders or lubricants according to the functions they serve in the plasticizing process. Binders give wet strength during formation and dry strength after drying (Jones and Berard, 1972). Lubricants lower the frictional forces present during formation, permitting formation at lower pressures. Water promotes the plasticizing abilities of these organic materials. Polyvinyl alcohols, MC, water-soluble gums, starches, proteins and other organic materials are binders that may be used in the presence of water. Waxes, stearates and water-soluble oils are examples of lubricants that may be used in the presence of water (Jones and Berard, 1972). In operations above the boiling point of water, the plasticizing abilities of natural and synthetic thermoplastic resins may be promoted with organic solvents, or plasticizers. The resin may act as a lubricant as well as a binder.

Organic lubricants and binders are used in dry formation, when ceramics are formed under high pressure with relatively low moisture content or sometimes with no moisture at all. The binder composition should be adjusted to the pressing pressure and the parts being pressed (Fig. 11.4). Wax binders are suitable for low pressures but harder, less tacky binders may be used at high pressures. In isostatic pressing, the binder system should be similar to that for dry pressing, but hard binders are generally avoided (Jones and Berard, 1972).

Tragacanth is used in ceramics production because of its emulsification and suspension abilities and because it is a thickening agent through the ordering of water molecules and the hydration of the water-swellable

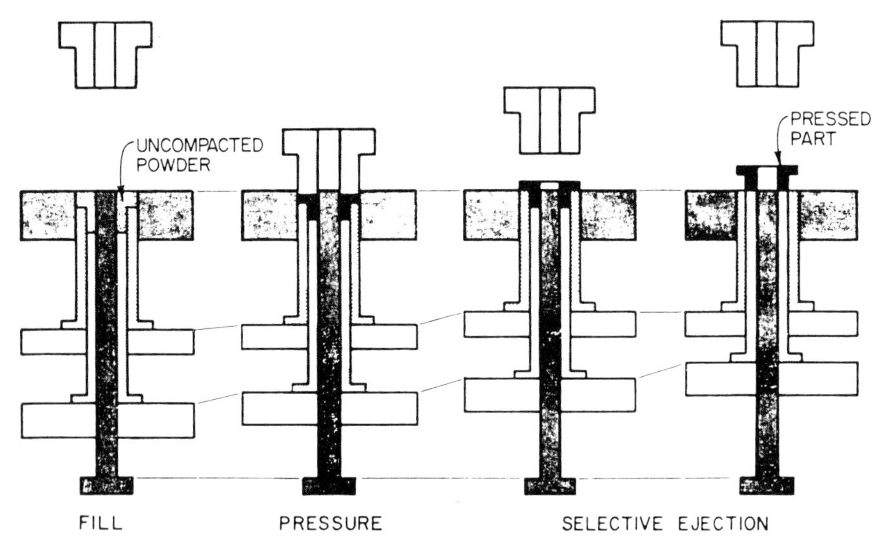

Figure 11.4 Details of dry pressing cycle for small parts. (Reproduced by permission of the Iowa State University Press.)

fraction. It is also used as an initial adhesive or suspension agent, which is later burned out in the furnace. Certain grades of tragacanth can be useful as binders because of their low ash content. The gum acts to suspend various ingredients in the mass before the ceramic is fired in the furnace.

HPC is used as a burnout binder for ceramics, ceramic glazes and electrical insulators. HPC (water and solvent soluble) is a strong binder that is readily burned out in nitrogen or air, leaving only negligible amounts of residual ash. It also contributes to green strength.

PEG is used in molded ceramics, foundry cores and ceramic glazes because of its water solubility and good lubricity, and because in the absence of air PEG chains undergo pyrolysis at temperatures above 300°C. Its decomposition is complete and leaves no residues. A key factor in choosing PEGs for such purposes is their ability to serve as an integral part of such products, in addition to contributing to green strength and the formation of the ceramic product. PEGs reduce die wear during extrusion operations.

PEOs (resins with high molecular weights, over 10^6, and concentrations of 0.1–2.0%) are used as binders in the ceramics industry. The resins improve binding applications where the combination of water solubility, almost no inorganic impurities and clean burnout are key factors in achieving satisfactory products. The green strength of clay and its refractory compositions is achieved by using PEO, which also lubricates the ceramic product during formation.

PVAs are water-soluble synthetic polymers that can be described as polyhydric alcohols with secondary hydroxyl groups on alternate carbon atoms. PVA is an efficient binder for ceramic materials. Only small amounts are required to improve the green strength of ceramic products. It is burned out when the ware is fired. The addition of 2–3% PVA (dry weight) to clay reduces possible damage (breakage) to the green ware; addition of 0.1% improves working conditions in the production of fine china flatware. Other desirable properties (color and transparency) are not affected (Pritchard, 1970; Finch, 1973). Most PVA-stabilized emulsions have good adhesion to glass and ceramics (Warson, 1972).

An anionic galactomannan polysaccharide gum from a newly isolated lactose-utilizing bacterium was reported to have an M_w value of $\sim 7 \times 10^6$. An aqueous solution of this so-called lactan gum had a stable viscosity over pH range 2–11, being particularly stable in alkaline environments. These gum solutions are thermostable and the flow properties indicate potential in areas of ceramics, paper, detergents and binders for building materials (Flatt *et al.*, 1992).

11.8.2 Adhesives

The innate adhesive and thickening properties of alkyl and hydroxyalkylalkylcellulose are used for many ceramics purposes, such as refractory mortars, glaze slips and high-temperature glaze slips, porcelain enamels, cements, tile mortars and plastic mixes. The wide range of multi-functional properties of cellulose derivatives can be used, among many other things, to provide green strength, water retention and lubrication in refractory mortars (thermal gelation ends with quick set following the incorporation of these products) and in cements and in glaze slips (such gums minimize the sagging that occurs during glazing). After firing, the amount of leftover ash in the product is minimal, a desirable property in the production of porcelain insulators for the electrical industry. Cellulose ethers can also be used in ceramic tile adhesives (Vaughan, 1985).

PVA is used as an adhesive for porous ceramic surfaces. The degree of PVA hydrolysis depends on the role the adhesive will play, e.g. completely hydrolysed for quick-setting water-resistent adhesives, partially hydrolysed for remoistenable adhesives.

11.8.3 Glazes

Alginates are gelling agents that can hold water, emulsify and stabilize (see Chapter 2). They are used in ceramics as stabilizers that import plasticity and suspend solids in ceramic glazes, they impart green strength and are burned out upon firing. In addition to improving plasticity of ceramic bodies, algin increases their wet strength. In glazes and slips, algin is used as a suspending, viscosity-controlling and drying agent. Penetration of the slips into porous ceramics is reduced; as a filming agent, it prevents pinholes and aids in the release of the casting (Thellmann, 1966; Rempes, 1980).

CMC exhibits binding, film-forming, thickening, suspending and water-holding properties. The last (control of water) is very important in the ceramics industry. Prior to firing, ceramic glazes require the incorporation of such a flow-control agent to prevent the glaze from sagging. This is successfully accomplished by CMC, which can thicken (increase viscosity) to various degrees depending on the particular glaze formulation. In addition, water-soluble cellulose ethers serve as auxiliary materials in the ceramics industry (Danielson, 1952; Stawitz, 1953).

Rice husks, produced in huge quantities in India, are mainly composed of cellulose and silica. Following pyrolysis, 'black ash' is obtained, comprising carbon and silica, which can serve as the starting materials for the production of silicon nitride and Sialon. Clay and coal, volcanic ash and rice husks cannot produce pure Sialons for sophisticated engineering applications such as the ceramic turbine, but they can provide useful refractory bricks, furnace linings and materials resistant to molten metals.

Carrageenans (Chapter 3) are used in ceramics as suspending agents for glazes. A carefully controlled calcium ion level in ι-carrageenan enables the suspension of insolubles in liquid media. Concentrations of 0.25–0.80% are satisfactory for achieving more uniformity and ease of handling in application.

Gum arabic (Chapter 7) is used in ceramics as a glaze binder, where its properties, including bonding, absence of slip and stability over a wide range of temperatures, are required. Gum arabic needs to meet specific demands so that it can be adjusted to its tasks, especially at elevated temperatures.

Hydroxyethylcellulose (Chapter 6) is used as a processing aid (thickener and water loss-control agent) in ceramic glazes. It also serves as a binder and behavior modifier (for glazes and refractory shapes), i.e. it is a promoter of workability and green strength that is later burned out.

Polyacrylic acid is used for thickening ceramic glazes (for use in patching compositions) and the thickener is burned away later. As a binder, polyacrylic acid imparts green strength to molded articles, thereby enabling them to maintain stability under wet conditions, before firing burns out the binder.

Xanthan gum (Chapter 9) serves as an ideal suspension stabilizer because it can maintain a viscosity that is almost independent of pH and temperature. With respect to most of the ingredients in a system, it is considered stable and has a yield value. All of these properties are important for the suspension of solids in ceramic glazes, enabling uniform application of the glaze to the ceramic body.

11.8.4 Special applications

PEG is used as a lubricant in the ceramics and tile industries, and as a binder for ceramic glazes and enamels. It is burned out cleanly during firing. PVA can be used for cementitious binding products such as thin-set mortars for ceramic-tile work. In this case a cold-water-soluble, partially hydrolysed grade is used. PVA improves the adhesion and resiliency related to joint cements and mortars. Polyacrylic acid can be used as a thickening agent for ceramic glazes, and thus can serve in patching compositions, with the gum being subsequently burned out. Non-ionic cellulose ethers are used in the electrical industry to help with the production of porcelain items. Since the gel has a thermal gelling point, quick set is achieved, minimizing sagging during glazing.

Hydrocolloids are not only connected to the production of ceramics. Sometimes gelling agents based on hydrocolloids are used as mediators between ceramic electrodes and the skin. A report on a NASICON-type ceramic (high-sodium-ion conductor) that does not need gel before its application is interesting. The principle of the measurements is based on sodium-ion exchange between the skin and the material. Electrical measurements performed in saline solutions show that the electrode is slightly polarizable. Skin–electrode impedance was also investigated (Gondran *et al.*, 1995).

References

Allen, W.H., Harmon, J.D. and Linvill, D.E. (1993) Evaluation of a ceramic roof coating. *Appl. Eng. Agric.*, **9**(3), 309–15.

Baczynskyj, W.M. and Yess, N.J. (1995) US Food and Drug Administration monitoring of lead in domestic and imported ceramic dinnerware. *J. AOAC Int.*, **78**(3), 610–14.

Beck, H. (1986) Notes on glazed stones: part I, glazed steatite, in *High Technology Ceramics, Past, Present and Future* (ed. W.D. Kingery), American Ceramic Society, Westerville, OH, p. 33.

Belle, J. (1961) *Uranium Dioxide, Properties and Nuclear Applications*. Naval Reactors Division, AEC.

Burke, T.J., Glatter, J., Hoge, H.R. *et al.* (1957) Fabrication of high density uranium dioxide fuel components for the first pressurized water reactor core, in *Nuclear Metallurgy*, vol. 4, Institute of Metals Division, AIME, p. 135.

Burrell, K., Gill, C., McKechnie, M. *et al.* (1994) Advances in separations technology for the brewer: ceramic crossflow microfilitration of rough beer. *Tech. Master Brew. Assoc. Am.* Wauwatosa, Wis., **31**(2), 42–50.

Butler, G.M. (1957) The past and future of silicon carbide. *J. Electrochem. Soc.*, **104**(10), 640–4.

Cadenhead, R.L. and DeCoursey, D.T. (1985) The history of microelectronics, part 1. *Int. J. Hybrid Microel.*, **8**(3), 14–30.

d'Albis, A. (1983) Steps in the manufacture of the soft paste porcelain of Vincennes, according to the books of Hellot, in *Ceramics and Civilization I: Ancient Technology to Modern Science* (ed. W.D. Kingery), American Ceramic Society, Columbus, OH, pp. 257–72.

Danielson, R.R. (1952) How to glaze structural clay products. *Brick Clay Rec.*, **121**(3), 53–5.

Dialdo, B., Vanhaelen, M. and Gosselain, O.P. (1995) Plant constituents involved in coating practices among traditional African potters. *Birkhauser Verlag Exp.*, Basel, **51**(1), p. 95–97.

Eisner, E.R. (1995) Restoring a tooth to form and function after endodonic treatment. *Vet. Med.*, **90**(7), 662–79 (Veterinary Medicine Publishing Group, Lenexa, KS).

Eriksson, T., Isaksson, J., Stahlberg, P. *et al.* (1993) Durability of ceramic filters in hot gas filtration. *Bioresour. Technol.*, **46**(1/2), 103–12.

Finch, C.A. (ed) (1973) *Polyvinyl Alcohol Properties and Applications*, Wiley, New York, pp. 338 and 534.

Flatt, J.H., Hardi, R.S. and Gonzalez, J.M. (1992) An anionic galactomannan polysaccharide gum from a newly-isolated lactose utilizing bacterium. I. Strain description and gum characterization. *Biotechnol. Prog.*, **8**(4), 327–34.

Fukang, Z. (ed.) (1985) Longquan celadon, in *Scientific and Technological Achievements in Ancient Chinese Pottery and Porcelain*, Shanghai Scientific & Technical Publishers, p. 172.

Givens, L. (1976) Spark plugs. *Automotive Eng.*, **84**(7), 27–31.

Godfrey, D.J. (1974) Silicon nitride ceramics for engineering applications. *Trans. SAE*, **83**, 1036–45.

Gondran, C., Siebert, E., Fabry, P. *et al.* (1995) Non-polarisable dry electrode based on NASICON ceramic. *Med. Biol. Eng. Comput.*, **33**(3), 452–7.

Guozhen, Li, Honming, Ye, Zhuhai, C. *et al.* (1985) An investigation of the Yue Kiln celadon of the past dynasties in Zhejiang. *Anthology 2nd Int. Conf. Ancient Chinese Pottery and Porcelain*, November 14, 1985, Beijing.

Hardie, D. and Jack, K.H. (1957) Crystal structures of silicon nitride. *Nature* (London), **180**, 332–3.

Jeffrey, B.A. (1932) *Method of and Apparatus for Shaping Articles*. US Patent No. 1,863 854, June 21.

Jones, J.T. and Berard, M.F. (1972) *Ceramics Industrial Processing and Testing*, The Iowa State University Press, Ames, IA.

Kahn, M. (1985) Structured microvoids in ceramic PZT, Paper 56-E-85. *Extended abstracts, Am. Ceram. Soc. 87th Annu. Meet.*, Cincinnati, OH, p. 115.

Kingery, W.D. (1986) The development of European Porcelain, in *High-technology Ceramics, Past, Present and Future*, vol. 3. (ed. W.D. Kingery), The American Ceramic Society, Westerville, OH, pp. 153–80.

Kingery, W.D. and Smith, D. (1985) The development of European soft-paste (Frit) porcelain, in *Ceramics and Civilization I: Ancient Technology to Modern Science* (ed. W.D. Kingery), American Ceramics Society, Columbus, OH, pp. 273–92.
Kingery, W.D. and Vandiver, P.B. (1984) Medici Porcelain, *Faenza*, **LXX** (5–6), 441–52.
Kingery, W.D. and Vandiver, P.B. (1986a) *Ceramic Masterpieces: Art, Structure, Technology*, ch. 2, *An Egyptian Faience Chalice*. Free Press MacMillan, New York, pp. 5–67.
Kingery, W.D. and Vandiver, P.B. (1986b) A Medici porcelain bottle, in *Ceramic Masterpieces: Art, Structure, Technology*, Free Press, Macmillan, New York.
Langton, C.A. and Roy, D.M. (1983) *Characterization of cement-based ancient building materials in support of repository seal materials studies*. Topical Report ONWI-523, Distribution Category UC-70, National Technical Information Service, US Department of Commerce, Washington, DC.
Langton, C.A. and Roy, D.M. (1984) Longevity of borehole and shaft sealing materials: characterization of ancient cement-based building materials. *Mat. Res. Soc. Symp. Proc.*, **26**, 543–9.
Lechtman, H.N. and Hobbs, L.W. (1986) Roman Concrete and the Roman Architectural Revolution, in *High-technology Ceramics, Past, Present and Future* (ed. W.D. Kingery), American Ceramic Society, Westerville, OH, pp. 81–128.
Lehtovaara, A. and Mojtahedi, W. (1993) Ceramic filter behavior in gasification. *Bioresour. Technol.*, **46**(1/2), 113–18.
Leverenz, H.W. (1940) Cathodoluminescence as applied to television. *RCA Rev.*, **5**(2), 1131–75.
Lloyd, D.E. and Howard, V.C. (1968) The fabrication and properties of large-diameter tubes in coatings in pyrolytic silicon carbide, in *Special Ceramics 4* (ed. P. Poper), British Ceramic Research Association, Stoke on Trent, UK, pp. 103–19.
Matte, T.D., Proops, D., Palazuelos, E. *et al.* (1994) Acute high-dose lead exposure from beverage contaminated by traditional Mexican pottery. *Lancet*, 344, 1064–5.
Owens, J.S., Hinton, J.W., Insley, R.H. *et al.* (1977) Development of ceramic insulators for spark plugs. *Am. Ceram. Soc. Bull.*, **56**(4), 437–40.
Parr, N.L., Martin, G.F. and May, E.R.W. (1959) Study of silicon nitride as a high-temperature material. *Admiralty Materials Report No.* A/75.
Pritchard, J.G. (1970) *Poly (Vinyl Alcohol) Basic Properties and Uses*, Gordon and Breach, New York.
Rempes, P.E., Jr (1980) US Patent No. 2,990,292 (*Chem. Abstr.*, **55**, 26310), in *Handbook of Water-soluble Gums and Resins*, ch. 2 (ed. R.L. Davidson), McGraw-Hill, New York, pp. 2.1–2.43.
Rohn, R.J., Adamantiades, A.G., Kenton, J.E. *et al.* (1985) *A Guide to Nuclear Power Technology*, Wiley-Interscience, New York.
Saito, T., Yoshida, Y., Kawashima, K. *et al.* (1994) Immobilization and characterization of a thermostable beta-galactosidase from a thermophilic anaerobe on a porous ceramic support. *Appl. Microbiol. Biotechnol.*, **40**(5), 618–21.
Schwartz, B. and Wilcox, D.L. (1967) Laminated ceramics. *Proc. Elect. Comp. Conf.*, pp. 17–26.
Smith, C.S. (1976) On art, invention and technology. *Technol. Rev.*, **6**, 36–40.
Stawitz, J. (1953) Water-soluble cellulose ethers as auxiliary materials in the ceramic industry. *Toind. Ztg.*, **77**, 14–15.
Thellmann, E.L. (1966) *Metal Slip Casting Composition*. US Patent No. 3,216,841 (1965) (*Chem. Abstr.*, **64**, 4739).
Trigg, M.B. and Jack, K.H. (1984) Silicon oxynitride and *O'*-Sialon ceramics, in *Ceramic Components for Engines* (eds S. Somiya, E. Kani and K. Ando), KTK Scientific, Tokyo, pp. 199–207.
Vandiver, P.B. (1983) Appendix A. The manufacture of faience, in *Ancient Egyptian Faience* (eds. A. Kaczmarczyk and R.E. Hedges), Aris and Phillips, Warminster, UK, p. 64.
Vandiver, P.B. and Kingery, W.D. (1986) Egyptian faience: the first high-tech ceramic, in *High-technology Ceramics, Past, Present and Future* (ed. W.D. Kingery), American Ceramic Society, Westerville, OH, pp. 19–34.
Vaughan, F. (1985) The use of cellulose ethers in ceramic tile adhesives, in *Cellulose and its Derivatives: Chemistry, Biochemistry and Applications* (ed. E. Horwood), Halsted Press, New York, pp. 311–18.

Walter, R.P., Kell, D.B., Morris, J.G. *et al.* (1989) Immobilization of *Candida cylindracea* lipase on a new range of ceramic support. *Biotechnol. Tech.*, **3**(5), 345–8.
Warson, H. (1972) *The Applications of Synthetic Resin Emulsions*, Ernest Benn, London, p. 295.
Watson, P.J. (1965) The chronology of North Syria and North Mesopotamia from 10 000 to 2000 BC, in *Chronologies in Old World Archaeology* (ed. R.W. Ehrich), University of Chicago Press, Chicago, pp. 61–100.
Yajima, S., Shishido, T. and Okamura, K. (1977) SiC bodies sintered with three-dimensional cross-linked polycarbosilane. *Am. Ceram. Soc. Bull.*, **56**(12), 1060–3.
Ziegler, G. and Wotting, G. (1985) Post-treatment of pre-sintered silicon nitride by hot isostatic pressing. *Int. J. High Technol. Ceram.*, **1**, 31–58.

12 Cosmetics

12.1 Introduction

The field of cosmetics combines knowledge and skills from various sciences, such as chemistry, physics, biology and medicine. After the late 1940s' revision of the Federal Food and Drug Act of 1906, regulations concerning 'cosmetics' in its broadest sense began to surface. Federal agencies are responsible for regulating the cosmetics industry and its products, and for providing guidelines for good manufacturing practices. The cosmetics industry attracts a large share of consumer spending each year. There are so many health and beauty aids on the market that they can only be listed by product families, such as hair preparations, oral-hygiene products, feminine hygiene, dieting aids, baby needs and medicaments, foot products, packaged medications, shaving preparations, fragrances, hand preparations, personal-cleanliness items, among many others (Greenburg and Lester, 1954; de Navarre, 1957; Wall, 1957).

The aim of this chapter is to screen the list of cosmetics for those in which hydrocolloids are found or can be used in the near and/or distant future. Toxicity and test methods are not reviewed. The Cosmetics, Toiletry and Fragrance Association (CTFA) set up a cosmetics ingredient review (CIR) to evaluate the safety of cosmetic ingredients. The CIR began with the 2800 ingredients listed in the CTFA's cosmetic ingredient dictionary. To date, 700 items are under FDA review and only 189 substances are used in more than 25 cosmetic products. The evaluation of these has been prioritized according to health aspects, pharmaceutical, manufacturing and targeting activity, and public response (FDA, 1976; CTFA, 1977).

12.2 Cosmetic emulsions and their properties

Cosmetic emulsions include lotions and creams. Both oil-in-water (o/w) and water-in-oil (w/o) systems exist. Important emulsion properties include viscosity, color, ease of dilution and stability. These depend on the properties of the continuous and discontinuous phases, the ratio between these phases, particle size and ionic charges, among other things. Of major importance is the type of emulsion (w/o or o/w), which is controlled by the emulsifier type and its amount, the ratio of ingredients and the order of ingredient addition during mixing.

The continuous phase of the emulsion influences its solubility, determining whether it can be diluted with water or oil. Ease of dilution depends on the viscosity of the emulsion. If the proportion of the internal phase increases to the point where the internal phase exceeds the volume of the external phase, emulsion-particle crowding causes the apparent viscosity to be partially structural. Emulsions are stable as long as particle coalescence in the internal phase is prevented. Stability depends on particle size, the different densities of the two phases, viscosity of the continuous phase and of the completed emulsion, particle charges, the nature of the emulsifier used and storage conditions.

12.3 Emulsifiers

Emulsifiers are classified as ionic or non-ionic according to their behavior. Hydrophilic–lipophilic balance (HLB) is often used to characterize emulsifiers and related surfactant materials. Ionic types are divided into anionic and cationic. The lipophilic (L) portion of the molecule is considered to be the surface-active portion. Non-ionic emulsifiers are completely covalent and show no apparent tendency to ionize. These emulsifiers are less susceptible to the action of electrolytes than the ionic surface-active agents. As surface-active agents, emulsifiers lower surface and interfacial tensions and increase the tendency of their solution to spread. Emulsifying agents such as o/w ones produce emulsions with a continuous hydrophilic phase. Such emulsions are generally dispersible in water and will conduct electricity. Such surfactants have an HLB value >6.0 and the hydrophilic portion of their molecules predominates. A few examples of these o/w emulsifiers are the non-ionic PEG 300 and 400 distearates with HLB values of 7.3 and 9.3, respectively; sorbitan (monolaurate), a non-ionic emulsifier with an HLB value of 8.6; PEG 6000 monolaurate (non-ionic emulsifier with an HLB of 19.2) and others. Emulsifiers of the w/o type produce emulsions in which the continuous phase is lipophilic (such as oil, wax, fat). They usually have HLB values <6.0 and preferably below 5. A few examples are lanolin alcohols, ethylene glycol monostearate (non-ionic), and sorbitan monooleate, with HLB values of *c.* 1.0, 2.0 and 4.3, respectively. Lanolin, or a lanolin substitute, mineral oil and white petrolatum are also used as ointment vehicles (Osborne, 1993). These mixtures were investigated by determining their binary and ternary phase behavior. Variability of the anhydrous lanolin from lot to lot caused significant shifts in the binary and ternary phase behavior, which determined the viscosity of the ointment base (Osborne, 1993).

12.4 Cosmetic and vanishing creams

Cosmetic creams make use of materials that can be prepared in w/o or o/w emulsions. Acceptance of a particular cosmetic cream depends to a great extent on the emulsion type as well as the emulsion composition (Barnett, 1972). A cream-cooling effect can be achieved by the more rapid water evaporation in o/w versus w/o emulsions.

A typical Unguentum Aquae Rosae cold cream contains 45.0% rose water, 40.2% sweet almond oil, 11.8% spermaceti and 3.0% beeswax. About 100 years ago, this basic formula was changed to include more beeswax and sweet almond oil, 0.5% borax was added and a reduction in rose water content to 19.4% was suggested. Now the almond oil has been replaced with mineral oil. Creams with considerable stability at acidic and alkaline pH can be achieved by non-ionic emulsifiers, whereas anionic emulsifiers (such as triethanolamine) are used to prepare slightly alkaline creams. Cationic emulsifiers are used to prepare emulsion systems with increased deposition on the negatively charged skin and hair among other things. A free fatty acid (e.g. stearic acid emulsion) in a non-alkaline medium is considered a vanishing cream.

PEGs are synthetic water-soluble polymers that have a special combination of properties including water solubility, lubricity, blandness, low toxicity, stability, solvent action and non-volatility (Powell, 1982b). They are, therefore, often found in pharmaceutical and cosmetic applications. These cosmetics include cleansing creams, toothpastes, hair dressings, stick deodorants and the like.

Many gums can be used in cream preparations. A partial list includes alkyl and hydroxyalkylcellulose, carrageenan, guar gum, gum arabic, tragacanth, hydroxyethylcellulose (HEC), HPC, PEG, PEO and PVA. (A comprehensive review of the cosmetic, pharmaceutical and technological applications of the most widely used tree gum exudates – gum arabic, tragacanth, karaya gum and ghatti gum – can be found in Wang and Anderson (1994).) Alkyl and hydroxyalkylcelluloses are the preferred agents in cream preparations because of their ability to control viscosity, emulsification and stabilization, their lubricity and feel, and surfactant compatibility in lotions, hand and face creams and other cosmetic applications such as hair dressings, deodorants, depilatory creams, shampoos and toothpastes. For creams and lotions, MHPC with a methoxyl content of 28–30% and 7–12% hydroxypropoxyl is utilized, respectively. Lambda-carrageenan can be used in hand lotions and creams to improve rub-out and provide slip, at concentrations of up to 1.0%. Agar can also be used in some cosmetic creams and lotions. Gum arabic stabilizes creams and lotions by increasing viscosity, which also contributes to their spreading properties as well as giving a smooth feel to the skin. Tragacanth is unique in its viscosity-creating ability, its stability at low pH and its unique emulsifying ability, which thickens aqueous systems and lowers interfacial tensions in o/w emulsions. Tragacanth is, therefore, used in cosmetic creams and lotions as well as in facial clays, toothpastes and hair-dressing preparations. PEG with a molecular weight of 4000 can be used as a major ingredient (45% by weight) in cosmetic creams, other ingredients being propylene glycol, zinc stearate and water, 10, 22 and 23% by weight, respectively. The cream is prepared by melting the PEG, adding the propylene glycol and then the zinc stearate. The water is added slowly with stirring to form a smooth cream (Powell, 1982b).

12.5 Manufacture

For manufacturing purposes, all oil-soluble components are heated (in a steam-jacketed kettle) to ~75°C. Water-soluble components are dissolved separately in the aqueous phase to the same temperature. Water (5%) is added to compensate for evaporation. The inner phase is then added to the outer phase followed by stirring and homogenization, the preferred procedure for creams and lotions. Stirring times and cooling rates are monitored to control lotion viscosity, cream consistency and emulsion stability. Also important are the type of equipment used, the cooling method and the temperature at which the perfume oils are added to the cream or lotion.

12.6 Cosmetic lotions

Lower concentrations of the oils and waxes found in emollient creams are included in cosmetic lotions. During the day, lotions with a lighter character (providing a less oily emollient film) are preferred, but the formulation can be designed to contain oil at the level regularly used in creams. Lotions should remain stable for at least 1 year, even at temperatures of 45–50°C. The lotion should also remain stable during a freeze–thaw test, i.e. it should be able to return to room temperature after being held at −5°C for 24 h without separating and without losing its pourability. All-purpose hand, face and body lotions contain deionized water, stearic acid, lanolin, mineral oil, alcohol, triethanolamine, preservatives and perfume, arranged in decreasing amounts in terms of their respective weight percentages in the formulation.

Alkylcellulose and hydroxyalkylalkylcellulose are used in lotions as well as in hand and face creams to provide viscosity control, emulsification and stability. Carrageenans are used to produce cosmetics, most requiring concentrations of about 1%. Carrageenans are preferred because of their stability at natural to alkaline pH. Carrageenan is a cost-effective agent and because it is a gelling agent as well, it can be utilized in hand lotions as an emulsion stabilizer. Carrageenans are also used in toothpaste as a binder and in lotions and creams as an additional bodying agent, and as a provider of slip and emollience. Other hydrocolloids used regularly in lotions are gum arabic, gum tragacanth, HEC, HPC, PEG and PEO.

It should be noted that suspensions are argued to act in multiple ways on the skin. Irrespective of the type of application, skin is nearly totally refractive to the penetration of (ordered) gel-phase vesicles (Cevc *et al.*, 1996). This is not the case for some lipid-vesicle formulations with fluid membranes (liposomes), which have already been shown to bring more drugs into the skin than the conventional hydrogels or ointments. Attempts to employ similar liposomes for systemic drug delivery across the skin, however, have been nearly always unsuccessful. Only the most modern self-optimizing aggregates with ultraflexible membranes (transfersomes) are able to deliver drugs reproducibly either into or through the skin, depending on the choice of administration or application, with a very high efficacy (Cevc *et al.*, 1996).

12.7 Antiperspirants and deodorants

Deodorants and antiperspirants are marketed via different commercial channels as lotions, gels, creams, powders, aerosols, soaps and sticks. These products include non-toxic colors and fragrance. FDA requirements state that such products must be at least 20% effective on at least one half of the persons tested, with a confidence rate of 95%. In order to claim extra effectiveness (>30%), special tests need to be carried out. Moreover, the capacity for skin irritation and compatibility with normal skin flora need to be retested each time new ingredients are added to the formulation (Mueller and Quatrale, 1975).

An example of an active ingredient used in deodorants and antiperspirants is aluminum chloride. This low-pH ingredient can damage fabrics and irritate the skin. Therefore, basic aluminum compounds have been developed. Examples are aluminum chlorohydroxide, basic aluminum bromide, iodide, nitrate and basic aluminum hydroxychloride–zirconyl hydroxyoxychloride with and without glycine. Zirconium salts have been banned for use in aerosol antiperspirants since the 1970s.

Deodorant-antiperspirant sticks are composed of 7–9% gelling agent (sodium stearate). Stearate derivatives are more often used in antiperspirant stick formulations. A few different stick formulations for antiperspirants exist: their stearamide wax content is 26–29%, and the active concentration 20–25%. Newer antiperspirant formulas include 40–50% volatile silicones to reduce cracking and crumbling of the sticks. If volatile silicones are used (e.g. tetrameric and pentameric cyclic silicones), then at application the silicones evaporate and the ingredients left on the skin produce a thin non-sticky layer. Deodorant stick formulations comprise alcohol, propylene glycol, sodium stearate, perfume and trichlorocarbanilide, in decreasing order of percentage by weight.

One of the major characteristics of silicone emulsions imparted by fluid silicones is a coating with good release, which prevents the adhesion of most types of tacky material. Polysiloxanes modified with diisocyanates dry to a tacky paste-like mass that is useful in cosmetics, as well as hydraulic fluids, cleaning compositions and dispersants.

Roll-on antiperspirant composed of water, HEC, pure glyceryl monostearate, emulsifier and aluminum chlorohydroxide can be prepared by adding the water to a mixing tank, agitating the HEC while the tank contents are being heated, and after its slow dissolution, the monostearate and emulsifer are added. Heating is to 80°C with constant stirring, followed by cooling to 35°C with agitation. Then the aluminum chlorohydroxide is added and the mix stirred well. HEC is chosen for this preparation, as well as for many specialized preparations in the cosmetics field, because it is a non-ionic polymer unaffected by dissolved salts or modifications. Also advantageous is its high thickening efficiency, complete solubility and tolerance for water-miscible organic solvents (Powell, 1982a). HPC is used in alcohol-based cosmetics because of its solubility in alcohol, as a thickener and film-former. In antiperspirants, HPC is used as a non-ionic polymer which serves as a thickener, stabilizer and binder (Butler and Klug, 1982).

Cream and lotion antiperspirants are generally composed of non-ionic systems, which are less irritating and more stable. Typical formulations include deionized water and an aluminum chlorohydroxide complex composed of magnesium aluminum silicate, glyceral monostearate, imidazolidinyl urea and perfume.

12.8 Sunscreens

Sunscreens are designed to protect the skin from excessive exposure to the sun's energy. Prolonged exposure produces most of the aging effects seen on the skin and can be a factor in basal skin cancer, squamous cell cancer and melanoma. Sunscreens contain physical screening agents such as titanium dioxide and zinc oxide. These are opaque materials that block and scatter light, physically blocking the radiation at all wavelengths. Chemical screening agents absorb UV light (i.e. they afford selective protection against certain wavelengths). These materials include anthranilates, cinnamates, benzyl and homomethyl salicylate, and *p*-aminobenzoic acid (PABA) and its ester derivatives. Sunscreens should be able to resist washing-off by swimming or sweating. The chemicals seem to be basically substantive although they can diffuse into the horny skin layer (Klarman, 1957).

12.9 Aerosols

Aerosol sprays are used as a packaging system for a variety of different cosmetic products such as shaving creams, hair sprays, deodorants, antiperspirants, colognes and sunscreens. The products are defined as aerosols if liquid, solid, gas or a mixture of these is discharged from a disposable container through a valve via the propellant force of liquified or non-liquified compressed gas. Compressed gases such as nitrogen and carbon dioxide, and liquified hydrocarbons are the primary propellants, rather than Freons (which are still used in pharmaceuticals such as asthma spray) (Chemical Specialities Manufacturers' Association, 1955, 1966; Sanders, 1970; Johansen *et al.*, 1972). HPC, which is sold as a powder, is available in a wide range of viscosities and has FDA clearance for use as a direct additive to foods. It is used in cosmetic emulsions packaged as aerosol products and serves as an emulsion stabilizer and foaming aid (Butler and Klug, 1982).

12.10 Make-up

Make-up preparations include colors, face powders, lipsticks, cream mascaras, eye shadows and nail products, to name a few. The principal primary colorants are red, yellow and blue. Via the proper blending of these colors, every hue can be produced. Colorants are certified for different uses and to indicate these initials are attached to the name of the color. Colorants designated FD&C may be used for foods, drugs and cosmetics. Colorants used in cosmetics must not contain more than 0.002% lead, 0.0002% arsenic and 0.003% of other heavy metals. Most cosmetic colorants are certified coal-tar colorants (Peacock, 1944). A few natural colors are also used, including alkanet, annatto, carotene, chlorophyll, cochineal, saffron and henna. Inorganic colorants include white pigments such as titanium dioxide, zinc oxide and talcum. Naturally colored minerals (including iron oxide) are ocher, umber and sienna. Many dyes are pH sensitive, a fact that needs to be taken into consideration. The hue varies with the percentage of moisture present. The color effects are themselves affected by the nature of the light source, degree of transparency and the opacity of the color (Peacock, 1948).

Gum arabic (acacia gum) is unique in its ability to form solutions of greater than 50% concentration. In cosmetics, gum arabic's main advantage is its non-toxicity and the lack of dermatological and allergic reactions associated with its use (Greenberg and Lester, 1954). It is also used in facial masks because of its adhesive ability, and its binding ability is useful in formulations of rouges and compact cakes. This hydrocolloid can also be used as a foam stabilizer in liquid soaps. In cream rouge preparations, PEG 4000 can be included at a level of 4% of the formulation. Except water

(~55%), stearic acid, cetyl alcohol, propylene glycol, isopropyl palmitate, potassium hydroxide, color lakes, preservative and perfume are used in decreasing order of weight percentage (Powell, 1982b). PEO can be used for its lubricant–emollient effect in personal care products such as creams and lotions and in other cosmetic applications, such as hand soaps, shaving preparations and toothpastes. Cold creams, cleansing creams, shaving creams and other cosmetic applications make use of the emulsifying, binding, film-forming and thickening properties of polyvinyl alcohol.

12.10.1 Face powders: formulations, manufacture and stability

Face powders, both loose and pressed (cakes), are generally composed of white pigments that have been tinted and perfumed. This powder is designed to impart a matte, smooth finish to the skin by masking skin secretions. These aims can be achieved with a few constituents which make up agents with covering power (of enlarged pores and shine), slip (to assist in spreading), absorbency (to absorb oily secretions), adherence (to improve clinging to the face) and bloom (to give a smooth appearance to the skin).

12.10.2 Lipstick

Lipstick is a solid fatty-acid product containing dissolved and suspended colorant materials. A good lipstick must have some thixotropic properties. It must become softer and create a thin film upon the application of minimal pressure.

12.10.3 Mascara, eye shadows, nail products and lacquer

Mascara is used to emphasize the lashes by making them appear longer. Because of its proximity to the eye, the ingredients comprise natural dyes and inorganic and carbon pigments (iron oxides, carbon black and ultramarine blue). In cream mascara, water is incorporated to provide a product which is ready to use without previous trituration. In eye shadows, up to 25% carbon black, ultramarine blue and iron oxide pigments (yellow, brown and red) are used. Nail products also contain pigment, in addition to resins, plasticizers and solvents.

12.11 Hair preparations

Hair preparations include shampoos (with soap and soapless, low pH, amphoteric and antidandruff preparations). Additives include foam builders (fatty acid alkanolamides), conditioning agents (amine oxides, fatty alcohols, lanolin derivatives, esters of fatty acids, silicones, cationic materials and polyacrylamides), opacity agents (higher fatty alcohols), sequestering agents (citric and tartaric acids, and EDTA salts), viscosity builders (cellulose derivatives, alkanolamides and carboxyvinyl polymers) and preservatives (hydroxybenzoates, formaldehyde, imidazolidinyl urea compounds and sorbates, among others). To evaluate the exposure of the general population to formaldehyde from the use of cosmetic products, as well as to monitor whether cosmetic products comply with national regulations, 285 shampoos, creams and the like were analysed for formaldehyde content (Rastogi, 1992). It was shown that 29.5% of the products investigated contained 0.001–0.147% total formaldehyde. In 10 products (3.5%), total formaldehyde content was $>0.05\%$. Eight of these products contained $>0.05\%$ free formaldehyde.

Hair straighteners based on thioglycolate are viscous products used to keep hair straight while it is being softened. They are composed of water, glyceryl monostearate, ammonium thioglycolate, ammonium hydroxide, Brij 35, mineral oil, perfume and preservatives.

Depilatories are used for the removal of unwanted hair by chemical agents. They can be based on calcium thioglycolate in a strongly basic medium (pH = 12.3). They also include small amounts of a wetting agent, alkali and alkali-stable perfume. This method is newer than the previous hair removal by wax (a mixture of rosin and beeswax), which is melted, applied to the area to be treated and allowed to solidify, after which it is peeled and the hair pulled out with the wax.

In the 1950s a long list of preparations, especially those that dealt with controlling women's hair and were based on shellac or oils, were replaced with PVP-based sprays. This substitution was effected because of the substantivity to hair, the formation of rewettable, transparent films and PVP's contribution to luster and smoothness. These qualities have led to PVP's use in other hair-grooming products as well.

12.11.1 Wave set

Karaya gum is used in typical formulations of wave-set alcohol concentrates. These formulations include a superfine grade of karaya gum, borax and alcohol, 19.5, 6.5 and 74% by weight, respectively. Preparation includes dry mixing of the gum and the borax, then adding it to the stirred alcohol, stirring the mixture for at least 0.5 h, then separation. Before use, the concentrate needs to be mixed well again; 170 g of this concentrate makes up $\sim$15 l of the finished wave-set product. A typical wave-set formula includes $\sim$1.0% karaya gum, colorant, distilled water, lanolin, borax, formalin and alcohol. The gum is added to the alcohol to prepare a slurry that is later mixed with the aqueous solution, which includes all the other ingredients. After the addition of the gum slurry, stirring takes at least 0.5 h (Meer, 1982).

12.12 Bath products

Bath products include bath oils and soaps. Bathing removes dirt, perspiration and microorganisms from the surface of the skin. Bath oils can be spread, or floated and dispersed in the bath water. Spreading is facilitated by including 5% of an oil-soluble surfactant, which lowers the surface tension of the bath water and enables the formation of a continuous oil film, rather than individual droplets. HEC's special properties make it useful in a range of bubble baths. Its advantages are its non-ionic nature and its tolerance for water-miscible solvents, in addition to its highly efficient thickening and stabilizing action (Powell, 1982a). Differences in the responses of distinct layers of the skin to surfactants were probed. Soap and other surfactant-containing cleansers were applied to the skin for 2 consecutive days (Simion *et al.*, 1991). Transepidermal water loss showed that the stratum corneum is readily damaged, even by a mild insult, when no erythema is induced. A more severe treatment, such as a 24 h exposure to a 5% soap solution, induced the maximal level of barrier damage but a submaximal level of erythema. Even 2 days of exposure to 5% soap did not elicit maximal erythema response. These results suggest that the stratum corneum is more readily damaged than the dermis, which is not unexpected because the stratum corneum is the initial point of contact between surfactants and skin (Simion *et al.*, 1991).

12.13 Shaving preparations

Shaving preparations include shaving soap, shaving cream, which produces a lather, brushless shaving cream, aerosol shaving cream, self-heating shaving cream and gel shaving cream. These products remove the complex mixture of lipids that are secreted by the sebaceous glands and absorbed by the beard, thus making shaving easier; they also reduce lubrication between the blade and the skin.

12.14 Toothpastes

Toothpastes are suspensions of polishing agents (calcium carbonate, magnesium carbonate, di- or tricalcium phosphate, talc), detergents (soap, anionic surface-active agents), suitable binders (excipients), humectants (glycerol, propylene glycol, sorbitol, mucilages or gums), sweeteners (saccharin, sorbitol), preservatives, flavor and water. Cellulose derivatives (e.g. MC) are responsible for the desired texture, and stannous fluoride or sodium monofluorophosphate may be added as anticavity agents. Since the early 1970s, there has been a decrease in the prevalence of caries in US schoolchildren, a change in the intraoral caries pattern and a slowing of the progress of lesions. The relative reduction in caries is attributed to fluoride mouth rinses and self-applied gels. Although inadvertent ingestion of fluoride can result from the use of mouth rinses and gels, there is little evidence to suggest that it has contributed to the current rise in fluorosis (Ripa, 1991). The greatest effect on the reduction of plaque and gingivitis can be expected from chlorohexidine, essential oils and triclosan-containing products. These chemical agents vary in form and include mouth rinses, gels and dentifrices (Ciancio, 1992).

12.15 Product shelf life

The shelf life of cosmetic products is extended by the addition of preservatives and antioxidants (Brannan, 1995). The preservatives should be active at low concentrations, be compatible with the product, effective against microorganisms, soluble and stable over a wide range of temperatures, lack odor, color and toxicity and not cause irritations at the levels used.

References

Barnett, G. (1972) *Emollient Creams and Lotions*, vol. 1, Wiley Interscience, New York, pp. 27–104.

Brannan, D.K. (1995) Cosmetic preservation. *J. Soc. Cosmetic Chemists*, **46**(4), 199–220.

Butler, R.W. and Klug, E.D. (1982) Hydroxypropylcellulose, in *Handbook of Water-soluble Gums and Resins*, ch. 13 (ed. R.L. Davidson), McGraw-Hill, New York, pp. 13.00–13.12.

Cevc, G., Blume, G., Schatzlein, A. *et al.* (1996) The skin – a pathway for systemic treatment with patches and lipid-based agent carriers. *Adv. Drug Delivery Rev.*, **18**(3), 349–78.

Chemical Specialities Manufacturers' Association (1955) *Glossary of Terms Used in the Aerosol Industry*, Aug. 25, New York.

Chemical Specialities Manufacturers' Association (1966) *Bulletin No. 211-66*, Aug. 25, New York.

Ciancio, S.G. (1992) Agents for the management of plaque and gingivitis. *J. Dental Res.*, **71**(7), 1450–4.

CTFA (1977) *Cosmetic Ingredient Dictionary*, 2nd edn, CTF Association, Washington DC.

de Navarre, M.G. (1957) *International Encyclopedia of Cosmetic Material Trade Names*, Moore Publishing, New York.

FDA (1976) *Federal Food, Drug and Cosmetic Act*, as amended Oct., Food and Drug Administration, Washington DC.

Greenburg, L.A. and Lester, D. (1954) *Handbook of Cosmetic Materials*, Interscience Publishers, New York, p. 19.

Johansen, M.A., Dorland, W.E. and Dorland, E.K. (1972) *The Aerosol Handbook*, 1st edn, Wayne E. Dorland, New York.

Klarman, E.G. (1957) *Suntan Preparations*, Interscience Publishers, New York, pp. 189–212.

Meer, W. (1982) in *Handbook of Water-soluble Gums and Resins*, ch. 10 (ed. R.L. Davidson), McGraw-Hill, New York, pp. 10.12–10.13.

Mueller, W.H. and Quatrale, R.P. (1975) Antiperspirants and deodorants, in *Chemistry and Manufacture of Cosmetics*, 2nd edn., vol. 3 (ed. M.G. de Navarre), Continental Press, Orlando, FL, pp. 205–28.

Osborne, D.W. (1993) Phase-behavior characterization of ointments containing lanolin or a lanolin substitute. *Drug Devel. Ind. Pharmacy*, **19**(11), 1283–302.

Peacock, W.H. (1944) The application of the certified coal-tar colors, *Calco Technical Bulletin, No. 715*, Calco Chemical Div., American Cynamid Co., Wayne, NJ, Dec. 1944.

Peacock, W.H. (1948) The practical art of color matching, *Calco Technical Bulletin, No. 573*, Calco Chemical Div., American Cynamid Co., Wayne, NJ, Dec. 1948.

Powell, G.M. (1982a) Hydroxyethylcellulose, in *Handbook of Water-soluble Gums and Resins*, ch. 12 (ed. R.L. Davidson), McGraw-Hill, New York, pp. 12.16–12.21.

Powell, G.M. (1982b) Polyethylene glycol, in *Handbook of Water-soluble Gums and Resins*, ch. 18 (ed. R.L. Davidson), McGraw-Hill, New York, p. 18.27.

Rastogi, S.C. (1992) A survey of formaldehyde in shampoos and skin creams on the Danish market. *Contact Dermatitis*, **27**(4), 235–40.

Ripa, L.W. (1991) A critique of topical fluoride methods (dentrifrices, mouth rinses, operator-applied, and self-applied gels) in an era of decreased caries and increased fluorosis prevalence. *J. Public Health Dentistry*, **51**(1), 23–41.

Sanders, P.A. (1970) *Principles of Aerosol Technology*, 1st edn, Van Nostrand Reinhold, Wayne E. Dorland, New York.

Simion, F.A., Rhein, L.D., Grove, G.L. *et al.* (1991) Sequential order of skin-response to surfactants during a soap chamber test. *Contact Dermatitis*, **25**(4), 242–9.

Wall, F.E. (1957) *Origin and Development of Cosmetics Science and Technology*, ch. 2, Interscience Publishers, New York.

Wang, W.P. and Anderson, D.M.W. (1994) Non-food applications of tree gum exudates. *Chem. Ind. Forest Prod.*, **14**(3), 67–76.

13 Explosives

13.1 Introduction

Hydrocolloids have played a large role in the development of the explosives industry. Taking a simplified view, the detonation process can be considered as a very rapid 'redox' reaction between oxidizers (which release oxygen) and fuels to produce large quantities of gases accompanied by the liberation of heat. The dramatic increase in volume and heat enables explosives to perform such tasks as crushing rocks and moving mountains (Sudweeks, 1985).

For many years, ammonium nitrate was used as an oxidizer to increase the explosive strength of dynamite, but it was not until after a series of accidents in which ammonium nitrate fires broke out that it was realized that ammonium nitrate itself could be the basis of cheap and effective industrial explosives. Fertilizer-grade ammonium nitrate in the form of porous round particles was found to absorb fuel oil, thereby forming a detonatable combination (Mark *et al.*, 1980).

Because of the low cost and ease of manufacture of ammonium nitrate–fuel oil (ANFO), these mixtures subsequently replaced dynamite in many applications and also expanded the explosives market, making previously uneconomical operations feasible. The maximum theoretical energy output for explosives occurs at zero oxygen balance (i.e. stoichiometrically enough oxygen to react with all the fuel). The optimum weight ratio for ANFO is 94.5:5.5 ammonium nitrate:fuel oil.

13.2 Historical perspective

The origin of explosives is lost in antiquity (Sudweeks, 1985) but is credited to the discovery of black powder, a mixture of charcoal, sulfur and potassium or sodium nitrate. Under burning (deflagration) and confinement, sufficient pressure builds up, reaching the considerable force needed for gun powder, blasting agents and pyrotechnic devices.

Once nitroglycerin had been synthesized, the invention of dynamite became possible, and this became the foundation for a worldwide explosives industry. Dynamite dominated the commercial explosives market for about 100 years after Alfred Nobel succeeded in marketing blasting caps as a safe and effective means of initiating dynamite. During this period, 'military-type' explosives such as trinitrotoluene (TNT), pentaerythritol tetranitrate (PETN) and cyclonite (RDX) were, and still are, used in boosters, blasting caps and detonation cords. Except for black powder, all of the explosives mentioned here are molecular in nature, since their basic ingredients consist of a single molecule (Sudweeks, 1985).

Water-gel explosives are a fairly recent invention, first introduced to the market in the latter half of the 1950s. The idea was to use a water gel to transform a high-velocity solid explosive to a plastic form. The first such explosive was called Securit: it had the properties of a high-density, cap-sensitive, oxygen-balanced explosive, as well as unique safety characteristics. At about the same time, a somewhat similar idea was being developed in the USA. This was called 'slurry explosives', developed in around 1958 (Cook, 1974). In the big open pit mines in the USA and Canada, drilling of large-diameter boreholes was well underway. The use of boosters was known from the ANFO technique. Under these circumstances and with the availability of cheap surplus TNT, a non-cap-sensitive explosive was developed with TNT as the sensitizer in a porridge-like mix thickened with hydrophilic colloids and supplemented with oxygen-delivering salts. Since the investment required to develop such explosives was small, and the explosives themselves were not expensive, they were rapidly introduced in big-hole mining. Water gels could also be sensitized in the same way that nitroglycerins are sensitized by air bubbles. Using the appropriate size and number of air bubbles in combination with very fine, flaky aluminum, or materials such as methyl amine nitrate, the water-gel-based explosives became cap sensitive. Explosions of this type are of low density and are remarkably different from the high-density Securit (Cook, 1974).

In the early 1970s, the main explosive used in criminal and terrorist activities in the USA was dynamite. Now dynamite is less frequently used in criminal bombing, having been replaced by slurry or emulsion explosives (Midkiff and Washington, 1974; Midkiff and Walters, 1993). This trend is not only present in North America but worldwide where, as previously mentioned, traditional explosives for blasting, mining and construction have been replaced with successors that have water as a significant component of their composition. Beginning in 1978, a fee-based crime laboratory proficiency testing program was launched in the USA. It now involves almost 400 laboratories worldwide. These laboratories are moderately successful in the identification and classification of common evidence types: flammables, explosives, and fibers (Peterson and Markham, 1995). Details of a comprehensive forensic analysis of debris from the fatal explosion of a 'cold-fusion' electrochemical cell can be found elsewhere (Grant *et al.*, 1995).

Slurry/gel or emulsion explosives (and the more modern two-phase water/oil explosives) are cap sensitive and were first used for blasting in Canada (Cook, 1958) and in Sweden (Wetterholm, 1981). The latter was sensitized with *N*,*N*-bis(trinitroethyl)urea and was capable of initiation by a no. 8 detonator. In the beginning, North American slurry explosives were sensitized with explosives such as TNT, smokeless powder or flake aluminum (Robinson, 1969).

Slurry sensitizers were used by the producers of explosives in the 1960s, but most were unsatisfactory because their ingredients tended to segregate at elevated temperatures or the oxidizer tended to salt out at lower temperatures. These problems were solved in the 1970s. In the mid 1960s, new sensitizers were introduced to the field of water explosives. They included nitrate salts of hexamine, ethanolamine or monomethylamine. In addition to chemical sensitizers, slurry-emulsion-type explosives are also sensitized by voids in the form of tiny bubbles. Since gas is not the ideal sensitizer (at low temperature or under hydrostatic pressure), microspheres of hollow glass, resin or ceramic spheres are added to control density, to detonate wave propagation and as a primary or secondary sensitizer. Partial information on the composition of sensitized slurries (products and producers) can be found elsewhere (Kaye, 1980; Midkiff and Walters, 1993).

13.2.1 Medicinal and biological aspects

From a medicinal point of view, many reports have been published discussing the effects of explosives and explosive-related health issues in our present and future lives. Blast injuries occur more often than previously, because of the wide use of explosives (Shuker, 1995). This is especially the case in wartime. More and more people lose their lives every day to blast injuries. The mechanism of the injury, pathophysiology of the trauma and clinical effects, as well as management of these, are presented by Shuker (1995). A limited list of biological–medicinal issues that have been described include the physical characteristics of gunfire impulse noise and its attenuation by hearing protectors (Ylikoski *et al.*, 1995); urogenital wounds during the war in Croatia in 1991–2 (Tucak *et al.*, 1995); a study of the toxic effects of organic explosives, propellants and related compounds (Nikolic *et al.*, 1994); microbial degradation of explosives and related compounds (Spiker *et al.*, 1992; Gorontzy *et al.*, 1994); the results of the simultaneous explosion of the Zaria and Plamia grenades (Putintsev and Kitaev, 1994); allergic contact dermatitis caused by nitroglycerin (Kanerva *et al.*, 1991); and blast injuries in humans and animals (Bruins and Cawood, 1991; Savic *et al.*, 1991; Tatic *et al.*, 1991).

13.3 More detailed anatomy of a slurry

Despite the popularity of the ANFOs, they do have certain drawbacks: ammonium nitrate is a highly water-soluble and hygroscopic salt. Therefore, ANFOs are not suitable for use in wet ground, since the ammonium nitrate

Table 13.1 Ingredients for slurry explosives

Ingredient	Function	Ingredient	Function
Ammonium nitrate	Oxidizer	Guar gum	Gelling agent
Sodium nitrate	Oxidizer	Starch	Gelling agent
Calcium nitrate	Oxidizer/density	$KCrO_4$	Cross-linking
Sodium perchlorate	Oxidizer/sensitizer	Sb compound	Cross-linking
Microspheres	Density/sensitizer	Bi compound	Cross-linking
Amine nitrate	Sensitizer	Boric acid	Cross-linking
Glycol nitrate	Sensitizer	Borax	Cross-linking
Flake TNT	Sensitizer	Aluminum	Fuel/sensitizer
Smokeless powder	Sensitizer		

tends to be leached from the salt explosive before the shot can be fired and the cartridges become incapable of initiation or detonation. Moreover, ammonium nitrate–gelatin explosives stiffen as a result of moisture absorption from the atmosphere. Therefore, wet conditions call for the use of more expensive explosives.

ANFOs have certain limitations, such as no water resistance and low density, restricting energy options. To overcome these problems, three approaches have been proposed, namely predissolving the ammonium in a small amount of water, thickening the solution with hydrocolloids (i.e. guar gum, starch) and, if possible, cross-linking the gum thickener to produce a gelled product (Sudweeks, 1985). Soluble or insoluble fuel components are added to this system. Dry oxidizers can also be suspended, and their presence, as well as that of ammonium nitrate crystals, has led to the creation of composite slurries. Slurry ingredients (Table 13.1) contain oxidizers, thickeners and soluble and insoluble fuels. The first slurries needed to be sensitized with additives such as TNT. Later improvements included sensitization with organic salts such as nitrates and perchlorates. In today's more advanced formulations, even small air or gas bubbles can be used as sensitizers. In general, slurries are much less sensitive to impact, friction or accidental detonation.

Slurry formulations are presented in Table 13.2. As with ANFO, the ratio of oxidizers and fuels is adjusted to give the desired oxygen balance. The energy output of the formulation can be increased by adding sulfur and aluminum. Cross-linking agents are included in the formulation for thickening and controlling product density. Density is also determined by the presence of the air bubbles or 'void spaces' required for efficient detonation. Sensitization by air bubbles (Fig. 13.1) occurs via the formation of hot spots. A wave of very high-pressure shock is introduced by the initiator blasting cap or booster into the slurry charge and compresses the air bubbles. The rapid rate causes adiabatic volume changes and heating to very high temperatures, and explosive decomposition is obtained in the surrounding

Table 13.2 Representative slurry formulations

	Slurry 1	Slurry 2
Ammonium nitrate	53.1	57.4
Sodium nitrate		14.4
Water	15.8	14.4
Thickeners	0.4	0.5
Ethylene glycol		0.4
Fuel oil	5.0	
Sulfur		2.0
Gilsonite (asphalt)		3.6
Aluminum		7.0
Dry prills	24.8	
Trace ingredients	0.9	0.3

material. The energy release contributes to the propagation of the shock wave through the entire column or charge of the explosive.

Data on typical slurry series properties can be found in Hansen (1983). Average borehole densities are 1.1 to 1.21 g cm^{-3}, energies 680–1145 cal g^{-1} (calculated by assuming the formation of the most probable highest energy products), weight strength (ANFO = 1.00 at density 0.82 g cm^{-3}) ranges between 0.83 and 1.34, bulk strengths are 1.11 to 1.98 (theoretical energy determined by comparison with ANFO), gas volume 45 and a reduction to 37 mol kg^{-1}, respectively, and velocity 4000 m s^{-1} for all mixes at

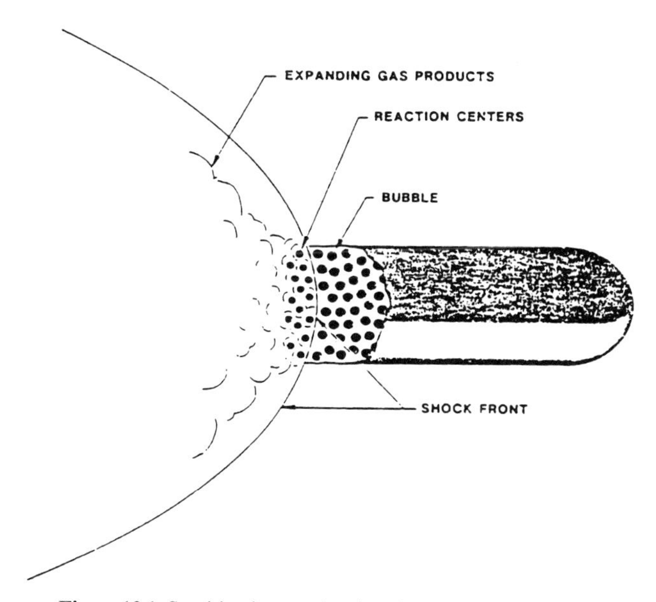

Figure 13.1 Sensitization mechanism. (From Sudweeks, 1985.)

5°C. Possible detonation products include calcium sulphate and carbonate, sodium carbonate, chromic oxide, carbon dioxide, carbon, water and nitrogen, of which water and nitrogen are produced at the highest mol kg^{-1} quantities: 25.44 and 7.55, respectively.

Recent developments include the commercialization of water-in-oil emulsion explosives and blasting agents. Here, a continuous oil phase constitutes the main component into which an aqueous oxidizer solution is dispersed as small droplets. One big advantage of such slurries is the water resistance of the product, which is higher than that of the cross-linked slurries. Emulsion explosives can be successfully used, even after being kept underwater for months. The inherent viscosity of such emulsion systems is 10000–20000 cP up to 10^6 cP. The main sensitization is provided by the direct contact between fuel and oxidizer. Additional sensitization is achieved by the presence of small air bubbles. Such products include various types of oil and/or wax. Emulsion detonation velocities tend to be in the range of 5000 to 6000 m s^{-1} with detonation pressures from 100 to 120 kbar. As a result of such advances, the sale of explosives in the USA increased fourfold between 1960 and 1980 (Mineral Industry Surveys, 1961–81).

13.4 Definitions of slurries and water gels

Slurries and water gels are explosives consisting of aqueous solutions of ammonium nitrate and sodium or calcium nitrate, gelled by the addition of guar gum or cross-linking agents. They are sensitized by nitro-explosives or organic amine nitrates. Combustible materials such as aluminum, urea, sugar or glycol are mixed with these solutions. Typical slurry explosives may contain 30–70% ammonium nitrate, 10–15% sodium nitrate, 15–20% calcium nitrate, up to 40% aliphatic amine nitrates, 15–25% aluminum, 5–15% TNT or other explosive sensitizer, 1–2% gellants, 0.1–2.0% stabilizers, 3–15% ethylene glycol and 10–20% water. Several commercial examples of the many such explosives are Sigmagel, Titagel, Gelsurite 2000, Tutagex 110, Explogel, Tovex, AXD 720, Slurmex 200 and Iremite (Gerhartz, 1987).

13.5 Role of hydrocolloids

More modern explosives resembling a 'gelatin dessert' are referred to as either water gels or slurries (Midkiff and Walters, 1993). The slurries are basically composed of one or more solid components suspended in a semi-solid continuous gel phase. The presence of hydrocolloids in these mixtures also prevents settling of the solid ingredients and this preserves the integrity of the explosive. The nature of the mixture, in which solids are dispersed in solution, has led to the designation of 'slurries' for these explosive formulations.

A certain degree of water resistance was found to be obtainable by predissolving ammonium nitrate in a small amount of water, thickening the solution with a hydrocolloid and, optionally, cross-linking. Fuel components could be added as soluble or finely divided insoluble materials. Thus, when the explosive mixture is immersed in water, the thickened solution or gel prevents water penetration into the mixture and the consequent leaching of salts. Gels are simply impermeable to water and in thickened solutions the gum absorbs the water.

Typical ingredients for slurry explosives have already been mentioned in this chapter and are given in Table 13.1 and in the literature (e.g. Kaye, 1980; Sudweeks, 1985; Midkiff and Walters, 1993). Generally, they are composed of an inorganic oxidizer (such as ammonium nitrate), sensitizers, density-formers, gelling agents such as the natural polysaccharide guar gum, cross-linking agents and fuel. (Another related water-containing explosive, an emulsion composed mainly of oil and water phases, is becoming increasingly popular as a tool for criminal and terrorist activities.) Water usually constitutes 10–20% of the mixture and hydrocolloids 1–2%. All other ingredients are present in varying amounts. Ingredients can be selected to vary the energy, sensitivity, oxygen balance, rheology and stability of the final product. This versatility makes it possible to adapt the fume and detonation characteristics as well as the physical properties of explosive slurries to meet the specific requirements of various applications. It is worth noting that these results can be achieved with ingredients that are themselves considerably non-explosive. Moreover, they are commodity items with relatively stable price parameters and are widely available. Thus slurries represent a very significant advance in the economy of commercial explosives manufacturing, shipping and application. Although they perform the same function as dynamite and molecular explosives, slurries are much less sensitive to impact, friction or accidental detonation. Since they can be made from non-explosive ingredients, a unique site-mixed system has been devised for bulk delivery that avoids transporting any explosive product. Upon arrival at the site, the ingredients are mixed in the proper proportions and pumped directly into the hole where the final gassing and gelling reactions occur. This way the mixture does not become detonatable until after it is loaded into the hole. Since they were first marketed, the commercial use of slurries has undergone continuous and rapid growth.

13.6 The use of hydrocolloids in slurry explosives

Any industrialist planning the use of hydrocolloids in products needs to consider a few important factors: the physical and chemical properties required, the cost of the hydrocolloid, its cost stability and the constancy of supply and composition (Whistler, 1973; Heckman, 1977).

13.6.1 Natural gums in slurry explosives

Guar gum is the most popular hydrocolloid used in slurry explosives, simply because it is most favorable in terms of these four factors. The structure of guar gum is described in Chapter 8. It is typically composed of 78–82% galactomannan, 10–15% moisture, 4–5% protein, 1.5–2.5% crude fiber and 0.5–1.0% ash. The most important structural feature of guar gum with respect to explosives is that it is non-ionic. This enables guar gum's full hydration in concentrated salt solutions, thereby producing maximum viscosities. As mentioned previously, slurry explosives consist of a high concentration of ammonium nitrate and other oxidizers. This property is, therefore, of great significance. The viscosity of guar gum solutions remains nearly constant over a wide range of pH (1.0–10.5), affording the manufacturer more flexibility in terms of including other ingredients within the slurry. Furthermore, the high viscosity obtained using small gum concentrations is advantageous. It enables a certain control over the pouring of slurry explosives into boreholes or into the polyethylene tubes in which they are often packaged (Davidson, 1980).

Borate ions are generally used as cross-linking agents to obtain a gelled explosive. The optimum pH for gel formation is 7.5–10.5, provided by the explosives solution. Alternatively, guar gum may be gelled by transition metal ions. Examples of commonly used cross-linking agents are potassium chromate, tin compounds and bismuth compounds. A typical blasting composition containing guar gum can be prepared by mixing 49% of a 60% ammonium nitrate solution at 120–180°F (50–80°C) for 2–3 min with a premixed preparation of 25% granular sodium nitrate. To this is added 0.75% finely ground guar gum flour and 25% pelleted or flaked TNT; the solution is mixed for a further 1–2 min. Sufficient zirconium tetraacetate is then added as a cross-linker to provide 0.7 mmol zirconium per gram guar gum. Magnesium oxide (0.25%) is then added and the thickened mixture is extruded at 90–100°F (32–38°C) into polyethylene tubes and sealed.

Guar gum is very useful in the production of slurry explosives or water gels. Guar gum and guar derivatives are capable of thickening nitrate salt solutions, the basic component of slurry explosive formulations. (Guar derivatives are created by etherification and esterification via the hydroxyl functionalities, of which guar has an average of three per sugar unit.) Guar gum's thickening capacity is unique under a variety of difficult conditions, in addition to its capability of being readily cross-linked or gelled. Gums are cross-linked to form stable water-resistant gels. A basic formulation includes, in addition to nitrate salts, sensitizing ingredients (organic and inorganic), water and water-soluble cross-linkable thickener to produce the slurries or gels. The advantage of these explosives is their safer use, compared with previous explosives. Slurries of different natures and textures can transform a loose, cohesive and pourable formulation to a rubbery, almost rigid solid (Davidson, 1980).

Starch has many applications in the production of slurry explosives, presumably because of its low cost and worldwide abundance, a trait particularly suited to this industry. The presence of electrolytes in slurry explosives has a marked effect on the swelling of starch. In general, cations and anions increase swelling and decrease gelatinization temperatures to variable extents. A saturated solution of ammonium nitrate will greatly increase swelling and decrease gelatinization temperature. The normal range of pH values for starches is 5.0–7.0, and cooking within this range has little effect on performance. Cooking at below pH 5.0 or above pH 7.0 tends to lower the gelatinization temperature and accelerate the whole cooking procedure. At highly acidic pH, hydrolysis of the glycosidic bonds may occur, decreasing gel viscosity. Therefore, the pH of slurry explosives needs to be carefully monitored. Syneresis and retrogradation are problems associated with starch solutions, but since slurry explosives are usually used shortly after preparation on site, this problem is seldom encountered. Starches are relatively inefficient thickeners and must be used at higher concentrations than galactomannan (4.0–6.0%, compared with 0.5–1.0% for the latter) (Davidson, 1980).

Guar gum, starch and LBG are the hydrocolloids (extracted from natural sources) most commonly encountered in slurry explosives. Other hydrocolloids that have been used include ghatti gum, psyllium seed gum, CMC and tamarind-kernel powder, to name a few. These are hydrocolloids that are either already suitable for use in slurry explosives, or conditions may be manipulated to make them suitable. In actual fact, the manufacture of slurry explosives is centered around manipulation, and explosive compositions are generally considered to be trade secrets.

13.6.2 Synthetic hydrocolloids and resins in explosives and pyrotechnics

The use of synthetic resins as binders in pyrotechnic compositions is one of their less obvious applications, but if one considers that the binding properties of many resins are extremely good, and an aqueous medium is an extremely safe one to handle these materials, it is scarcely surprising. There seems to be no reason why plasticized polyvinyl acetate should not be in pyrotechnic compositions, and it is probably used in this way (Warson, 1972). For the most serious type of composition, i.e. flares, it has been claimed that one part of acrylic polymer and between 1 and 3.3 parts of strontium perchlorate should be added to an acrylic monomer such as methyl methacrylate (Douda, 1966). A process of this type ought to work quite well with the direct use of an emulsion, particularly if the various ingredients can be first wetted out with the monomers, although one would presume that oxidizing and reducing agents should be treated separately. It should be noted that magnesium for some incendiary compositions is desensitized by the addition of the small quantity of linseed oil with which it is wetted out (Warson, 1972). In this connection, the use of polyvinyl acetate as a binder for match beads has been claimed, in conjunction with PVA (Watkins *et al.*, 1968). As far as coatings for explosives are concerned, PETN crystals may be bound into friable aggregates by means of brittle binding agents such as polyvinyl acetate, in emulsion form. The particles aggregate in a way that renders them particularly suitable for enclosure within a protective sheath to form a fuse cord (Gow and McAuslan, 1967; Warson, 1972). A cartridge case, itself very readily combustible, may be prepared from a mixture of cellulose fibers, nitrocellulose fibers and a compatible resin binder. The latter can conveniently consist of a self-reactive copolymer consisting of a monomer (probably acrylic) with a functional group selected from glycidyl, methylol, ureido, hydroxyl, amine and carboxyl (Warson, 1972).

Polyacrylamide and its modified forms are used in many commercial applications. Its gelling or thickening ability is essential for many formulations. Cross-linking of polyacrylamide can be accomplished by methylolation or by ionic bridging. Methylolated polyacrylamide is dried for convenient storage and shipping, but methylolated polymer solution can be used directly, without the drying step. Gelled explosives are prepared by making use of the properties of acrylamide copolymers, where methylolated acrylamide and methylenebisacrylamide combine to form three-dimensional structures which can be used to gel explosives. An example of such a formulation is 9.5% nitroglycerin, 71.5% ammonium nitrate, 1.2% sodium nitrate and 4.0% of a mixture of a dried copolymer of methylolacrylamide (10%), acrylamide (75%) and methylmethacrylate (15%) with oxalic acid (3% of the mixture) and the usual combustibles (Scalera and Bender, 1958).

Another kind of explosive gelled via ionic bonding is obtained by polyacrylamide gelation through the generation of hydrous oxides of multivalent cations such as aluminum (3+) and iron (3+). An example of such a formulation consists of mixing the dry ingredients ammonium nitrate, uncoated, whole prill (18.5%), urea (5%) and sodium chloride (2.8%) in a suitable bowl, adding TNT (35%) and polyacrylamide (0.62%, which contains 1.5% carboxy groups and has a molecular weight of $\sim$5 million) and mixing the ingredients for 2 min. This is followed by the addition of a sodium chloride solution (5.2% in 14.2% hot water) for 5 min of mixing. Aluminum sulfate (0.1%), dissolved in 0.8% water, is mixed in for 2 min to provide the final gel product.

PEO resins are used in a wide variety of binding applications. Such applications take advantage of the water solubility of the resin, its low level of inorganic impurities and its clean burnout characteristics. These binding applications can be used in diverse products, from explosives to pharmaceutical tablets and tobacco filters.

13.6.3 Less commonly used natural gums

Many natural hydrocolloids have been used in the field of water-based explosives. Agar has been used as a flash inhibitor in sulfur-mining explosives. The gelling ability of alginate was useful for the production of explosives formulations. In this case, rubbery stable elastic gels of alginate were formed by reaction with borates in cold-water systems. Sodium alginate has also been used for suspending solids in aqueous gelled explosives (Davidson, 1954). Powdered ghatti gum was used in ammonium nitrate–semi-gelatin mixtures and powdered explosives to improve their resistance to water damage (Meer Corp., 1980). Tamarind gum powder has been used in the explosives industry as a thickener in slurry explosives and as a water barrier in blasting explosives because of its moisture-absorbing capabilities (Ghose and Krishna, 1943; Boyd, 1960). Xanthan and guar gum blends have been used to suspend and gel slurried explosives systems containing saturated ammonium nitrate solutions. Dextran nitrate or nitrate sulfate (dextran derivatives) have been used as gelled propellants and explosives.

References

Boyd, G. (1960) Australian Patent No. 229,190 (*Chem. Abstr.*, **55**, 25255 (1961)).

Bruins, W.R. and Cawood, R.H. (1991) Blast injuries of the ear as a result of the Peterborough lorry explosion: 22 March 1989. *J. Laryngol. Otol.*, **105**(11), 890–5.

Cook, M.A. (1958) Water-compatible explosives, in *The Science of Explosives*, Reinhold, New York, pp. 316–21.

Cook, M.A. (1974) *The Science of Industrial Explosives*, IrecoChem, Salt Lake City, UT, pp. 1–26.

Davidson, R.L. (ed.) (1980) *Handbook of Water-soluble Gums and Resins*, McGraw-Hill, New York.

Davidson, S.H. (1954) *Blasting Explosive Containing a Water Soluble Polysaccharide Ether.* US Patent 2,680,067 (*Chem. Abstr.*, **48**, 11062).

Douda, B.E. (1966) *Process for Producing Microporous Polymeric Structures by Freeze-coagulation of Lattices.* US Patent No. 3,376,158.

Gerhartz, W. (ed.) (1987) *Ullmann's Encyclopedia of Industrial Chemistry*, 5th edn, vol. A10, VCH, Weinheim.

Ghose, T.P. and Krishna, H. (1943) *Ind. Text. J.*, **53**, 236.

Gorontzy, T., Drzyzga, O., Kahl, M.W. *et al.* (1994) Microbial degradation of explosives and related compounds. *Crit. Rev. Microbiol.*, **20**(4), 265–84.

Gow, R.S. and McAuslan, J.H.L. (1967) British Patent No. 1,070,660.

Grant, P.M., Whipple, R.E. and Andresen, B.D. (1995) Comprehensive forensic analyses of debris from the fatal explosion of a 'cold fusion' electrochemical cell. *J. Foren. Sci.*, **40**(1), 18–26.

Hansen, B.L. (1983) *Technical Data Sheets*, IRECO Inc., Salt Lake City, UT.

Heckmam, E. (1977) Starch and its modifications for the food industries, in *Food Colloids*, ch. 12 (ed. H. Grahams), Avi Publishing, Westport, CT.

Kanerva, L., Laine, R., Jolanki, R. *et al.* (1991) Occupational allergic contact dermatitis caused by nitroglycerin. *Contact Dermatitis*, **24**(5), 356–62.

Kaye, S.M. (1980) Slurry explosives, in *Encyclopedia of Explosives and Related Items*, vol. 9, PATR 2700, USAADRC Dover, NJ, pp. S121–47.

Mark, H.F., Othmer, D.F., Overberger, C.G. *et al.* (1980) *Kirk-Othmer Encyclopedia of Chemical Technology*, 3rd edn, vol. 9, Wiley, Interscience, New York.

Meer Corp. (1980) *The Use of Gums in Wax Emulsions*. Technical Bulletin No. 1-11.

Midkiff, C.R. and Walters, A.N. (1993) Slurry and emulsion explosives: new tools for terrorists, new challenges for detection and identification, in *Advances in Analysis and Detection of Explosives* (ed. J. Yinons), Kluwer Academic, the Netherlands, pp. 77–90.

Midkiff, C.R. and Washington, W.D. (1974) Systematic approach to the detection of explosives residues. III. Commercial dynamite. *J. Assoc. Offic. Anal. Chem.*, **57**(5), 1092–7.

Mineral Industry Surveys (1961–81) *Apparent Consumption of Industrial Explosives and Blasting Agents in the United States*, US Department of the Interior, Bureau of Mines, Washington, DC (annual).

Nikolic, S., Medic Saric, M., Rendic, S. *et al.* (1994) Toxic effects and a structure–property study of organic explosives, propellants and related compounds. *Drug Metab. Rev.*, **26**(4), 713–38.

Peterson, J.L. and Markham, P.N. (1995) Crime laboratory proficiency testing results 1978–1991. I. Identification and classification of physical evidence. *J. Foren. Sci.*, **40**(6), 994–1008.

Putintsev, A.V. and Kitaev, N.N. (1994) The results of the simultaneous explosion of the Zaria and Plamia special grenades. *Sud. Med. Ekspert*, **37**(2), 16–17.

Robinson, R.V. (1969) Water gel explosives – three generations. *Can. Mining Meta. Bull.*, **72**, 348–56.

Savic, J., Tatic, V., Ignjatovic, D. *et al.* (1991) Pathophysiologic reactions in sheep to blast waves from detonation of aerosol explosives. *Vojnosanit Pregl.*, **48**(6), 499–506.

Scalera, M. and Bender, M. (1958) *Water-resistant Explosive Compositions*. US Patent No. 2, 826,485.

Shuker, S.T. (1995) Maxillofacial blast injuries. *J. Craniomaxillofac. Surg.*, **23**(2), 91–8.

Spiker, J.K., Crawford, D.L. and Crawford, R.L. (1992) Influence of 2,4,6-trinitrotoluene (TNT) concentration on the degradation of TNT in explosive-contaminated soils by the white-rot fungus *Phanerochaete chrysosporium*. *Appl. Environ. Microbiol.*, **58**(9), 3199–202.

Sudweeks, W.B. (1985) Physical and chemical properties of industrial slurry explosives. *Ind. Eng. Chem. Prod. Res. Dev.*, **24**(3), 432–6.

Tatic, V., Ignjatovic, D., Mrda, V. *et al.* (1991) Morphologic changes in the tissues and organs of sheep in blast injuries caused by detonation of aerosol explosives. *Vojnosanit Pregl.*, **48**(6), 541–5.

Tucak, A., Lukacevic, T., Kuvezdic, H. *et al.* (1995) Urogenital wounds during the war in Croatia in 1991/92. *J. Urol.*, **153**(1), 121–2.

Warson, H. (1972) *The Application of Synthetic Resin Emulsions*, Ernest Benn, London, p. 1000.

Watkins, T.F., Cackett, J.C., Hall, R.G. (1968) *Chemical Warfare, Pyrotechnics and the Fireworks Industry*, Pergamon Press, London.

Wetterholm, A. (1981) The first commercial water gel explosives. *Polish J. Chem.*, **55**, 1445–57.

Whistler, R.L. (1973) *Industrial Gums*, 2nd edn, Academic Press, New York.

Ylikoski, M., Pekkarinen, J.O., Starck, J.P. *et al.* (1995) Physical characteristics of gunfire impulse noise and its attenuation by hearing protectors. *Scan. Audiol.*, **24**(1), 3–11.

14 Glues

14.1 Introduction

An adhesive is a substrate capable of holding materials together by surface attachment. The bond (joint) is the location at which the two materials (adherents) are held together by a layer of adhesive (ASTM, 1982). Materials are held together by forces of attraction, which are the result of chemical bond formation, physical interactions (e.g. dispersion forces) and mechanical interlocking (Patrick, 1966). The last is a consequence of adhesive penetration into pores (microscopic scale) and other surface irregularities to produce mechanical interlocking. It is not sufficient, in analysing interactions between an adhesive and an adherent, to consider the bulk properties of the materials involved: major concern should be given to the zone between bulk adhesive and bulk adherent 'interphase'. The level of performance required of the adhesive will depend on the strength of the adherent and the load that it is expected to carry (Bolger, 1983).

Although the use of adhesives can be traced back many centuries, their production on an industrial scale started about 300 years ago. Modern structural adhesives can be dated from about 1910, when resins of phenol-formaldehyde were introduced (Bruno, 1970). To appreciate how structural adhesives 'stick', a knowledge of surface chemistry and wetting is necessary.

14.2 Hydrocolloids as adhesives

The adhesive properties of many hydrocolloids (gums) have been known for centuries. The word 'gum' means a sticky substance and was previously defined as such by the Egyptian term qemai or kami, referring to the exudate of the acanthus plant and its adhesive capacity (Glicksman, 1982). Hydrocolloid glues are distinguished from most organic-based adhesives by their hydrophilic, non-toxic nature and are used at many different concentrations, viscosities and molecular weights. Gums are inflammable and possess good wettability properties, which enhance their penetration into porous substrates. The development of synthetic hydrocolloids during this century has broadened their uses with such commodities as paper, wood, textiles, leather, food, cosmetics and medicine.

Bioadhesive applications of hydrocolloids in medicine and cosmetics have been widely explored. A few examples are adhesive biodelivery systems (Smart *et al.*, 1984; Robinson *et al.*, 1987; Bottenberg *et al.*, 1991; Smart, 1991; Bouckaert and Remon, 1993; Irons and Robinson, 1994), adhesive bioelectrodes (Keusch and Essmyer, 1987), cosmetic preparations (Toulmin, 1956), pressure-sensitive adhesives (Piglowski and Kozlowski, 1985; Kiyosi and Yasuo, 1986), ostomy rings (Glicksman, 1982), adhesive ointments (Kanig and Manago-Ulgado, 1965), and dental adhesives (Shay, 1991).

A large number of hydrocolloids have been mentioned in the literature as adhesive agents (Chen and Cyr, 1970; Bauman and Conner, 1994). They include gum talha (similar to gum arabic), gum ghatti, gum karaya, gum tragacanth, arabinogalactan (AG), dextran, pectin, tapioca–dextrin, CMC, MC, HPC, MHPC, carbopol, PEO, PVP, PVA, retene, pullulan and chitosan. Some lesser-known water-soluble gums that possess wet-adhesive bonding properties include gum angaco, brea gum and psyllium seed gum (Mantell, 1947), gum cashew (Howes, 1949), gum damson, jeol, myrrh (Smith and Montgomery, 1959) and scleroglucan (Glicksman, 1982).

14.3 Hydrocolloid glues: industrial uses

14.3.1 Paper

The use of hydrocolloids in the paper industry began in ancient Egypt when starch was used to adhere papyrus sheets (Lucas, 1962). Now starch, as well as dextrin, gelatin, PVA, gum arabic and cellulose derivatives make up the major portion of the glue market for the paper industry. Synthetic glues are based on polyvinyl acetate, ethylene vinyl acetate, acrylic acid polyurethane and latex, among many others (Brief, 1990). Hydrocolloid glues are used for paper lamination, to prepare cardboard (Torrey, 1980). Special methods (e.g. the ‘Stain Hall’ procedure) are used to prepare starch adhesives for paperboards (Sadle *et al.*, 1979; Baumann and Conner, 1994). For envelopes, stamps, wallpapers and similar products, special dry hydrocolloid glues that regain their adhesive properties upon wetting are used. They are based on gum arabic, PVA, dextrins and other gums (Kirby, 1967; Sharkey, 1987). For paper bags, mostly starch or modified starch and dextrin are used (Baumann and Conner, 1994). Glue’s resistance to water is increased by including PVA and/or polyvinyl acetate in its composition, as well as urea-formaldehyde at 5–15% of the gum weight in the formulation (Caesar, 1962). Paper glues can be used to attach paperboard to gypsum sheets. Such glue can be composed of calcium carbonate, mica, attapulgite clay, MHPC, ethylene glycol, preservative, anti-foaming agent and latex. Tobacco is sometimes packaged in paper cylinders that are glued together with starch or gum tragacanth (Brief, 1990).

14.3.2 Wood

In ancient times, gum arabic and gelatin were used by the Egyptians in furniture manufacturing (Keimel, 1994). The use of hydrocolloid glues in the modern wood industry began at the turn of the century. Until 1912, gelatin and starch were the most commonly used adhesives. Today, many other glues are also in use. They include polyamide, neoprene, polyvinyl acetate, urea-formaldehyde, phenol-formaldehyde, melamine-urea-formaldehyde, recorcinol-phenol-formaldehyde, epoxy ethylene vinyl acetate and others. Gelatin is still used, especially for gluing cloth to wood (Krage and Wootton, 1987). CMC, at ~30% of the dry solids in the glue, is used to retain better moisture within the recipe, without influencing its adhesive strength. Synthetic polyvinyl alcohol is used as an additive in urea-formaldehyde glues to improve the strength of the wet glue and its stability over time. Chitosan has been used to glue pine-tree plates, in a mixture with water and acetic acid (Rigby, 1936). HEC, as part of a glue composed of bentonite, asbestos, coquina, and glycerol, was used to fill cracks in wooden boats. Another hydrocolloid used in the preparation of elastic and strong glues for woods is fonuran, combined with carbon disulfide. The preparation involves mixing with sulfuric acid and later adding sodium hydroxide, heating to 50°C for 10 min, followed by heating for 3 h at the same temperature after addition of the carbon disulfide (Schacat and Glicksman, 1959).

14.3.3 Leather and textiles

Hydrocolloid glues are used as a tool in drying leather. The leather is smeared with a polymeric glue and attached to glass, aluminum, stainless steel or porcelain plates. The plates move perpendicularly through a drying tunnel at ~200°F (93°C) for a few hours. The better the adherence of the leather to the plate the less shrinkage occurs. The glues are composed mainly of MC, CMC and polyacrylic acid (PAA). MC has been found to be the most efficient hydrocolloidal ingredient because of its tendency to gel under heating and to maintain its elasticity during drying, resulting in less leather shrinkage during the process. In addition, the gel can be easily removed at the end of drying without damaging the leather fiber structure. CMC can be used as a substitute for MC (Paist, 1959; Baird and Speich, 1962). PAA can also be used in combination with starch or CMC (Glavis, 1962). In ancient times, mummies were wrapped in bandages glued with gum arabic (Lucas, 1962). Starch and dextrins are used for fiber strengthening as are HEC and CMC. A typical formula for such glues includes cornstarch, sulfonated oil, kerosene and water. Special starches after acetate modification are also applicable. Oxidized starch is also used for printing and finishing (Kruger and Lacourse, 1990).

14.3.4 Food

A simple way of achieving different textures and tastes in the same bite is to build a food product composed of layers with different properties. A few multi-layered food products are already available, e.g. crunchy wafers, which include a sweet vegetable-fat-based chocolate or vanilla taste filling between brittle wafers. For children, a multi-layered, sweetened agar–or starch-based–three-color confection can also be found, in which the texture of the layers is similar, but the tastes and colors differ. In the Orient, where the awareness of different gel textures is much more developed than in the West, a curdlan-based, sweetened, multi-layered gel has been developed (Harada, 1977, 1979; Morris, 1991). In this case all layers are built from the same hydrocolloid, because two types of curdlan gel can easily be prepared from its powder by heating the suspension to different temperatures. Multi-layered foods based on hydrocolloids are important in the framework of foods of the future.

Recently, multi-layered gels (composed of different combinations of agar, four galactomannans, xanthan, carrageenan and konjac mannan), and gelled texturized fruits (based on banana, apple, kiwi and strawberry pulps) and agar–LBG combinations, glued together by three different adhesion techniques, were studied (Ben-Zion and Nussinovitch, 1996). The gluing techniques consisted of pouring hot hydrocolloid solution on a gelled layer, using melted agar as a glue between already gelled layers, or simultaneously pouring pregelled (gum solution before setting) hydrocolloid solutions. The compressive deformabilities of these gels were predicted. Two assumptions were made: that the normal force in the layers is the same and that the deformations are additive. The effects of lateral stresses were considered negligible. The calculation was performed using a mathematical model previously developed for double-layered curdlan gels (Nussinovitch *et al.*, 1991). The model constants were determined from the behavior of the individual layers. Good agreement was found between experimental and fitted results over a considerable range of strains. Thus the model's applicability to a given gel system was demonstrated, suggesting a very convenient tool for analysing and predicting the compressive behavior of any number of arrays with different layer combinations (Ben-Zion and Nussinovitch, 1996).

14.3.5 Biomedicine

Hydrocolloids are used in biodelivery adhesive systems to carry fluoride (Irons and Robinson, 1994), benzydamine hydrochloride or other drugs to cure gingivitis. The hydrocolloids are part of both the biodelivery and protective layers. The first is prepared, for example, from HPC, HPMC, gum karaya, propylene glycol alginate 400 and the drug. The second is composed of ethyl cellulose, sodium CMC and HPC. This latter layer is not adhesive. The whole preparation is dried and compressed to a thickness of $\sim 100\,\mu m$ (Nagai and Machida, 1993). Many other adhesive biodelivery systems have been described (Anders and Merkel, 1989; Ishida *et al.*, 1982). Other additional gums are used in these preparations, such as PVP, PVA and carbopol. Agar is frequently used for the intermediate, non-adhesive layer of the preparation.

Many hydrocolloids have been used as biological glues to the mucus, with PC and HPC being preferred. One of the earliest discussions on hydrocolloids' ability to serve as glues in biological systems was published 26 years ago (Chen and Cyr, 1970). These authors also presented results of *in vitro* tests comparing adhesion characteristics. Almost 15 years later, Smart *et al.* (1984) examined the mean adhesive forces of many hydrocolloids to mucus. They found that sodium CMC produces the strongest adhesive force, followed by carbopol and gum tragacanth. Similar results were obtained by Robinson *et al.* (1987). Other research (Anders and Merkle, 1989) estimated the minimal amount of polymer needed for the glue to stay in position for about the same duration. The following order of efficiency was found: HEC $>$ HPC $>$ PVP and PVA. Other scientific approaches can be found in the literature, calculating adhesion energy of different formulations attached to biological membranes (Bottenberg *et al.*, 1991). It is interesting to note that although LBG–xanthan systems were evaluated as mucosal glues and found unsatisfactory, they are nevertheless used because of good mouthfeel and for health safety reasons (Nagai and Machida, 1993). Biological hydrocolloid glues for the vaginal area and eyes, and in sprays for the nose and other mucosal areas, have been developed (Nagai and Machida, 1993; Irons and Robinson, 1994). Chitosan, pullulan, gum tragacanth, HPC, starch and carbopol have been suggested for these preparations. Adhesive dressings contain many hydrocolloids in their compositions, such as PVP, PEO and PVA. Adhesive bioelectrodes use PVA, hydroxyethyl methacrylate, gum karaya, agar, MC or CMC in their formulations (Keusch and Essmyer, 1987). Other biological applications of hydrocolloids are the use of dextran in cosmetic preparations, pullulan and CMC as part of medicinal adhesive tapes, gum karaya for ostomy rings, and a wide range of hydrocolloids such as gum karaya, gum arabic, cellulose derivatives, polyox and others for dentistry.

14.4 Hydrocolloid adhesion tests

A relatively large number of tests have been proposed to evaluate adhesive-bonding strength. They include peeling at 90°, and tensile-bond and lap-shear tests. The bond-strength value measured by a specific test is not only an inherently fundamental property of the type of adhesive, it also depends on other factors. Many experimental procedures, using biological or other hydrocolloid adhesives, have been conducted to test different important variables such as crosshead speed at debonding (Smart, 1991), adhesive-layer thickness (Smart *et al.*, 1984), water-holding capacity of a specimen (Kanig and Manago-Ulgado, 1965; Chen and Cyr, 1970), length of contact (Chen and Cyr, 1970; Smart *et al.*, 1984), the effect of molecular weight (Smart *et al.*, 1984) and the type of adhesive (Chen and Cyr, 1970).

14.5 Hydrocolloids as wet glues

Recently, 26 hydrocolloids were tested for their ability to produce wet glues, or, in other words, they were studied for their ability to create very thick suspensions with 'good' adhesive properties at predetermined gum concentrations ranging from 10 to 75%. The hydrocolloids were gum talha, ghatti gum, AG, karaya gum, tragacanth, dextran, apple pectin, CMC, MHPC, tapioca–dextrin, carbopol-934, HPC, MC, gelatin, casein, starch, LBG, guar gum, alginate, κ-carrageenan, tara gum, fenugreek gum, konjacmannan, xanthan gum, gellan and curdlan. The hydrocolloids (at different concentrations) were added in powdered form to double-distilled water and mixed with a standard dough mixer for at least 15 min until a thick, uniform and smooth paste–wet glue was obtained (Ben-Zion, 1995; Ben-Zion and Nussinovitch, 1997).

Preliminary tests revealed that only 13 of these, namely gum talha, ghatti gum, karaya gum, tragacanth, AG, dextran, pectin, tapioca–dextrin, CMC, MC, HPC, MHPC and carbopol, can serve as bioadhesives in hydrophilic systems. Wet glues were produced from these hydrocolloids and tested over a wide range of concentrations, i.e. 10 to 75%, and the color and pH of each glue was determined.

Table 14.1 gives some of their physical properties. All preparations were tested immediately following their production. Paste temperature was taken at the end of mixing, and 5 min later had risen by 0.97 ± 0.14°C. The pH values of the wet glues ranged from 1.2 with carbopol to ~9.6 with AG. pH may be an important factor in the utilization of bioadhesive materials. Studies have shown (Shay, 1991) that karaya gum (pH ~3.6) may cause allergic reactions such as hives and angioneurotic edema. Paste color may be a factor in choosing an ointment or bioadhesive for a particular application. A variety of different colors could be found among the wet glues. The higher the *L* value, the lighter the paste. Wet glues with *L* values of ~60 were found to be the lightest of all the glues tested. People tend to prefer pastes with a 'whitish' appearance, i.e. a higher *L* value. The *a* value (hue) spans the green–red axis of the color system and the *b* value (chroma), the yellow–blue axis. Many of the wet glues in Table 14.1 had a yellowish appearance (higher *b* values), and only a few were dark brown.

Table 14.1 Physical properties of hydrocolloid pastes

Hydrocolloid	Conc. (g 100 g^{-1})	pH	Color parameters *L*	*a*	*b*	Hue
Tree and shrub exudates						
Gum talha	65	4.02	22.16	0.00	0.43	Dark brown
	70	3.98	21.69	0.59	1.08	Dark brown
	75	3.95	24.21	3.07	4.03	Dark brown
Gum karaya	25	3.68	27.98	2.00	4.95	Light brown
	35	3.61	29.17	2.09	4.82	Light brown
	45	3.52	26.94	2.51	5.40	Light brown
Gum tragacanth	25	4.65	43.94	−0.92	6.26	Yellowish
	35	4.61	47.76	−0.34	8.73	Yellowish
	45	4.56	52.51	−0.11	11.67	Yellowish
Gum ghatti	55	4.22	23.65	1.18	2.55	Dark brown
	65	4.17	23.69	1.45	2.64	Dark brown
	75	4.12	24.82	1.68	2.58	Dark brown
Tree extracts						
Arabinogalactan	65	9.57	69.08	−2.35	12.45	Yellow
	75	9.08	60.13	−0.29	25.19	Yellow
Fruit extracts						
Apple pectin	15	2.32	62.7	−1.5	6.84	Yellowish-white
	25	2.28	64.55	−0.72	9.77	Yellowish-white
Grains						
Tapioca	55	3.11	62.68	−0.11	8.76	Yellowish
	65	2.98	65.34	0.67	11.2	Yellowish
Exocellular polysaccharides						
Dextran	65	4.72	46.09	−0.29	1.05	White
	70	4.89	61.8	−0.14	0.54	White
Cellulose derivatives						
CMC	10	6.87	52.77	−0.90	3.64	Yellowish-transparent
	15	7.06	34.05	−0.49	5.73	Yellowish-transparent
	20	7.08	35.05	−0.25	5.45	Yellowish-transparent
	25	7.12	39.57	−0.13	6.24	Yellowish-transparent
MC	10	5.18	32.86	−0.34	0.81	Light white
	15	5.11	45.73	−0.65	1.58	Light white
	20	5.07	54.78	−0.69	2.07	Light white
	25	4.85	59.29	−0.75	2.70	Light white
MHPC	10	4.49	31.25	−0.58	2.02	Semi-transparent
	15	4.48	29.17	−0.44	1.74	Semi-transparent
	20	4.42	31.22	−0.43	1.99	Semi-transparent
	25	4.37	34.38	−0.42	2.26	Semi-transparent
HPC	10	4.10	30.3	−0.52	1.40	Semi-transparent
	15	4.08	29.71	−0.40	1.41	Semi-transparent
	20	4.02	31.46	−0.44	1.35	Semi-transparent
	25	3.99	31.08	−0.51	1.83	Semi-transparent
Petrochemicals						
Carbopol-934	15	1.90	36.05	−0.60	−0.39	White
	25	1.42	32.88	−0.55	0.83	White
	35	1.24	38.48	−0.87	3.19	White

14.5.1 Properties of hydrocolloidal wet glues

Once in very thick paste form, the 13 hydrocolloids were smeared homogeneously onto two different substrates: (a) cellulose acetate film, normally used for dialysis, and (b) a skin-surface model (SSM), previously proposed and adapted (Charkoudian, 1988, 1989) for testing the adhesion of medical adhesives. The water content in the SSM was approximately 1%.

Three mechanical tests were performed to check the properties of the wet glues (Ben-Zion, 1995): peel, tensile and lap-shear tests (Table 14.2). (These data are given in g force where 1 unit = 4.4482 N.) Seven typical hydrocolloids (ghatti gum, AG, pectin, tapioca–dextrin, dextran, MHPC and carbopol-934) were chosen for an evaluation of their physical properties as representatives of tree and shrub exudates, tree extracts, fruit extracts, grains, exocellular polysaccharides, cellulose derivatives and petrochemicals, respectively (Ben-Zion, 1995). Typical curves for the 90° peeling, tensile-load and lap-shear tests are presented in Figs. 14.1–14.3, respectively.

Two common types of curves were obtained from the peel test. When a sample is pulled apart at a constant crosshead speed, the measured force should ideally be constant after reaching a steady-state condition. In practice, however, this is not always the case (Fig. 14.1). There is evidence in the literature that when such results are reported, the mean values (averages) of the curve's ruggedness (deviations from a smooth line after reaching a steady-state condition) can be calculated and observed. In some instances, as observed macroscopically during testing, the rupture process occurs abruptly, sample failure propagates faster than the rate at which the sample is pulled apart, and failure is initiated periodically. The force has been claimed to go through well-defined maxima and minima, and the distance between two minima or maxima is independent of testing rate (Gardon, 1966). This author also mentions that the variability of the force is caused by sample imperfections, that all the points of such a curve can be considered a statistical population and that their frequency distribution is Gaussian.

Tensile load (Fig. 14.2) and lap shear (Fig. 14.3) were applied and plotted as load (g force cm^{-2}) versus displacement (cm) curves (Ben-Zion and Nussinovitch, 1997). Tensile-bond strength increased in parallel to increases in deformation, until the beginning of failure. Tensile-bond tests are commonly used to analyse various adhesives, ranging from those for wood to those for metal (ASTM D-897 and ASTM D-2094, respectively (ASTM, 1982)).

Lap-shear strength decreased linearly as deformation increased. Lap-shear tests are used to examine adhesion when the samples are relatively easy to construct and closely resemble the geometry of many practical joints (ASTM D-1002; ASTM D-3528 (ASTM, 1982)). Data were collected continuously as the specimens were stretched uniaxially until rupture. The maximal force was recorded in both cases (Ben-Zion and Nussinovitch, 1997).

Table 14.2 Peel-bond strengths for 13 hydrocolloid pastes tested on cellulose acetate film and SSM

Hydrocolloid	Conc. (g 100 g^{-1})	Peel bond strength (g force cm^{-1})	
		Cellulose acetate membrane (SD)	SSM (SD)
Tree and shrub exudates			
Gum talha	65	5.9 (1.1)	3.6 (0.4)
	70	18.8 (1.6)	7.5 (0.6)
	75	45.2 (2.5)	15.4 (1.3)
Gum karaya	25	4.9 (0.7)	3.4 (0.5)
	35	12.2 (0.7)	10.7 (0.8)
	45	31.4 (2.4)	26.3 (2.1)
Gum tragacanth	25	4.4 (0.6)	3.4 (0.4)
	35	11.9 (0.6)	9.4 (0.3)
	45	30.9 (1.1)	23.2 (0.9)
Gum ghatti	55	7.1 (0.9)	6.0 (1.0)
	65	22.0 (1.6)	11.5 (0.6)
	75	62.6 (1.4)	0.0 (0.0)
Tree extracts			
Larch arabinogalactan	65	4.2 (0.6)	2.6 (0.4)
	75	14.3 (2.2)	6.4 (0.9)
Fruit extracts			
Apple pectin	15	0.3 (0.5)	0.7 (1.0)
	25	6.0 (0.6)	4.0 (0.6)
Grains			
Tapioca–dextrin	55	8.5 (0.8)	5.2 (1.3)
	65	16.8 (1.2)	14.7 (1.9)
Exocellular polysaccharides			
Dextran	65	7.4 (0.9)	4.5 (0.5)
	70	62.5 (1.9)	49.6 (1.8)
Cellulose derivatives			
CMC	10	2.5 (0.3)	1.6 (0.4)
	15	3.3 (0.4)	2.3 (0.5)
	20	6.3 (0.7)	2.7 (0.2)
	25	4.8 (0.6)	2.6 (0.4)
MC	10	9.2 (0.6)	9.4 (0.6)
	15	7.3 (0.7)	7.2 (0.5)
	20	6.0 (0.4)	4.9 (0.5)
	25	3.8 (0.5)	1.8 (0.3)
MHPC	10	2.7 (0.5)	1.4 (0.1)
	15	4.8 (0.4)	3.0 (0.7)
	20	9.1 (0.7)	6.3 (1.0)
	25	7.0 (0.6)	5.9 (0.5)
HPC	10	1.7 (0.4)	2.0 (0.5)
	15	2.5 (0.4)	3.3 (0.3)
	20	3.9 (0.5)	3.9 (0.5)
	25	7.7 (0.4)	5.0 (0.7)
Petrochemicals			
Carbopol-934	15	4.1 (0.8)	4.1 (1.1)
	25	13.7 (1.0)	14.8 (1.5)
	35	38.8 (1.3)	40.4 (2.7)

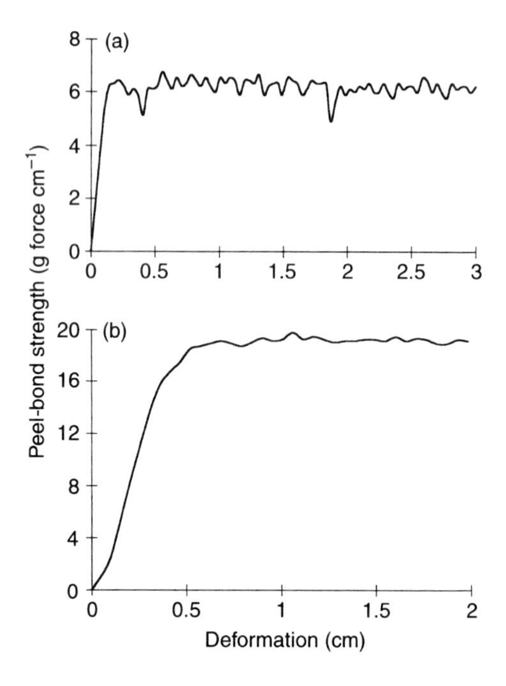

Figure 14.1 Typical curves for the 90° peel test: (a) pectin, 25%; (b) gum talha, 70%. (From Ben-Zion and Nussinovitch, 1997; by permission of Oxford University Press.)

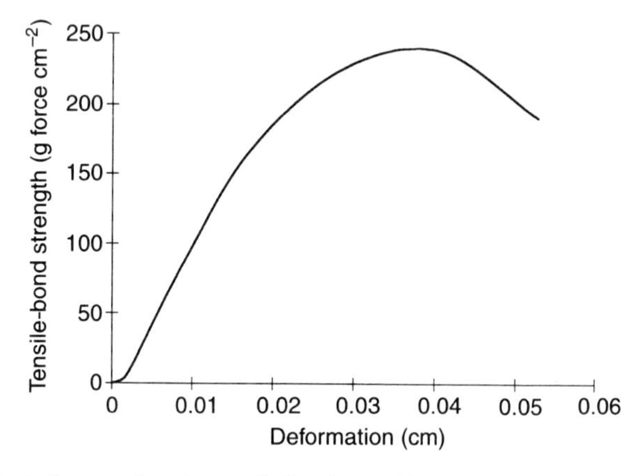

Figure 14.2 Typical curve for the tensile-bond test (dextran, 65%). (From Ben-Zion and Nussinovitch, 1977; by permission of Oxford University Press.)

It is difficult to make comparative economic or practical analyses of the various hydrocolloids tested in this study, since thc concentrations and viscosities differ, and formation of the desired suspension does not necessarily require the same time or energy. Therefore, in choosing a hydrocolloid for a particular application, such as wet biological adhesives or ointments,

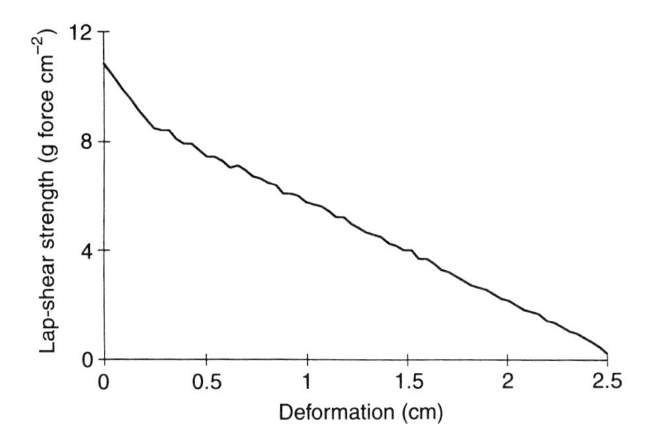

Figure 14.3 Typical curve for the lap-shear test (HPMC, 20%).

one needs to consider, in addition to these factors, clinical or medicinal aspects such as the ability of the gum to perform as a medium for different electrolytes that could cause changes in viscosity and thus changes in adhesion potential (e.g. bioelectrodes). Also important are the wet glue's potential to cause dermatological irritation or allergic reactions, and its cost (Ben-Zion and Nussinovitch, 1997).

Table 14.3 gives the experimental values of tensile-bond and lap-shear strengths for seven hydrocolloids (representing seven different origins of natural and synthetic gums). Under the test conditions, carbopol exhibited the highest tensile-bond strength, dextran the lowest. It is important to note that the concentrations studied were not the same: the gum concentrations chosen were designed to produce the highest bond strength based on the previously performed peel test. Dextran differed from the other gums in its ability to dry fast after preparation, making it difficult to use at higher concentrations. In the lap-shear test, tapioca was found to produce the highest lap-shear strength, dextran the lowest. This test was performed to

Table 14.3 Tensile-bond and lap-shear test values for seven hydrocolloid adhesive pastes

Hydrocolloid	Conc. (g 100 g^{-1})	Tensile-bond strength (g force cm^{-2})	COV	Lap-shear strength (g force cm^{-2})	COV
Dextran	65	232.8	2.0	8.1	13.8
Carbopol-934	35	1112.7	1.5	46.1	2.3
Ghatti	65	452.8	4.9	27.1	1.3
Tapioca–dextrin	65	884.6	2.2	49.7	5.7
Arabinogalactan	75	568.7	2.3	29.5	2.7
Pectin	15	334.7	11.5	20.4	1.3
MHPC	20	259.2	5.8	12.4	17.4

COV is the coefficient of variance, i.e. $100\,SD/\bar{X}$.

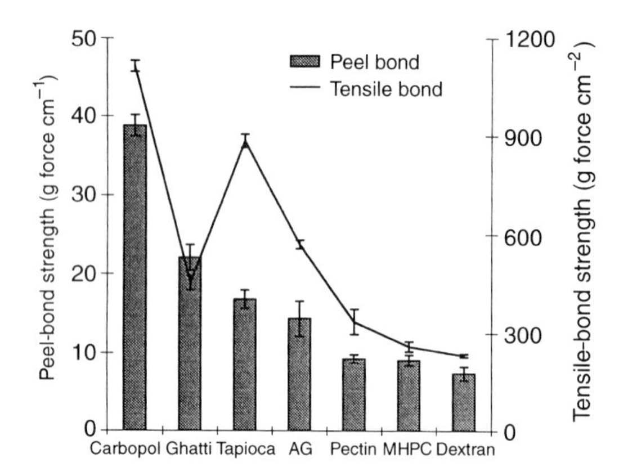

Figure 14.4 Peel-bond strength versus tensile-bond strength for seven hydrocolloid adhesive pastes. (From Ben-Zion and Nussinovitch, 1997; by permission of Oxford University Press.)

examine the cohesiveness of the glue, because the adhesive pressed between two sliding glass plates undergoes shearing in the medium itself, and not at the interface between the medium and the adhered substrate.

Figure 14.4 shows tensile-bond versus 90°-peel adhesive strength. Since both tests estimate the adhesiveness of the various gums, the order in which the gums fell was expected to be the same. Four of the seven hydrocolloids behaved as expected (carbopol, pectin, MHPC and dextran). The other three gums (ghatti, tapioca, AG) yielded different results, perhaps because of the distinct patterns of stress distribution through the bond interface with each test (Kutai, 1994). In the tensile-bond test, the entire area involved is stressed simultaneously. In the 90°-peel test, only a separation line exists, and this line is extended as the hydrocolloid paste is stretched and pulled away. These findings agree with those of other works performed on resilient liners (Kutai, 1994).

The relatively large standard deviations occurring with the tensile-bond test (Fig. 14.4) may be a consequence of the procedure followed. We found that although samples were produced very carefully and appropriate testing machine alignment was established, it was still difficult to apply a truly centric load. As a result, the experiments often exhibited quite a lot of data scattering. The same phenomenon was observed with the lap-shear test. Even under the best circumstances, one would not anticipate the stress distribution in such a case to be very uniform. The exact stress distribution is highly dependent on the relative flexibilities of both the parallel beams and the adhesive (Ben-Zion and Nussinovitch, 1997).

Figure 14.5 shows the influence of deformation rate on the peel-bond-strength values. The adhesive strength increase paralleled that of deformation rate, suggesting that the adhesive bond is of a viscoelastic nature, i.e. that a faster rate of stress application to the adhesive bond gives it less time

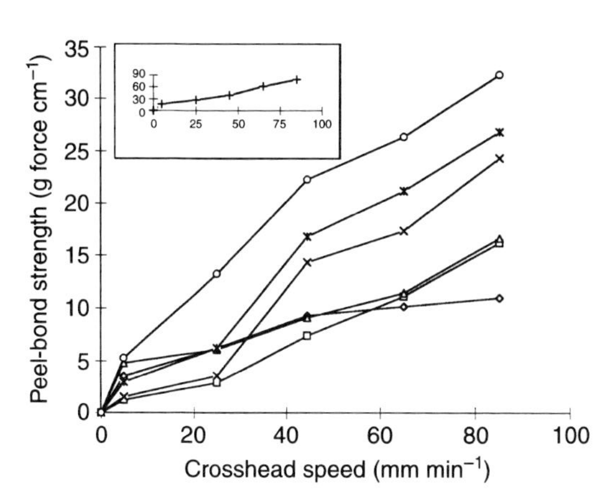

Figure 14.5 Peel-bond strength versus deformation rate for seven hydrocolloid adhesive pastes. Main graph: pectin (◇), dextran (□) MHPC (△), AG (×), tapioca (*), ghatti (○); inset: carbopol. (From Ben-Zion and Nussinovitch, 1997; by permission of Oxford University Press.)

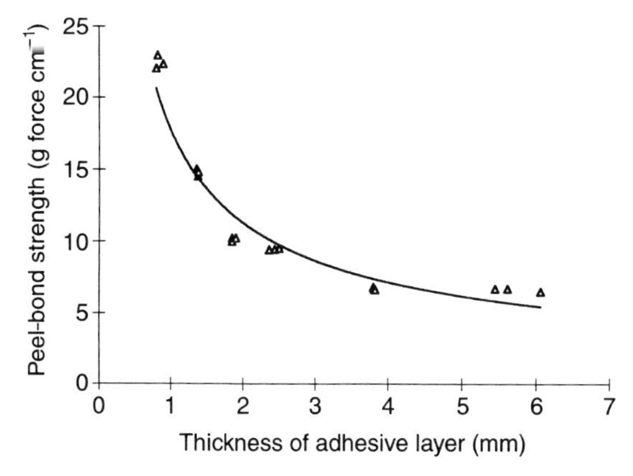

Figure 14.6 Peel-bond strength versus gum ghatti adhesive paste thickness. (From Ben-Zion and Nussinovitch, 1997; by permission of Oxford University Press.)

to deform and flow (Smart, 1991). At higher crosshead speeds, the peel-bond strength tends to be rate independent because the interactive forces at the interface are characterized by their short range. Pectin is a good example of this behavior.

Figure 14.6 represents the dependence of peel-bond strength on adhesive-layer thickness. Ghatti gum was chosen for this time-consuming experiment since it is the easiest gum with which to produce various thicknesses of wet glue with maximal accuracy (thickness uniformity). In addition, the peel-bond strength is relatively high, making it a worthwhile substance for experimentation. There was a reduction in peel-bond strength as the

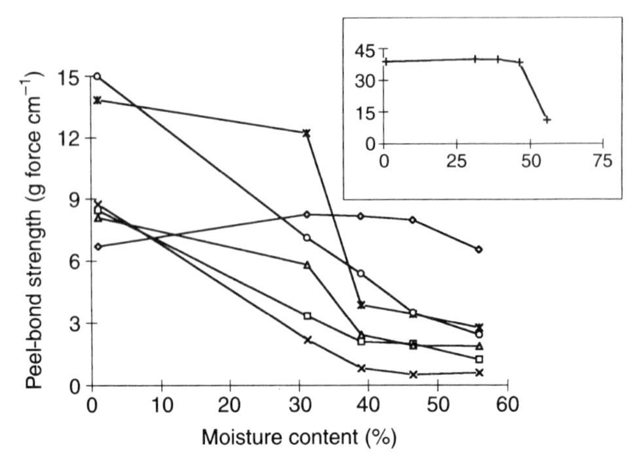

Figure 14.7 Peel-bond strength versus water content of SSM films for seven hydrocolloid adhesive pastes. Main graph: pectin (◇), dextran (□), MHPC (△), AG (×), tapioca (∗), ghatti (○); inset: carbopol. (From Ben-Zion and Nussinovitch, 1997; by permission of Oxford University Press.)

thickness of the wet glue decreased (Fig. 14.6). When the wet glue was thicker than ~3.8 mm, the peel-bond strength reached an asymptotic value of ~7 g force cm^{-1}. This may be explained by a cohesive type of failure (i.e. inside the glue layer), whereas for thicknesses of <3.8 mm the failure was adhesive. We do not know whether this observation, which is typical for ghatti gum, is common with other gums, and this, therefore, needs to be studied. Data found in the literature for non-gummy materials (such as cured adhesives) provide evidence for different behaviors (Gardon, 1966).

To study the influence of water absorbance on peel-bond strength, we produced SSM films with different moisture contents by immersing them in water for 0, 2, 5, 9 and 13 min. This produced films with water contents of ~1, 31, 39, 46 and 56%, respectively. Figure 14.7 shows the relationship between this parameter and 90°-peel-bond strength. AG, dextran, MHPC, ghatti gum and tapioca–dextrin exhibited decreases in peel-bond strength as water content increased. Hydrocolloids develop their wet adhesive properties at various degrees of hydration, reaching maximum adhesion at an optimum degree of hydration (Chen and Cyr, 1970). Some hydrocolloids exhibit wet adhesiveness in the presence of only very little water, whereas excessive water causes the deformation of a slippery, non-adhesive mucilage. A wet surface used as a substrate for adhesives is different from a dry one. The first state is not static and water diffuses from the surface of the wet substrate (SSM) into the hydrocolloid interface. The rate and capacity of water absorbance of the hydrocolloids affect the amount of water present near the interface, between the adhesive and the substrate. These properties appear to be important in determining the time required for initial wet adhesion, hydration time and the duration of adhesion. Rapid water absorbance may shorten the duration of adhesion because erosion proceeds

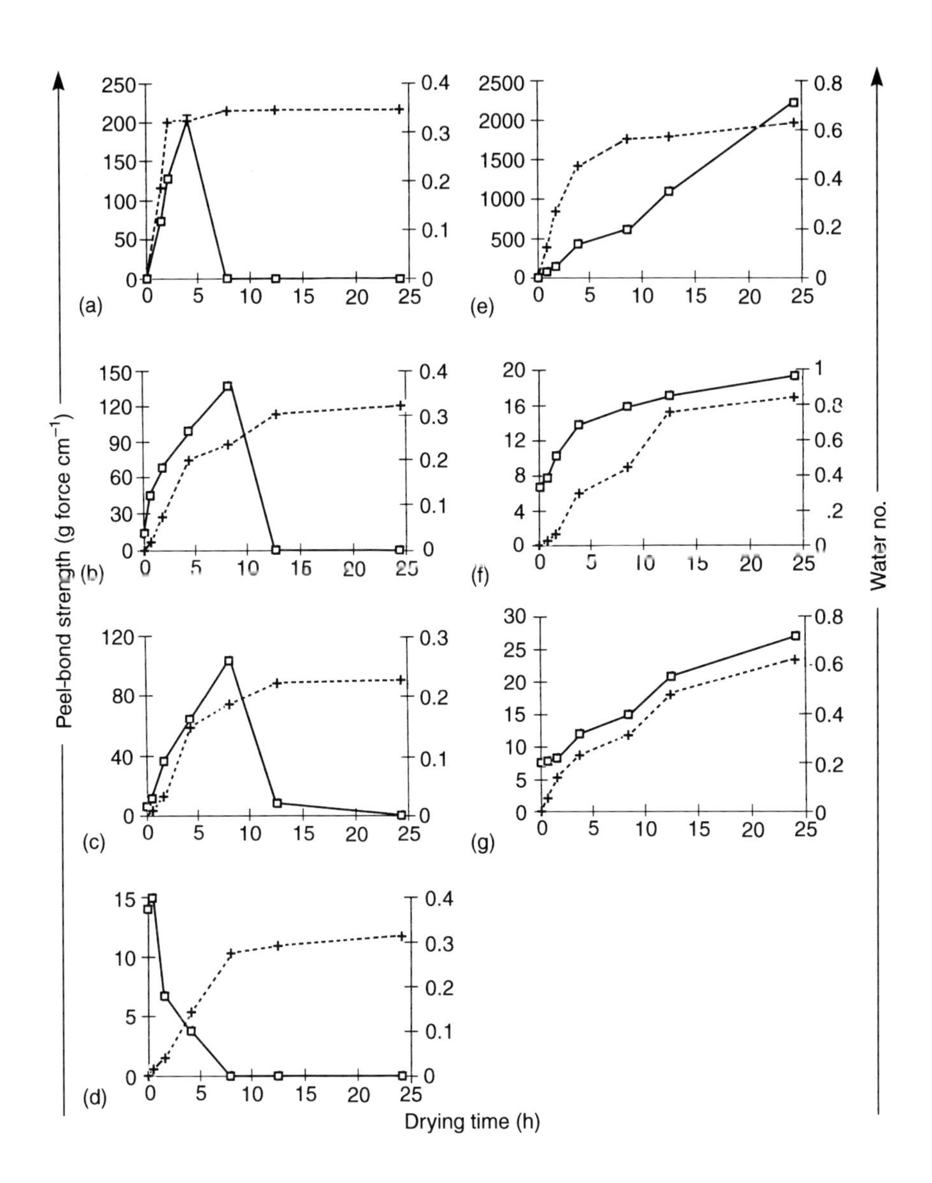

Figure 14.8 Relationship between contact time of the covering cellulose acetate film adhered to SSM substrate with a 1% water content by hydrocolloid adhesive, and the 90°-peel-bond strength effected by water loss from the paste, expressed as water number. The dashed line represents the water number; the solid line represents the peel-bond strength. Gum concentrations given as (g per 100 g): (a) dextran (65 g), (b) ghatti (65 g), (c) AG (75 g), (d) tapioca (65 g), (e) carbopol (35 g), (f) pectin (15 g), (g) MHPC (20 g). (From Ben-Zion and Nussinovitch, 1997; by permission of Oxford University Press.)

rapidly. An excessive amount of water at the interface causes overextensions of the hydrogen bonds and other adhesive forces, leading to a weakening in adhesive bond strength (Chen and Cyr, 1970). The mechanism of wet adhesion is based on the following concept, explained by Chen and Cyr (1970). They state that when a dry hydrocolloid is hydrated, the long chains of the polymer are liberated to a freely moving state. The stretched, entangled, or twisted molecules are able to match their active adhesive sites with those on the hydrophilic substrate to form an adhesion interaction, or match with each other to form cohesive bonds. When an optimum amount of water is present, at or near the interface, perfect matching is attained. A hydrocolloid that is hydrated by insufficient amounts of water will not possess completely liberated and exposed adhesive sites.

In the case of carbopol-934 and pectin (Fig. 14.7), no significant change in peel-bond strength was found in films with over $\sim 50\%$ water present in the SSM. This suggests that adhesion is not affected by such degrees of hydration and that the constant peel-bond strength is a result of the cohesive force being measured.

Figure 14.8 shows the relationship between peel-bond strength and water number (amount of water evaporated from a standard sample) and the drying time of the wet glues produced from the seven gums representing the various natural and synthetic hydrocolloids in this study. The wet glue served as an adhesive between the covering cellulose acetate membrane and the SSM layer with $\sim 1\%$ water content. Each of the seven hydrocolloids tested exhibited different behaviors, in terms of the aforementioned factors. For dextran, ghatti gum and AG (Fig. 14.8a–c) the peel-bond strength increased with time up to ~ 8 h and then decreased, as a result of the gradual drying of the wet glues. Tapioca–dextrin's behavior (Fig. 14.8d) was somewhat similar. For carbopol, pectin and MHPC (Fig. 14.8e–g), a gradual increase in peel-bond strength was observed over the entire test period. For pectin and MHPC, adhesion strength values were lower than those of the other adhesives. After 8–12 h, peel-bond strength for all hydrocolloid glues except carbopol, pectin and MHPC decreased to a value of ~ 0. This is because the wet glue dried and adhered strongly to the SSM only, whereas no attachment existed with the cellulose acetate film. The water number for all tested hydrocolloids increased with drying time, except in the case of MHPC and pectin where the initial water content of the glue was higher than in other systems.

The influence of molecular weight on peel-bond strength was investigated using HPC with molecular weights of 10^5, 3.7×10^5 and 10^6. We found that the higher the molecular weight, the higher the peel-bond strength, exhibiting values of 1.7, 5.9 and 7.7 g force cm^{-1}, respectively. These results concur with previous findings for other hydrocolloids (Chen and Cyr, 1970; Smart *et al.*, 1984). Adhesion and bioadhesion studies on hydrocolloids reveal that compounds capable of inducing good wet adhesion have high molecular weights. Therefore, chain length may contribute considerably to adhesive strength.

14.6 Future prospects

Around 1930, water-based glues comprised ~90% of the adhesives market. From 1930 to 1986, this figure decreased to ~60%. However, an average annual increase of ~3.9% since the mid-1980s in such glue types, owing to renewed interest by the packaging, construction and medicinal industries, indicates that nature-based products will still compose 44% of the market in 1996, whereas early synthetic adhesives, including PVA, PVP and PAA, will compose 18% of the market and new synthetic adhesives ~38% (Hagquist *et al.*, 1990; Darnay and Redd, 1994).

References

Anders, R. and Merkle, N. (1989) Evaluation of laminated mucoadhesive patches for buccul drug delivery. *Int. J. Pharm.*, **49**, 231–40.

ASTM (1982) *Annual Book of ASTM Standards*, part 22, American Society for Testing and Materials, Philadelphia, PA.

Baird, G.S. and Speich, J.K. (1962) Carboxymethylcellulose, in *Water Soluble Resin* (ed. R.L. Davidson), Reinhold, London, pp. 110–32.

Bauman, M.G.D. and Conner, A.H. (1994) Carbohydrate polymers as adhesives, in *Handbook of Adhesive Technology* (ed. A. Pizzi), Marcel Dekker, New York, pp. 299–313.

Ben-Zion, O. (1995) Physical properties of adhesed hydrocolloid systems. M.Sc. thesis, The Hebrew University of Jerusalem, pp. 92.

Ben-Zion, O. and Nussinovitch, A. (1996) Predicting the deformability modulus of multi-layered texturized fruits and gels. *Lebensm. Wiss. Technol.*, **29**, 129–34.

Ben-Zion, O. and Nussinovitch, A. (1997) Hydrocolloid wet glues. *Food Hydrocolloids* (in press).

Bolger, J.C. (1983) in *Adhesives in Manufacturing* (ed. G.L. Schneberger), Marcel Dekker, New York, p. 133.

Bottenberg, P., Cleymaet, R., De-Muynck, C. *et al.* (1991) Development and testing of bioadhesive fluoride containing slow-release tablets for oral use. *J. Pharm. Pharmacol.*, **43**, 457–64.

Bouckaert, S. and Remon, J.P. (1993) In vitro bioadhesion of a buccal, miconazole slow-release tablet. *J. Pharm. Pharmacol.*, **45**, 504–7.

Brief, A. (1990) The role of adhesives in the economy, in *Handbook of Adhesives* (ed. I. Skeist), Van Nostrand Reinhold, New York, pp. 21–38.

Bruno, E.J. (1970) *Adhesives in Modern Manufacturing*, Society for Manufacturing Engineers, Dearborn, MI.

Caesar, G.V. (1962) Starch and its derivatives, in *Handbook of Adhesives* (ed. I. Skeist), Van Nostrand Reinhold, New York, pp. 170–86.

Charkoudian, J.C. (1988) A model skin surface for testing adhesion to skin. *J. Cosmet. Chem.*, **39**, 225–34.

Charkoudian, J.C. (1989) *Model Human Skin.* US Patent No. 11,877,454.

Chen, J.L. and Cyr, G.N. (1970) Compositions producing adhesion through hydration, in *Adhesion in Biological Systems* (ed. R.S. Manly), Academic Press, New York, pp. 163–81.

Darnay, A.J. and Redd, M.A. (1994) *Adhesives and Sealants*, Market Share Reporter, Gale Research Inc. Staff, Detroit, MI, pp. 232–3.

Gardon, J.L. (1966) Some destructive cohesion and adhesion tests, in *Treatise on Adhesion and Adhesives*, Vol. I (ed. R.L. Patrick), Marcel Dekker, New York, pp. 286–323.

Glavis, F.J. (1962) in *Water Soluble Resin* (ed. R.L. Davidson), Reinhold, London, pp. 133–52.

Glicksman, M. (1982) *Food Hydrocolloids*, vol. 3, CRC Press, Boca Raton, FL, p. 176.

Hagqist, J., Meyer, K.F. and Sandra, K.M. (1990) Adhesives market and applications, in *Adhesives and Sealants* (ed. C.A. Dostal), ASM International, London, pp. 87–8.

Harada, T. (1977) Production, properties and application of curdlan, in *Extracellular Microbial Polysaccharides* (ed. A. Sanford), ASC Symp. Ser., Washington DC, 265–83.

Harada, T. (1979) Curdlan: a gel forming *p*-1,3-glucan, in *Polysaccharides in Food* (ed. J.M.V. Blanshard), Butterworths, London, p. 298.

Howes, F.N. (1949) *Vegetable Gums and Resins*, Chronica Botanica Comp., Waltham, MA, pp. 56–8, 61–2.
Irons, B.K. and Robinson, J.R. (1994) Bioadhesives in drug delivery, in *Handbook of Adhesive Technology* (ed. A. Pizzi), Marcel Dekker, New York, pp. 615–27.
Ishida, M., Namubu, N. and Nagai, T. (1982) Mucosal dosage form of lidocaine for toothcare using HPC and carbopol. *Chem. Pharm. Bull.*, **30**, 980–4.
Kanig, J.L. and Manago-Ulgado, P. (1965) The in vitro evaluation of orolingual adhesives. *J. Oral Therap. Pharm.*, **4**, 413–20.
Keimel, F.A. (1994) Historical development of adhesive binding, in *Handbook of Adhesive Technology* (ed. A. Pizzi), Marcel Dekker, New York, pp. 3–15.
Keusch, P. and Essmyer, J.L. (1987) *Adhesive Polyethylene Oxide Hydrogel Sheet and its Production*. US Patent No. 4,684,558.
Kirby, K.W. (1967) Vegetable adhesives, in *Adhesion and Adhesives* (ed. R. Howink), Elsevier, Amsterdam, pp. 167–85.
Kiyosi, O. and Yasuo, S. (1986) Japanese Patent No. 86,250,080.
Krage and Wootton (1987) In *Process for Flavouring Dry Vegetable Matter* (C.H. Koene, C. Vos and J. Brasser. European Patent 0,109,968B1.
Kruger, L. and Lacourse, N. (1990) Starch-based adhesives, in *Handbook of Adhesives* (ed. I. Skeist), Van Nostrand Reinhold, New York, pp. 153–66.
Kutai, O. (1994) Comparison of tensile and peel-bond strength of resiliant liners. *J. Prosthet. Dent.*, **71**, 525–31.
Lucas, A. (1962) Adhesives, in *Ancient Egyptian Materials and Industries*, Edward Arnold, London, pp. 8–14.
Mantell, C.L. (1947) *Water-soluble Gums*, Reinhold, New York, pp. 48, 71, 72.
Morris, O. (1991) *Dextrin-based Food Grade Adhesive including Xanthan or Carboxymethylcellulose or Mixture Thereof*. US Patent 4,981,707.
Nagai, T. and Machida, Y. (1993) Buccal delivery systems using hydrogels. *Adv. Drug Deliv. Rev.*, **11**, 179–81.
Nussinovitch, A., Lee, S.J., Kaletunc, G. *et al.* (1991) Model for calculating the compressive deformability of double layered curdlan gels. *Biotechnol. Prog.*, **7**, 272–4.
Paist, W.D. (1959) Cellulosics, in *Industrial Gums* (ed. R.L. Whistler), Academic Press, New York, p. 181.
Patrick, R.L. (1966) in *Treatise on Adhesion and Adhesives*, vol. 1 (ed. R.L. Patrick), Marcel Dekker, New York, p. 4.
Piglowski, J. and Kozlowski, M. (1985) Rheological properties of pressure sensitive adhesives: polyisobutylene/sodium carboxymethylcellulose. *Rheol. Acta*, **24**, 519–24.
Rigby, G.W. (1936) *Chemical Process and Chemical Compounds Derived Therefrom*. US No. Patent 2,047,226.
Robinson, J.R., Longer, M.A. and Veillard, M. (1987) Bioadhesive polymers for controlled drug delivery, in *Controlled Delivery of Drugs* (ed. R.L. Juliano), Ann. N.Y. Acad. Sci., **507**, 307–14.
Sadle, A., Norristown, N.J. and Pratt, T.J. (1979) *Starch Carrier Composition for Adhesive Containing Urea as a Gelatinizing Agent*. US Patent No. 4,157,318.
Schacat, R.E. and Glicksman, M. (1959) Some lesser known seaweed extracts, in *Industrial Gums* (ed. R.L. Whistler), Academic Press, New York, p. 396.
Sharkey, J.B. (1987) Chemistry of stamps: dyes, phosphors, adhesives. *J. Chem. Educ.*, **64**, 195–200.
Shay, K. (1991) Denture adhesives – choosing the right powders and pastes. *J. Am. Dent. Assoc.*, **122**, 70–6.
Smart, J.D. (1991) An in vitro assessment of some mucosa dosage forms. *Int. J. Pharm.*, **73**, 69–74.
Smart, J.D., Kellaway, I.W. and Orthington, H.E.C. (1984) An in vitro investigation of mucosa-adhesive materials for use in controlled drug delivery. *J. Pharm. Pharmacol.*, **36**, 295–9.
Smith, F. and Montgomery, R. (1959) *The Chemistry of Plant Gums and Mucilages*, Reinhold, New York, pp. 15–20, 199, 404–5.
Torrey, S. (1980) *Adhesive Technology Developments since 1977*, Noyes Data Corporation, NJ, pp. 197–200.
Toulmin, H.A. (1956) *Dextran Compound Coated Body Powder*. US Patent No. 2,749,277.

15 Immobilization and encapsulation

15.1 Introduction

Flavoring materials can be encapsulated using edible films (Reineccius, 1991). Encapsulation enables the creation of a dry, free-flowing powdered flavor. The coating protects the flavoring from interaction with the food, inhibits oxidation and can enable controlled flavor release. A variety of processes can be used to encapsulate the flavoring within the film, with the latter's properties being dependent upon processing as well as composition. Of the many processes described for flavor encapsulation, spray-drying and extrusion are the most commercially advantageous (Reineccius, 1989).

15.2 Spray-drying and extrusion processes for flavor encapsulation

More than 90% of the encapsulated flavoring on the market today is produced by spray-drying. This process generally involves producing an emulsion of the flavoring in an encapsulation matrix (at a ratio of about 1:4, respectively, on a dry weight basis), taking into consideration that a minimal amount of water, preferably 40% (wet weight), needs to be included in the formulation. Homogenization is used to prepare the emulsion, with a particle size averaging $\sim 1\,\mu m$. The emulsion is fed into a spray dryer; it is passed through atomization into hot-air streams where evaporative cooling brings the inlet air temperature down from 200–325°C to 80–90°C. During drying, the temperature of the flavoring particle never exceeds that of the exit air. The matrix materials need to be water soluble, making hydrocolloids a good choice. The water solubility of the matrix is important because many flavoring materials are designed to be released by contact with water. Since continuous pumping and atomization of the feed material is essential for processing, the two main conditions that need to be fulfilled are low viscosity and a high concentration of solids (50–70%). The emulsion should be kept stable until its atomization. The encapsulation matrix should not become sticky at high temperatures, so that good yield can be achieved; after manufacturing, the product should not be hygroscopic. Of course once all of these requirements have been fulfilled, the encapsulated flavoring needs to have retained its volatile flavor constituents; the capsule should be capable of resisting evaporation and degradation during storage and of providing appropriate release of the finished product. All of these requirements limit the use of matrix materials. Maltodextrins, corn-syrup solids, modified starches and acacia gum are examples of suitable matrix builders and encapsulators.

The extrusion process involves an emulsion that is composed of a low-moisture ($\sim 15\%$) carbohydrate melt, which includes 10–20% flavoring (dry weight basis). The melt is extruded through a die under pressure and the formed flavoring drops are introduced into a cold isopropanol bath where an amorphous glass structure is formed; a mixer is used to break these structures into small pieces, followed by drying in hot air to yield the finished flavoring material (Risch, 1986). Although the process is different, requirements similar to those for the spray-drying process need to be met. The encapsulating material should be soluble with 85% solids at 110°C. Synthetic emulsifiers are added to eliminate phase separation during processing, although this is not likely to happen because of the very high viscosity of the medium, paralleling the short time between emulsion formation and solidification. Other requirements are the creation of a non-hygroscopic amorphous medium. The retention capacity of the flavoring material during production and storage and its release in accordance to requirements are of the utmost importance.

Most flavors are encapculated in water-soluble polymers: acacia gum, modified starches, maltodextrin and corn-syrup solids. Acacia gum is the traditional choice for encapsulation (Thevenet, 1986). The production of the gum, and its physical and chemical properties are described in Chapter 7. Only modified starches exhibit emulsification properties in encapsulation matrices (Trubiano and Lacourse, 1986), obtained via the chemical addition of octenyl succinate to partially hydrolysed starch. Octenyl succinate can be used at no more than 0.02 DS on the starch polymer and is still awaiting approval for food purposes. After substitution, the modified starches serve as excellent emulsifiers and, at a high solid content (40–50%), still exhibit low viscosity. If starches are to be treated with acid or an acid–enzyme combination, then corn-syrup solids or maltodextrins can be used. Products with a DE of less than 20 are classified as maltodextrins. Products having a DE that is equal to or higher than 20 are corn-syrup solids. Both are used at levels that produce high viscosities, to maintain a temporary emulsion but not confer emulsion stability to the finished product. Maltodextrins and corn-syrup solids are inexpensive compared with other hydrocolloids and can also be used as fillers.

15.3 Hydrocolloid performance in flavor encapsulation

In general, good performance of the encapsulating materials includes stabilization and formation of emulsions, flavor retention during processing and later during storage and controlled release at its destination. In some food products, no significant emulsification properties are required. In dry-beverage mixes, 1 week of shelf life (i.e. no visible ring of separation) is required, followed by reconstitution. In other words, the edible film used here as an encapsulator must provide the product with good emulsification properties. Since maltodextrins and corn-syrup solids offer no emulsion stability to the finished product, they cannot be used for dry-beverage flavor applications, unless a secondary emulsifier is incorporated. Secondary emulsifiers are not commonly used with spray-dried flavoring but are essential for the manufacture and stability of extruded flavorings. In a comparison of the stabilization of orange-oil emulsions by modified starch, corn-syrup solids or acacia gum (Reineccius, 1991), the modified food starch formed the cloudiest emulsion, i.e. the most stable flavor emulsion. This relative stability remained even after centrifugation, although all emulsions cleared to some extent.

Flavor retention during the drying process is essential to the success of a preparation. It should be noted that the more volatile constituents can easily be lost. The volatile organic compounds that are carried away by the drying air need to be removed by scrubbing, and thus poor retention creates complications during processing. In a comparison of the encapsulating efficiencies of acacia gum versus starch carriers (Trubiano and Lacourse, 1986), the true encapsulated flavor was highest when the low-viscosity starch octenyl succinate was used (29.4% versus 23.9% for acacia gum and 17.4% for dextrin). It should be noted that modified food starches can typically be used at a concentration of 50–55% solids, whereas acacia gum is used at 30–35% solids. A higher level of solids results in better flavor retention (Reineccius, 1989). The poor emulsification properties obtained using maltodextrins result in poor flavor retention during drying.

Decreases in the quality and shelf life of flavoring materials are caused by the oxidation of flavoring agents and the consequent development of off-flavors. Maltodextrins provide varying amounts of protection, depending upon their DE (Anandaraman and Reineccius, 1986): the higher the DE, the better the protection. This poses a problem since higher DE materials tend to be difficult to dry, are very hygroscopic and result in poor flavor retention, leaving open the possibility of achieving an unsuitable product with an excellent shelf life. Modified food starches are used to achieve encapsulated flavoring with good retention, but they have poor protection against oxidation.

A preparation's shelf life is related to the permeability of the encapsulating matrix to oxygen (Baisier and Reineccius, 1989). The rate of molecular diffusion is dependent upon the state of the polymer composing the capsule. At low water activity and temperature, the related polymers are in a glassy state, limiting molecular diffusion; at higher water activity and temperature the material changes to a rubbery state, permitting molecular diffusion (Levine and Slade, 1989; Nelson and Labuza, 1992). In other words, when the encapsulating film remains in a glassy state, oxygen diffusion is limited or non-existent, whereas it occurs when the film changes to a rubbery state. It should be noted that this relationship was not found for the oxidation of orange oil that had been spray-dried using maltodextrin as the encapsulating matrix. Nevertheless, as water activity increased, the product was less stable.

To prepare flavorings, water-soluble edible films are prepared. These preparations release flavor upon contact with water. Slow release is necessary in products such as chewing gums or baked and microwaved products. Such controlled properties are achieved by secondary coatings (with, for example, fats, oils, edible shellacs, modified cellulose) around an already encapsulated flavoring. Since this coating dilutes flavor strength and adds to the cost, better techniques need to be sought.

The fundamental equations governing the controlled release of active ingredients and the application of controlled-release technology in food systems have been recently reviewed (Pothakamury and Barbosa Canovas, 1995). The method of microencapsulation, among others, can be applied to achieve controlled release in foods. Some of the release mechanisms employed in the food industry involve one or a combination of the following stimuli: a change in temperature, moisture or pH, the application of pressure or shear, and the addition of surfactants. Encapsulation is a method of protecting food ingredients that are sensitive to temperature, moisture, microorganisms or other components of the food system. Such food ingredients include flavors, sweeteners, enzymes, food preservatives and antioxidants, and they are encapsulated using carbohydrates, gums, lipids and/or proteins. With a properly designed controlled-release delivery system, the food ingredients are released at the desired site and time, at the desired rate (Pothakamury and Barbosa Canovas, 1995).

15.4 Immobilization of cells

Immobilization relates to the prevention of free cell movement (by natural or artificial means), independent of its neighbors, to any part of the aqueous phase of the system under study (Bucke, 1983). At one extreme, the immobilized cells can be fully capable of division; at the other, they may possess only one type of enzymatic activity (Mattiasson, 1983). There are several techniques by which cells can be immobilized, including adsorption to neutral or charged supports, flocculation, entrapment by natural or synthetic polymers, covalent coupling and containment. The support materials for immobilization are divided into two types: organic and inorganic. The discussion in this chapter is limited to organic support materials, which fall into three categories: polysaccharides, proteins and synthetic polymers. Polysaccharides include agar, agarose, alginate, carrageenan, cellulose, dextran, pectate and lignin (within wood); proteins include collagen, egg white and gelatin, and the synthetic polymers include polyacrylamide, phenolic resins, polystyrene and polyurethane. In general, the desirable features for immobilized cell preparations are high biocatalytic activity, long-term stability of the biocatalyst, possibility of regenerating the biocatalyst, low loss of activity during immobilization, low leakage of cells, non-compressible particles, high resistance to abrasion, resistance to microbial degradation, low diffusional limitations, spherical shape, high surface area, appropriate density for the reactor type, technique simplicity, inexpensive support materials and non-toxicity of those materials (Tampion and Tampion, 1987).

15.4.1 Entrapment

Cells are often entrapped within a gel matrix. A wide range of characteristics are attributed to gels as an entrapment medium. On the one hand, they include macromolecules held together by relatively weak intermolecular forces, such as hydrogen bonding or ionic cross-bonding by divalent or multivalent cations. On the other hand, strong covalent bonding, where the lattice in which the cells are entrapped is considered as one vast macromolecule, is limited only by the particle size in the immobilized cell preparation (Tampion and Tampion, 1987; Nussinovitch *et al.*, 1994).

The major categories of entrapment have been reviewed (Cheetham, 1980; Bucke, 1983; Mattiasson, 1983; Nussinovitch *et al.*, 1994). They include some commonly used, single-step entrapment methods, such as the simple gelation of macromolecules by lowering or raising temperatures using hydrocolloids such as agar, agarose, κ-carrageenan, chitosan, gelatin and egg whites, among others (Fig. 15.1). These preparations regularly suffer from low mechanical strength and possible heat damage. Another simple single-step entrapment method is the iontropic gelation of macromolecules by di- and multi-valent cations, alginate (Fig. 15.1) and LMP being good examples of this. The limitations of such a system are low mechanical strength and breakdown in the presence of chelating agents. Other single-step entrapment methods include the use of synthetic polymers produced by chemical or photochemical reaction. Typically used materials include epoxy resins, polyacrylamide and polyurethane. In such cases, gel precursors are often toxic and some gels have low mechanical strength. Another single-step entrapment method is by precipitation from an immiscible solvent, as in the case of cellulose triacetate and polystyrene. Here the solvents are often toxic. An overview of hydrogels for cell immobilization has been recently published (Jen *et al.*, 1996). Current developments in the immobilization of mammalian cells in hydrogels are also reviewed. Discussions cover hydrogel requirements for use in adhesion, matrix entrapment and microencapsulation, the respective processing methods, as well as the current applications (Jen *et al.*, 1996).

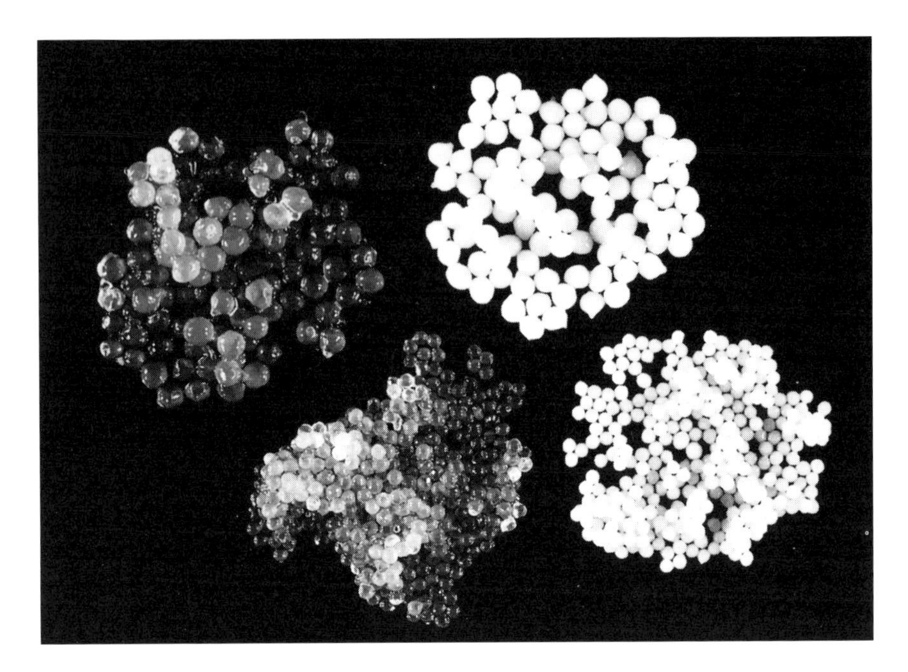

Figure 15.1 (a) Hydrocolloid beads of chitosan (top right), alginate (bottom right) and colored beads of the same (left) (actual size). (b) Entrapped alginate beads (magnification ×5) within a ruptered-compressed cylindrical gel: a convenient way of producing multi-textured food, composite material or complex entrapment.

When agar is used regularly, a dilute solution, usually 2–4% agar, is prepared in a medium suitable for the particular cells being immobilized. The agar is dissolved at temperatures of ~ 90°C before any cells are added to the solution. This addition is effected at a temperature a bit higher than the setting point to minimize heat damage. Usually a 1:1 ratio of cell suspension to agar solution is used (Brodelius and Nilsson, 1980). The gel can be formed in sheets or slabs of any desired thickness, to be cut later into smaller pieces. Perforated molds are used to produce short cylindrical beads (Brodelius and Nilsson, 1980). Spherical beads can be obtained by dropping the hot solution of the preparation into ice-cold buffer. Another technique is to add the hydrocolloid solution to a heated oil bath to produce a hot emulsion that is cooled later to achieve the gel beads (Wiksstrom *et al.*, 1982). Modifications of such preparations can be found in the literature (Banerjee *et al.*, 1982). Other reported entrapments include the immobilization of *Bacillus licheniformis* in agar and alginate gels. The optimal immobilization parameters (gel concentration, initial cell quantity, biomass age, bead size and solidification prolongation) were determined. The immobilization procedure was most effective at a gel concentration of 4% using cells from a 12 h culture. The optimal initial cell quantity was found to be 0.6–3.0% in the agar gel and 0.4% in the calcium alginate gel with bead sizes of 3 and 5 mm, respectively. An enzyme yield of $1.1 \times 10^6\,\mathrm{U\,l^{-1}}$ culture medium was reached in batch fermentation with agar-entrapped cells under optimal conditions. This activity represented 135% of the corresponding yield obtained with free cells. A significant increase (2.2-fold) was observed in the enzyme yields in the fourth cycle of repeated batch runs with cells entrapped in agar gel pellets with a bead size of 5 mm (Dobreva *et al.*, 1996).

The mechanical properties of entrapped bacteria, yeast and mold spores within an agar matrix have been recently discussed (Nussinovitch *et al.*, 1994). Agarose, the gelling agent within agar, was used (Khachatourians *et al.*, 1982) to entrap whole cells of *Escherichia coli* and anucleate minicells produced by a mutant with defective cell division at 50°C. Agarose entrapment is one of the mildest methods available and viability is retained. Special grades of agarose that have lower gelling temperatures are supplied by some manufacturers. One of these agarose grades was used (Brodelius and Nilsson, 1980) to immobilize *Catharanthus roseus.* These cells retained cellular integrity, as evidenced by their respiratory activity and susceptibility to plasmolysis, and were capable of growth. Many other applications, including the entrapment of algae and blue–green bacteria, have been reported (Wiksstrom *et al.*, 1982).

The use of κ-carrageenan as an immobilization medium was reported by Chibata (1981). It is more readily used than agar as a growth substrate by microorganisms. A *Saccharomyces cerevisiae*–carrageenan mixture was pumped for gelation into a 2% potassium chloride solution (Wang and Hettwer, 1982). A high number of cells were found to be loaded within the beads. Ten times more cells were observed in a batch reactor holding the immobilized cells than with free cells. Although both followed the same typical growth curve, the immobilized cells reached the stationary-phase plateau at a higher cell density, taking about twice the time of the free yeast cells. Cell leakage could be reduced from carrageenan beads by increasing the potassium chloride concentration to 4%, but with a concomitant loss in cell viability (Wang *et al.*, 1982). Eighteen species (Mattiasson, 1983) were entrapped by such beads. The stability of κ-carrageenan gels in the presence of potassium chloride was demonstrated experimentally by Krouwel *et al.* (1982). These authors claim carrageenan gels (beads) to be superior to agar and inferior to calcium alginate. A comparison between κ-carrageenan and calcium alginate as entrapment media for *Zymomonas mobilis* (Grote *et al.*, 1980) was reported. These authors found a 30% reduction in activity after 1 month of a continuous process. The *Z. mobilis* was added to 2% κ-carrageenan at 47–50°C and used to coat Raschig rings in a rotating flask. These rings were then packed into a column and stabilized with 0.75% potassium chloride in 15% glucose solution. The bacterial cells continued to reproduce, and the void space in the column reactor was, therefore, reduced. A small advantage of κ-carrageenan over calcium alginate was reported. Kappa-carrageenan and supplemental LBG were used to immobilize fungal spores of *Penicillium urticae* for the production of the antibiotic patulin (Deo and Gaucher, 1983). The production half-life was also extended from 6 days (free cells) to 16 days for the immobilized preparation. Biotransformation of progesterone by entrapped *Aspergillus phoenicus* in matrices of carrageenan and calcium alginate has also been reported (Kim *et al.*, 1982). In a study of *C. roseus* cells immobilized in κ-carrageenan, some evidence of heat damage was reported (Brodelius and Nilsson, 1980). The rate of bioconversion of tryptamine to ajmalicine was lower than in free cells or in those immobilized in agarose or calcium alginate.

Collagen is an animal protein derived from connective tissues. To stabilize its structure, it is usually tanned using glutaraldehyde, which cross-links the protein molecules. Cell immobilization also makes use of this reaction for matrix stabilization. However, excessive exposure to glutaraldehyde damages cell function, and optimal conditions need to be chosen carefully before preparation. An interesting overview of the entrapment of eight types of bacteria within a membrane cast from tanned collagen can be found in Venkatsubramanian *et al.* (1983). However, the method described by these authors is less favored because gelatin produced from the same source is of higher quality. Studies of *C. roseus* entrapment within gelatin, gelatin–agarose and gelatin–alginate were reported by Brodelius and Nilsson (1980). Cell growth and respiration were adversely affected by all methods involving glutaraldehyde treatment. Although the structural integrity of the plasma membrane was retained, its role as a selectively permeable membrane was assumed to have been affected. In fact, other available proteins such as egg white can be used for immobilization. Glutaraldehyde can be used as a cross-linking agent.

The most studied method of entrapment is sodium alginate gelation by cross-linking via di- or multi-valent ions. Such gelation is easy to perform: cells are mixed with a sodium alginate solution and added dropwise into dilute baths of calcium chloride. The method retains cell viability, but cell decomposition can occur via the introduction of chelating agents such as phosphate and/or citrate buffers. The strength and porosity of the beads can be somewhat controlled by choosing an alginate with a specific composition. The higher the L-guluronic acid content, the stronger the gel. If alginate with a higher proportion of D-mannuronic acid is chosen for immobilization, beads with larger internal pore sizes are produced. As the concentration of alginate in the beads increases, so does mechanical strength. An increase in cell mass has the reverse effect. Diffusion coefficients of glucose and ethanol in alginic membranes were studied (Hannoun and Stephanopoulos, 1986). Diffusivity decreased with increasing concentrations of alginate. In a work describing alginate-immobilized *Chlorella* (Dainty *et al.*, 1986), the importance of providing suitable environmental conditions, especially if prolonged use of alginate is desired, was discussed. Trivalent ions were beneficial in achieving greater alginate bead strengths (Rochefort *et al.*, 1986). Immobilization of *Bacillus acidocaldarius*, whole-cell rhodanese in polysaccharide matrices of calcium alginate, κ-carrageenan and chitosan (Fig. 15.2) and in

Figure 15.2 Chitosan beads (magnification ×7).

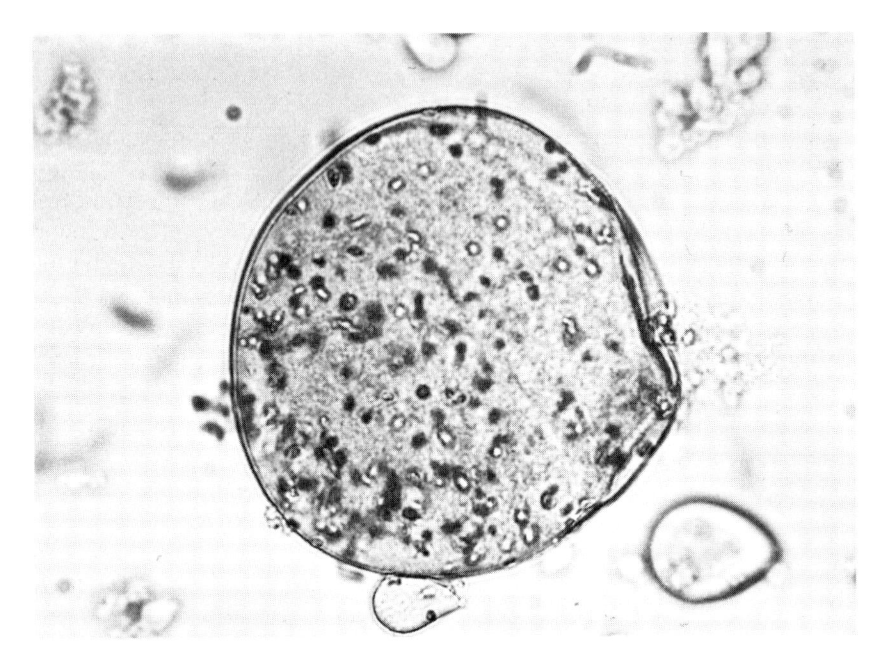

Figure 15.3 Yeasts immobilized within an alginate bead.

an insoluble gelatin gel was recently reported (Deriso *et al.*, 1996). The results obtained with the different immobilizates in terms of activity yield, possibility of regeneration and operative stability were evaluated with the aim of setting up a continuous system. This was achieved with a system consisting of *B. acidocaldarius* cells entrapped in an insolubilized gelatin matrix. The latter, in the form of a thin membrane, was employed in a custom-conceived reactor operating as a plug flow reactor (Deriso *et al.*, 1996). Another study described the immobilization of *Acetobacter aceti* cells in a calcium alginate gel and adsorption onto preformed cellulose beads. The cell number within the supports showed no significant alterations with changing temperature or pH, whereas acetic acid production was slightly increased by immobilization (Krisch and Szajani, 1996). When *S. cerevisiae* cells were inoculated at low density in alginate gel beads (Fig.15.3) and cylinders, cells grew in the form of distinct microcolonies throughout the gel matrix. Alginate-gel beads gave rise to microcolonies that became elongated and lens-shaped with their major axes aligned with the gel surface. The aspect ratio (major axis:minor axis length) of the microcolonies and the local concentration of alginate increased with increasing distance from the center of the gel particles. In contrast, spherical microcolonies were observed in alginate cylinders formed by internal gelation and no significant local concentration gradients of alginate were detected in these gels. Non-spherical microcolonies were also observed in carrageenan gel beads. However, the colonies were irregularly shaped, and their major axes demonstrated no preferential alignment (Walsh *et al.*, 1996).

Enzymes and enzymatic activity can be obtained easily following entrapment. The enzymatic oxidation of cephalosporin C to glutaryl-7-amino-cephalosporanic acid (glutaryl-7-ACA) was carried out utilizing permeabilized whole cells of the yeast *Trigonopsis variabilis* entrapped in calcium alginate beads. The biomass, cultured in a rich medium containing methionine and harvested after 72 h of growth, exhibited high levels of D-aminooxidase activity. Prior to use, the whole cells were permeabilized with four freeze–thawing cycles and immobilized in polysaccharide matrices, such as calcium alginate and κ-carrageenan, and in an insolubilized gelatin gel. The best results in terms of activity yield and storage stability were obtained with cells entrapped in calcium alginate beads (Dacunzo *et al.*, 1996). Cathepsin B, purified from goat brain, was immobilized in calcium alginate beads in the presence of bovine serum albumin. The immobilized enzyme retained close to 63% of its original activity and could be used for seven successive batch reactions with a retention of 22–30% of its initial activity (Kamboj *et al.*, 1996). Screening of matrices for the immobilization of naringinase demonstrated 2% sodium alginate to be the optimal matrix. Upon immobilization, 30 U naringinase effected 82% naringin hydrolysis in 3 h; a broadening of pH optima was attributed to imparting desirable flexibility for debittering kinnow juice of various pHs, and temperature profiles indicated improved thermostability, which could be useful for reducing the cost of debittering. Alginate permitted the attainment of equilibrium readily with no hindrance in the inflow of naringin and outflow of naringenin/prunin, and adequate mechanical stability, cumulatively indicating the feasibility of its commercial exploitation (Puri *et al.*, 1996).

Many industrial and semi-industrial processes to produce large masses of beads have been described. A rotating nozzle-ring was used to spray the alginate with the cells into a cross-linking solution. An alternative approach was to use a dual-fluid atomizer to shear sodium alginate beads off the tip of hypodermic needles by air stream into a calcium chloride solution. Many reports on the use of this popular entrapment method can be found in the literature (Grote *et al.*, 1980; Margaritis *et al.*, 1981; Klein and Kressdorf, 1983; Grizeau and Navarro, 1986; Nussinovitch *et al.*, 1994).

Chitosan is a polycation that can be used for cross-linking by multivalent anions. Many reports describing cell entrapment within chitosan beads can be found (Vorlop and Klein, 1981; Kluge *et al.*, 1982; Stocklein *et al.*, 1983). The higher stability of chitosan beads over alignate was reported by Stocklein *et al.* (1983). Chitosan beads dissolve in the absence of phosphate and swell under acidic conditions but are stable in the presence of chelating agents and monovalent cations (Fig. 15.4).

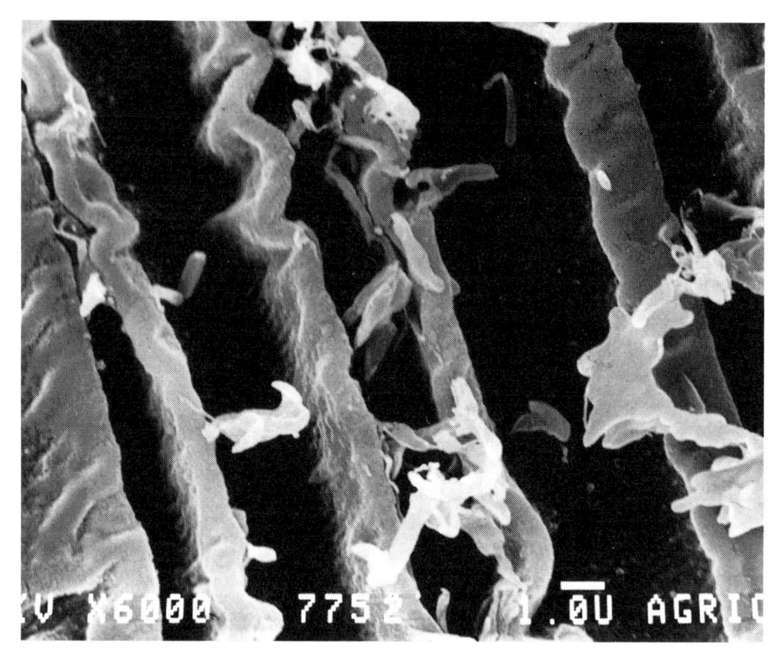

Figure 15.4 Denitrification bacteria on formed grooves within chitosan beads 14 days after the beginning of processing. (From Nussinovitch *et al.*, 1996; by permission of the American Chemical Society.)

Polyacrylamide is a synthetic polymer that exhibits different properties depending on the reagents and conditions used during its preparation. An early study of cell immobilization by entrapment in polyacrylamide was reported by Mosbach and Mosbach (1966). The mechanical strength of the gel increases in proportion to the square root of the concentration of the acrylamide monomer used with a given fixed concentration of cross-linking agent. This is followed by a decrease in the gel's pore size, which imposes some diffusional restrictions. The least cell damage is achieved by using relatively high concentrations of tetramethylethylenediamine and ammonium persulfate. The mechanical properties of the polyacrylamide gels were compared with those of alginate, agar and κ-carrageenan (Krouwel *et al.*, 1982). The achievement of maximal mechanical strength was also reported (Klein *et al.*, 1978). The diffusional properties of polyacrylamide gels containing immobilized *Alcaligenes faecalis* were studied (Whcatley and Phillips, 1983). These properties are clearly of paramount operational significance. In fact, the unsuitability of polyacrylamide gels for particular cell types was demonstrated. Positive results on the entrapment of *E. coli* in polyacrylamide were reported (Bang *et al.*, 1983). Polyacrylamide was found to be superior to polymethacrylamide or polyepoxide in terms of the production of tryptophan from serine and indole. Many other synthetic polymeric gel systems used for entrapment are mentioned in the literature. They include copoly(styrene–maleic acid), polyethylene glycol methacrylate, polyisocyanates and polyurethane, among many others. The most commonly used thermoplastic membrane material is probably acrylonitrile vinyl chloride copolymer (PAN–PVC), which was used in studies of encapsulated insulin-producing tissue in animal models of diabetes.

Cellulose itself is not soluble in water, but it can be dissolved in organic liquids to form beads entrapping bacteria and enzymes (Linko *et al.*, 1977). Cellulose acetate has been used for entrapment, but leakage and diffusional restrictions were reported. A detailed discussion (Dinelli, 1972) of the entrapment of both enzymes and cells in solid fibers made from a variety of polymers including celluloses was published in terms of the advantage of using standard equipment for the wet spinning of fibers, a method that can be easily scaled up to any desired quantity.

A different approach for achieving predetermined properties of entrapment media consists of two-step methods. Combinations of different methods could provide a means of overcoming evident shortcomings. Alginate-bead stabilization using a combination of three methods was discussed by Birnbaum *et al.* (1981). The exact mechanism of the three methods is not fully understood, but they all probably involve some ionic or covalent binding of the cells themselves. Even an element of direct coupling of the cells to the activated alginate has been considered. A mixed alginate–gelatin system for yeast encapsulation was studied. Following bead formation, the alginate was leached out in phosphate buffer and the porous matrix produced by the gelatin was stabilized by cross-linking with glutaraldehyde (SivaRaman *et al.*, 1982). Kappa-carrageenan beads are stabilized by treatment with hexamethylenediamine and glutaraldehyde (Chibata, 1981). As previously described, alginate alone can form a capsule, but more sophisticated applications have been proposed by combining alginate with the negatively charged poly-amino acid polylysine. In the latter case, the barrier is a weak polyelectrolyte complex formed by the reaction of alginate and polylysine. In many cases, an extra layer of alginate on the preformed membrane is applied to achieve better compatibility.

To modify the structure of phosphorylated PVA gel beads in order to improve their gas permeability, some additives such as soluble starch, partially saponified PVA and calcium alginate were used. A small amount of calcium alginate was added to the gel solution during gelation. The calcium alginate was later destroyed by treating the beads with phosphate solution, thereby effectively promoting gas permeation. The modification process with alginate did not cause a significant decrease in the nitrate reduction rate of immobilized beads. The gas permeability of the beads increased by as much as 62% (Chen *et al.*, 1996). Stable, semi-permeable polyamide microcapsules were prepared by interfacial polymerization from a mixture of 1,6-hexanediamine and poly(allylamine) cross-linked with diacid chlorides and were used to encapsulate baker's yeast. The encapsulated cells were studied as a biocatalyst for the model reduction of 1-phenyl-1,2-propanedione to 2-hydroxy-1-phenyl-1-propanone in a number of organic solvents. Reduction by the microencapsulated yeast cells was carried out more efficiently than by free cells or by those immobilized in alginate and κ-carrageenan beads (Green *et al.*, 1996).

Although immobilization in gel matrices is common practice, there seems to be little or no information on how the properties (number, size, etc.) of the entrapped microorganisms influence the mechanical characteristics of the matrix. Recently, some insight has been gained into the influence of bacteria, yeast and fungal spores (Nussinovitch *et al.*, 1994). *L. lactis* bacteria, *S. cerevisiae* yeasts and *Trichoderma viride* fungal spores were immobilized in agar and alginate gels and the mechanical properties of these gels were tested by compression using a universal testing machine. Data were collected as volts versus time and later converted to stress versus strain. Both gel systems exhibited similar behavior after incorporation of the microorganisms. In most cases the addition of microorganisms up to $10^8\,CFU\,l^{-1}$ of gel (CFU, colony-forming unit) did not change the gel's strength (stress at failure), or affect its deformability modulus or Hencky's strain at failure. Different behavior was observed for lactic acid bacteria in an alginate gel, when 10^{10}–10^{12} cells l^{-1} gel were immobilized. Both agar and alginate gels suffered weakened textures (even though their mechanical properties, gelation mechanisms and immobilization methods were very different). Gel strengths and deformability moduli decreased by a factor of 1.2–4.4, depending on the type of gel, the number of microorganisms per unit volume and the diameter of the embedded microorganisms. Following immobilization, gels became more brittle, as demonstrated by a decrease in their Hencky's strain at failure (Nussinovitch *et al.*, 1994). Because the situation of particles embedded within a continuous matrix very much resembles that of a composite material, the question was raised as to whether mathematical models designed for composite materials could also describe gels with entrapped microorganisms. To answer this, *T. viride* fungal spores were immobilized in alginate gels. The mechanical properties of these gels were tested by compression in the universal testing machine. A simple mathematical model for composite materials was found to properly describe the relative stress at failure versus the phase volume of the fungal spores. Random loose packing of the spores was proposed (Nussinovitch, 1994). The stability of alginate and agarose gels used in cell encapsulation has been recently studied (Shoichet *et al.*, 1996). The gel strength of agarose diminished in the presence of cells, because they interfered with the hydrogen-bond formation required for agarose gelatin. The gel strength of calcium- or barium-cross-linked alginate, in this case, decreased over 90 days, with an equilibrium gel strength being achieved after 30 days. The stability of calcium-cross-linked sodium alginate gels over a 60-day period was monitored by diffusion of proteins ranging from 14.5 to 155 kDa. From these diffusion measurements, the average pore size of the calcium-cross-linked alginate gels was estimated, using a semi-empirical model, to increase by 17.6 to 28.9 nm over a period of 60 days (Shoichet *et al.*, 1996).

The abilities of two grapevine cell suspensions to biotransform geraniol into nerol in a biphasic system based on culture medium and Miglyol 812 were compared. The Gamay grape-cell suspension was able to transform a higher concentration of geraniol into nerol than the Monastrell one. Immobilization proved to be advantageous in protecting cells against the toxicity of the substrate. Furthermore, immobilization seemed to have an effect on the secondary metabolism: the cells immobilized in polyurethane foam were more efficient at performing the isomerization process than either the freely suspended or calcium alginate immobilized cells (Guardiola *et al.*, 1996).

Current membrane-based bioartificial organs consist of three basic components: a synthetic membrane, cells that secrete the product of interest and encapsulated matrix material. Alginate and agarose have been widely used to encapsulate cells for artificial organ applications. It is important to understand the degree of transport resistance imparted by these matrices in cell encapsulation to determine if adequate nutrient and product fluxes can be obtained (Li *et al.*, 1996). In general, 2–4% agarose gels offered little transport resistance for solutes up to 150 kDa, whereas 1.5–3.0% alginate gels offered significant transport resistance for solutes in the range 44–155 kDa, lowering their diffusion rates by 10- to 100-fold compared with their diffusion in water. Doubling the alginate concentration had a more significant effect on hindering the diffusion of larger-molecular-weight species than did doubling the agarose concentration. Average pore diameters were approximately 17.0 and 14.7 nm for 0.5 and 3% alginate gels, respectively, and 48.0 and 36.0 nm for 2 and 4% agarose gels, respectively. These values were estimated using a semi-empirical correlation based on diffusional transport of different-size solutes. This method, developed for measuring diffusion in these gels, is highly reproducible and is useful for gels cross-linked in the cylindrical geometry relevant for studying transport through matrices used in cell immobilization in the hollow-fiber configuration (Li *et al.*, 1996).

References

Anandaraman, S. and Reineccius, G.A. (1986) Stability of spray-dried orange peel oil. *Food Technol.*, **40**(11), 88.

Baisier, W. and Reineccius, G.A. (1989) Spray drying of food flavors: factors influencing shelf-life of encapsulated orange peel oil. *Perfum. Flav.*, **14**, 48–53.

Banerjee, M., Chakrabarty, A. and Majumdor, S.K. (1982) Immobilization of yeast cells containing β-galactosidose. *Biotech. Bioeng.*, **24**, 1839–50.

Bang, W.G., Behrendt, U., Lang, S. *et al.* (1983) Continuous production of L-tryptophan from indole and L-serine by immobilized *Escherichia coli* cells. *Biotechnol. Bioeng.*, **25**, 1013–25.
Birnbaum, S., Pendleton, R., Larsson, P. *et al.* (1981) Covalent stabilisation of alginate gel for entrapment of living whole cells. *Biotechnol. Lett.*, **3**, 393–400.
Brodelius, P. and Nilsson, K. (1980) Entrapment of plant cells in different matrices. *FEBS Lett.*, **122**, 312–16.
Bucke, C. (1983) Immobilized cells. *Philos. Trans. R. Soc. London, Series B*, **300**, 369–89.
Cheetham, P.S.J. (1980) Developments in the immobilization of microbial cells and their applications, in *Topics in Enzyme and Fermentation Biotechnology*, vol. 4 (ed. A. Wiseman), Ellis Horwood, Chichester, NJ, pp. 189–238.
Chen, K.C., Chen, S.J. and Houng, J.Y. (1996) Improvement of gas permeability of denitrifying PVA gel beads. *Enz. Microb. Technol.*, **18**(7), 502–6.
Chibata, I. (1981) Immobilized microbial cells with polyacrylamide gel and carrageenan and their industrial applications, in *Immobilized Microbial Cells* (ed. K. Venkatsubramanian) *Am. Chem. Soc. Symp. Ser.*, **106**, 187–202.
Dacunzo, A., Dealteriis, E., Maurano, F. *et al.* (1996) D-Amino-acid oxidase from *Trigonopsis variabilis* immobilization of whole cells in natural polymeric gels for glutaryl-7-aminocephalosporanic acid production. *J. Ferm. Bioeng.*, **81**(2), 138–42.
Dainty, A.L., Goulding, K.H., Robinson, P.K., Simpkins, I. and Trevan, M.D. (1986) Stability of alginate-immobilized algal cells. *Biotechnol. Bioeng.*, **28**, 210–216.
Deo, Y.M. and Gaucher, G.M. (1983) Semi-continuous production of the antibiotic patulin by immobilized cells of *Pennicillium urticae. Biotechnol. Lett.*, **5**, 125–30.
Deriso, L., Dealteriis, E., Lacara, F. *et al.* (1996) Immobilization of *Bacillus acidocaldarius* whole-cell Rhodanese in polysaccharide and insolubilized gelatin gels. *Biotechnol. Appl. Biochem.*, **23**, 127–31.
Dinelli, D. (1972) Entrapment in solid fibres. *Process Biochem.*, **7**(8), 9–12.
Dobreva, E., Ivanova, V., Tonkova, A. *et al.* (1996) Influence of the immobilization conditions on the efficiency of α-amylase production by *Bacillus licheniformis*, **31**(3), 229–34.
Green, K.D., Gill, I. S., Khan, J.A. *et al.* (1996) Microencapsulation of yeast cells and their use as a biocatalyst in organic solvents. *Biotechnol. Bioeng.*, **49**(5), 535–43.
Grizeau, D. and Navarro, J.M. (1986) Glycerol production by *Dunaliella tertiolecta* immobilized with Ca alginate beads. *Biotechnol. Lett.*, **8**, 261–4.
Grote, W., Lee, K.J., and Rogers, P.L. (1980) Continuous ethanol production by immobilized cells of *Zymomonas mobilis. Biotechnol. Lett.*, **2**, 481–6.
Guardiola, J., Iborra, J.L., Rodenas, L. *et al.* (1996) Biotransformation from geraniol to nerol by immobilized grapevine cells (*V. vinifera*). *Appl. Biochem. Biotechnol.*, **56**(2), 169–80.
Hannoun, B.J.M. and Stephanopoulos, G. (1986) Diffusion coefficients of glucose and ethanol in cell-free and cell-occupied calcium alginate membranes. *Biotechnol. Bioeng.*, **28**, 829–35.
Jen, A.C., Wake, M.C. and Mikos, A.G. (1996) Review – hydrogels for cell immobilization. *Biotechnol. Bioeng.*, **50**(4), 357–64.
Kamboj, R.C., Raghav, N., Nandal, A. *et al.* (1996) Properties of cathepsin-B immobilized in calcium alginate beads. *J. Chem. Technol. Biotechnol.*, **65**(2), 149–55.
Khachatourians, G.G., Brosseau, J.D. and Child, J.J. (1982) Thymidine phosphorylase activity of anucleate minicells of *E. coli* immobilized in an agarose gel matrix. *Biotechnol. Lett.*, **4**, 735–40.
Kim, M.N., Ergan, F., Dhulster, P. *et al.* (1982) Steroid modification with immobilized mycelium of *Aspergillus phoenics. Biotechnol. Lett.*, **4**, 233–8.
Klein, J., Hackel, U., Schara, P., Washausen, P. *et al.* (1978) Polymer entrapment of microbial cells: preparation and reactivity of catalytic systems, in *Enzyme Engineering*, vol. 4, (eds G.B. Broun, G. Manecke and L.M. Wingard), Plenum Press, New York, pp. 339–41.
Klein, J. and Kressdorf, B. (1983) Improvement of productivity and efficiency in ethanol production with Ca-alginate immobilized *Zymomonas mobilis. Biotechnol. Lett.*, **5**, 497–502.
Kluge, M., Klein, J. and Wagner, F. (1982) Production of 6-aminopenicillanic acid by immobilized *Pleurotus ostreatus. Biotechnol. Lett.*, **4**, 293–6.
Krisch, J. and Szajani, B. (1996) Acetic acid fermentation of *Acetobacter aceti* as a function of temperature and pH. *Biotechnol. Lett.*, **18**(4), 393–6.

Krouwel, P.G., Harder, A. and Kossen, N.W.F. (1982) Tensile stress–strain measurements of materials used for immobilization. *Biotechnol. Lett.*, **4**, 103–8.
Levine, H. and Slade, L. (1989) Interpreting the behavior of low moisture foods, in *Fundamental Aspects of the Dehydration of Foodstuffs* (ed. T.M. Hardmann), Elsevier Applied Science, New York, pp. 71–134.
Li, R.H., Altreuter, D.H. and Gentile, F.T. (1996) Transport characterization of hydrogen matrices for cell encapsulation. *Biotechnol. Bioeng.*, **50**(4), 365–73.
Linko, Y.Y., Pohjola, L. and Linko, P. (1977) Entrapped glucose isomerase for high fructose syrup production. *Process Biochem.*, **12**(6), 14–16.
Margaritis, A., Bajpai, P.K. and Wallace, J.B. (1981) High ethanol productivity using small Ca-alginate beads of immobilized cells of *Zymomonas mobilis*. *Biotechnol. Lett.*, **3**, 613–18.
Mattiasson, B. (1983) *Immobilized Cells and Organelles*, vols 1 and 2, CRC Press, Boca Raton, FL.
Mosbach, K. and Mosbach, R. (1966) Entrapment of enzymes and micro-organisms in synthetic cross-linked polymers and their application in column techniques. *Acta Chem. Scand.*, **20**, 2807–10.
Nelson, K. and Labuza, T.P. (1992) Understanding the relationships between water and lipid oxidation rates using water activity and glass transition theory, in *Lipid Oxidation in Foods* (ed. A. St. Angelo), American Chemical Society, Washington DC, pp. 91–101.
Nussinovitch, A. (1994) Resemblance of immobilized *Trichoderma viride* fungal spores in an alginate matrix to a composite material. *Biotechnol. Prog.*, **10**, 551–4.
Nussinovitch, A., Nussinovitch, M., Shapira, R. *et al.* (1994) Influence of immobilization of bacteria, yeasts and fungal spores on the mechanical properties of agar and alginate gels. *Food Hydrocolloids*, **8**, 361–72.
Pothakamury, U.R. and Barbosa Canovas, G.V. (1995) Fundamental aspects of controlled release in foods. *Trends Food Sci. Technol.*, **6**(12), 397–406.
Puri, M., Marwaha, S.S. and Kothari, R.M. (1996) Studies on the applicability of alginate-entrapped naringinase for the debittering of kinnow juice. *Enz. Microb. Technol.*, **18**(4), 281–5.
Reineccius, G.A. (1989) Flavor encapsulation. *Food Rev. Int.*, **5**, 147.
Reineccius, G.A. (1991) Carbohydrate for flavor encapsulation. *Food Technol.*, **45**, 144.
Risch, S.J. (1986) Encapsulation of flavors by extrusion, in *Flavor Encapsulation* (eds S.J. Risch and G.A. Reineccius), American Chemical Society, Washington, DC, pp. 103–9.
Rochefort, W.E., Rehg, T. and Chau, P.C. (1986) Trivalent cation stabilization of alginate gel for cell immobilization. *Biotechnol. Lett.*, **8**, 115–20.
Shoichet, M.S., Li, R.H., White, M.L. *et al.* (1996) Stability of hydrogels used in cell encapsulation – an *in vitro* comparison of alginate and agarose. *Biotechnol. Bioeng.*, **50**(4), 374–81.
SivaRaman, H., Rao, B.S., Pundle, A.V. *et al.* (1982) Continuous ethanol production by yeast cells immobilized in open pore gelatin matrix. *Biotechnol. Lett.*, **4**, 359–64.
Stocklein, W., Eisgruber, A. and Schmidt, H.L. (1983) Conversion of L-phenylalanine to L-tyrosine by immobilized bacteria. *Biotechnol. Lett.*, **5**, 703–8.
Tampion, J. and Tampion, M.D. (1987) *Immobilized Cells: Principles and Applications*, Cambridge University Press, Cambridge.
Thevenet, F. (1986) Acacia gums: stabilizers for flavor encapsulation, in *Flavor Encapsulation* (eds S.J. Risch and G.A. Reineccius), American Chemical Society, Washington, DC, pp. 37–44.
Trubiano, P.C. and Lacourse, L. (1986) Emulsion-stabilizing starches: use in flavor encapsulation, in *Flavor Encapsulation* (eds S.J. Risch and G.A. Reineccius), American Chemical Society, Washington, DC, pp. 45–54.
Venkatsubramanian, K., Karkare, S.B. and Vieth, W.R. (1983) Chemical engineering analysis of immobilized cell systems, in *Applied Biochemistry and Bioengineering*, vol. 4, *Immobilized Microbial Cells* (ed. I. Chibata and L.B. Wingard, Jr.), Academic Press, New York, pp. 312–50.
Vorlop, K.D. and Klein, J. (1981) Formation of spherical chitosan biocatalysts by iontropic gelation. *Biotechnol. Lett.*, **3**, 9–14.

Walsh, P.K., Isdell, F.V., Noone, S.M. *et al.* (1996) Microcolonies in alginate and carrageenan gel particles – effect of physical and chemical properties of gels. *Enz. Microb. Technol.*, **18**(5), 366–72.

Wang, H.Y. and Hettwer, D.J. (1982) Cell immobilization in κ-carrageenan with tricalcium phosphate. *Biotechnol. Bioeng.*, **24**, 1827–38.

Wang, H.Y., Lee, S.S., Takach, Y. *et al.* (1982) Maximizing microbial cell loading in immobilized cell systems, in *Biotechnology and Bioengineering, Symposium 12* (ed. E.L. Garden, Jr.), John Wiley, New York, pp. 139–46.

Wheatley, M.A. and Phillips, C.R. (1983) The influence of internal and external diffusional limitations on the observed kinetics of immobilized whole bacterial cells with cell-associated β-glucosidase activity. *Biotechnol. Lett.*, **5**, 79–84.

Wiksstrom, P., Szwajcer, E., Brodelius, P. *et al.* (1982) Formation of α-keto acids from amino acids using immobilized bacteria and algae. *Biotechnol. Lett.*, **4**, 153–8.

16 Inks

16.1 Introduction

Evidence for the first uses of ink in ancient Egyptian and Chinese cultures can be traced back to 2600 BC. These inks are thought to have been composed of a mixture of soot or lampblack (carbonaceous materials) and glue based on animal sources or oil derived from vegetable sources. Solid ink blocks and pellets are a Chinese invention, which originated somewhere between 220 and 419 AD. The development of writing inks became an art in China 400 years before Gutenberg introduced the moveable type in Europe in the 15th century. The difference between printing and writing inks is that the former are introduced to a substrate by means of a printing press. Different categories of inks (printing and writing) and various ways of distributing them are popular (Bourachinsky *et al.*, 1982).

16.2 Printing inks

Printing ink is defined as a mixture of coloring matter, dispersed or dissolved in a carrier to form a fluid or paste, which can be applied (printed) to a medium and later dried. Colorants are chosen from a list of pigments, toners, dyes or their combinations to provide the desired color contrast between the ink and the background. The ink carrier enables ease of operation along with adhesion (gluing capacity) between the substrate and the ink (Bourachinsky *et al.*, 1982).

Wet ink color formulations contain a higher percentage of color pigments than previous formulations. This enables the designed image layer to be a more faithful reproduction of the art work (Belsito, 1994).

Media used for printing include paper and paperboard, metal sheets and metallic foil, plastic films and molded plastic, glass and textiles. The inks have several potential functions, e.g. decorative, protective and communicative, which can be involved solely or in combination. Four different classes of printing ink exist: letterpress (LP) and lithographic (litho) (also called oil or paste inks) and flexographic (flexo) and rotogravure (gravure) inks (also referred to as solvent or liquid inks) (Benkreira and Davie, 1994). The inks are ranked according to their dying method, printability, color and use.

Ink is dried by absorption (penetration into the pores of the substrate, possible separation of the pigment from the solvent and the latter's diffusion throughout the paper), solvent evaporation (by heat supply), precipitation (reducing the solubility of the resin in the ink, causing it to precipitate and dry), oxidation (catalysing the absorption of oxygen by the drying oil), polymerization, cold setting (ink in a molten state that undergoes film formation and drying via cooling), gelation (elevation of temperatures which causes dispersed polymer particles to undergo gelation and dry), radiation curing (polymerization via a free-radical mechanism), infrared drying (heat set and solvent evaporation) or radiofrequency and microwave drying (30–100 MHz and 1000–3000 MHz, respectively). This transforms the ink from a fluid state to a dry form that does not stick or transfer to another substrate (Bourachinsky *et al.*, 1982).

16.3 Physical properties of ink

Inks are generally non-Newtonian fluids. Their resistance to flow is described in terms of viscosity (resistance to flow), yield value (the point at which a liquid starts to flow under stress), thixotropy (decreasing viscosity with increasing agitation) and dilatancy (the opposite of thixotropy). Their rheological properties determine the fidelity of printing, drying speed and the holdout and trapping properties on the substrate. Flexo and gravure systems have a viscosity range of 50–100 cP (mPa s) and low yield values to permit easy pumping of the ink from the reservoir to the fountain. By definition then, such systems are not thixotropic (Bourachinsky *et al.*, 1982).

Offset inks and LP have higher viscosities than these ink-distribution systems. Letter-type news ink has viscosity values below 500 cP varying to values above 500 cP for special litho ink formulations. Since these inks have higher viscosities, special equipment, including a multitude of rollers in the ink-distribution unit, are used to permit their fine transfer to the plates where printing is carried out. Higher press speeds require inks with lower viscosities. Such inks are also preferred for fine-line flexography and shallow-cell gravure printing. Printing smooth, dense solids can best be achieved with higher viscosity ink. The higher the ink's pigment content, the higher its yield value and prevalence of thixotropic properties. Therefore, attention should be paid to the compromise between color intensity and rheological properties, as determined by appropriate instrumentation (Bourachinsky *et al.*, 1982).

16.4 Color versatility

Many properties are achieved with pigments in addition to color, including bulk, opacity, specific gravity, viscosity, yield value and printing quality, and this needs to be taken into account before choosing a pigment for a specific application. The colors also differ in their permanencies to light, heat and chemicals, and in their bleeding or staining properties in water, oil, alcohol, fat, grease, acid or alkali. Insoluble inorganic pigments and dyes are used in inks. Dye families that are of interest to the manufacturer include azo, triphenylmethane, anthraquinone, vat and phthalocyanine (Bourachinsky *et al.*, 1982).

Flushing is a pigment-preparation process that is designed to overcome the problem of not being wetted properly by oil phases. The pigment presscake is introduced with a water-immiscible agent into a dough mixer, separating and removing most of the water. Full dehydration is achieved by heating under agitation for several hours in a vacuum. After flushing, colors are glossier and stronger than their dry-pigment counterparts (Bourachinsky *et al.*, 1982). A theory explaining the mechanism by which pigments affect the gloss of paint films was strengthened and expanded by Braun and Fields (1994). Immediately after paint is applied, the surface tension of the wet film maintains its surface and molecular smoothness, as well as its gloss and perfect image distinctness. While the film dries or cures, it shrinks from loss of solvent or condensation products. Pigment particles and, if present, resin droplets come together, interact and develop a structure within the wet film. Eventually, the compressive strength of this structure overcomes the surface forces. As the film continues to shrink, its surface becomes rough, degrading distinctness-of-image gloss. Experimental data and theoretical arguments are presented by the authors showing that the effects are proportional to pigment particle size (Braun and Fields, 1994).

For easier use, concentrated colors in paste or liquid form containing 35–65% pigment are used. They are composed of monomeric and resinous dispersants, in addition to a solvent, which defines their classification. Such pigments are very well dispersed to achieve the desired color values. To produce dispersable pigments (called stir-ins), precipitated pigment is coated with a suitable dispersant before drying and pulverization. Toners are defined as full-strength, undiluted pigments for use with tinctorially weak ones. Occasionally dyes are used as toners (Bourachinsky *et al.*, 1982).

The ink industry makes use of black pigment (carbon black), white pigments (titanium dioxide, zinc sulfide, zinc oxide), inorganic and organic color pigments, blue, purple and green pigments, daylight-fluorescent pigments and silver powder. If white is mixed with other pigments, opacity is added and hue lightened.

Transparent pigments (extenders), such as aluminum hydrate, magnesium carbonate, calcium carbonate, precipitated barium sulfate, natural barium sulfate and clay, are used to reduce color strength and change the ink's flow properties. Inorganic color pigments such as chrome yellow, chrome orange, and molybdate orange can be toxic and are restricted to non-related food products and items that are not used by children. Other inorganic yellow, red and orange colors contain mercury and are, therefore, considered toxic and their use restricted. Iron blue and ultramarine have poor alkali and acid resistance, respectively.

16.5 Other ingredients

Inks can also contain driers, waxes, antioxidants and miscellaneous additives. The latter include lubricants, surface energy-reducing agents, thickeners, gelling agents, defoamers, wetting agents and shorteners. Some of these tasks can be performed by gums and resins at various levels.

Ink vehicles can also be prepared via the polymerization of vegetable oils (Erhan and Bagby, 1994). By controlling the polymerization conditions, the desired viscosity, color and molecular weight can be achieved with a variety of vegetable oils having a broad range of iodine values and fatty acid compositions. Of the oils tested, the polymerization rate constant of safflower oil is the highest, followed by soybean, sunflower, cottonseed and canola oils, in decreasing order (Erhan and Bagby, 1994).

16.6 Inks for different purposes

Liquid inks were developed after paste inks. Paste inks utilize different solvents and have different applications to the liquid (gravure) inks. Dry offset (DO) inks are not treated separately because of their similarity to LP inks in all properties except the former's higher viscosity and an up to 30% increase in color strength. Ink properties can be divided into three classes: LP, DO and litho. LP printing can be used for printing news, publications, advertisements, cartons, containers and boxes destined for folding, books, bags and wrappers. DO is used for folding cartons, books, bags, metal containers and plastics. Litho inks are used for news publications, commercial documents, business documents, folding cartons, books, metal containers and plastic. Printing speeds of ~ 80–$180\,\mathrm{m\,min^{-1}}$ are usual for paper, cloth and plastic bags, where the typical vehicle is composed of resin oil, resin glycol or resin solvent oil. Much higher printing speeds are used in printing papers and comics (300–$450\,\mathrm{m\,min^{-1}}$) or publications such as magazines and periodicals (300–$370\,\mathrm{m\,min^{-1}}$). Typicial vehicle compositions in such cases are mineral oil and resin solvent oil for newspapers, and acrylate oligomers and resin solvent oil for other publications (Bourachinsky *et al.*, 1982).

LP and litho newsprint inks are based mainly on mineral oil, sometimes combined with a resin. Drying is achieved by penetration into the printed medium. Litho inks include resins such as rosin, fossil and hydrocarbon dissolved in aliphatic hydrocarbon solvents. Mineral oil with its higher boiling point relative to other solvents can be used in non-drying versions.

Litho newsprint can be heat-dried by high-velocity hot air, infrared and gas flame.

Rotary heat-set publications and commercial inks are based on solvents (vehicles) composed of natural or semi-synthetic resins dissolved in fractions of aliphatic hydrocarbons that boil at 200–280°C. These dry in 1 s through evaporation or solubilization of dispersed resins at the web temperatures of up to 180°C achieved by high-velocity hot air, or infrared- and gas-flame-type dryers (Bourachinsky *et al.*, 1982).

In sheet-fed presses, mainly using the litho process, inks based on resins (phenolic, maleic and rosin ester) dissolved in vegetable drying oils and diluted with hydrocarbon solvents are regularly used. Document reproduction inks contain drying-oil alkyds or plain-bodied oils, along with litho-type (non-polar) resins and high-boiling hydrocarbon solvents. Folding-carton inks are based on quick-set vehicles. If better gloss is desired, then glossy oleo-resinous vehicles are used. They dry by oxidation to yield tough glossy films. Ink vehicles for corrugated and kraft-liner containers are based on oleoresinous materials. Book, bag and wrapper inks are similar to those used for publication and commercial purposes. Metal-container inks are based on blends of oleoresinous and heat-set varnishes containing resins, alkyds and oils. They require 10 min at 150–200°C in a gas-fired oven for drying and hardening. Ink vehicles for preformed aluminum or steel containers are based on polyester vehicles used in conjunction with melamine cross-linkers. Inks for plastics are based on hard-drying oleoresinous varnishes sometimes diluted with hydrocarbon solvents (Bourachinsky *et al.*, 1982).

A novel use of ballpoint-pen-ink coating on nucleation enhancement of low-pressure diamond was reported by Peng (1994). It was found that coating the silicon with ballpoint pen ink can greatly increase the diamond nucleation density when the coating was followed by heat treatment at 100 to 400°C. The optimum enhancement effect was reached when the ink coating was first heat treated at 300°C for 30 min (Peng, 1994).

16.7 Ink manufacture and analysis

Paste inks can be produced by either mixing predispersed or flushed pigment concentrates with vehicles, solvents, oils or compounds, or mixing dry pigments or resin-coated pigments with vehicles or compounds and then grinding them on ink mills. The mixing and milling are performed with different machines in accordance with predetermined demands. Finished inks are packed in metal cans, metal or plastic pails, or metal or fiber drums or bins. Ink vehicles are produced in separate resin-varnish plants. Quality control of produced inks includes an examination of their color, strength, hue, tack, rheology, drying rate, stability and product proof (Bourachinsky *et al.*, 1982).

The usefulness of time-of-flight secondary ion mass spectrometry for the analysis of colorants on paper surfaces was investigated (Pachuta and Staral, 1994) with 21 pen inks and 16 printed-paper specimens. The use of static analysis conditions had no detectable effect on the paper and made secondary ion mass spectrometry suitable for non-destructive analyses. In cases in which the specimen is too large to fit in the instrument, a single paper fiber can be unobtrusively removed and analysed separately. This method permits rapid differentiation of the colorants and provides qualitative identification of some components (Pachuta and Staral, 1994).

16.8 Hydrocolloids and resins in the ink industry

16.8.1 Carrageenans

Carrageenans are used to gel, thicken or suspend. They are used in emulsion stabilization, for syneresis control, and for bodying, binding and dispersion. Carrageenans are utilized as pigment dispersion agents in ink products.

16.8.2 Polyethylene glycols

PEGs are synthetic water-soluble polymers that are very low in toxicity and are non-irritating. They possess good stability and lubricity, broad compatibility and good solvent action and can, therefore, be used in many commercial applications. Ballpoint pens require thixotropic inks, which can be obtained by using liquid or solid PEGs as the internal phase. A typical formulation of such ink is 56% oleic acid salt of basic dyestuff, PEG 1000 (15%), 27% pigment and 2% surfactant. The oleic salt and PEG serve as the external and internal phases, respectively. The viscosity can be controlled and regulated by changing the PEG molecular weight. The age of ballpoint-pen ink can be determined by gas and densitometric thin layer chromatography (TLC) (Aginsky, 1994). Densitometric TLC is also a very effective tool for the comparative TLC examination of similarly colored inks, paint fibers and other materials of forensic interest (Aginsky, 1994).

Solid PEG solutions are used in preparations of water-based stencil inks to obtain the desired viscosity and flow control. The unique humectant and solvent properties of PEG help in the production of steam-set printing inks. Liquid PEGs also help in producing stamp-pad inks. The water-soluble polymers help to maintain the stamp's characteristics over the required periods of time. When high-speed printing is required, the addition of these water-soluble polymers minimizes the hygroscopicity and premature ink setup at high levels of humidity (Gaylord, 1963).

16.8.3 Gum arabic

Gum arabic is the most widely used gum in the industry. It is of major importance in various applications of inks and related products. Gum arabic is unique in its ability to form solutions at concentrations of over 50%, and it is non-toxic, odorless, colorless and tasteless. It is used as a suspending agent for soluble inks, water colors, quick-drying inks and typographic and hectographic inks (as a fountain solution) (Meer, 1982). Typographic inks (emulsions) make use of gum arabic. Emulsion inks are oil-in-water emulsions with the pigment dissolved in the oil phase. Typically, such ink contains lampblack, mineral oil, rosin, catechu black, formalin, sodium silicate, sodium carbonate, gum arabic, aluminum resinate, a coloring agent and water. Hectographic inks have been used for many years in hectograph duplication equipment. They are prepared by mixing methyl violet dye with water or ethanol. Some formulae include hydrochloric acid, oxalic acid, lactic acid and tannic acid. The gum arabic is used to yield and control the desired property (Meer, 1982).

Gum arabic serves as a constituent of many special-purpose inks (Ellis, 1940). Its protective colloidal properties are of major importance. The development of the ink industry began with the preparation of lampblack dispersions in water. In a short time, gum arabic found a major role in this industry as a suspending agent. Sometimes the lampblack and gum were mixed into a thick paste and allowed to harden in molds or into ink sticks (Meer, 1982). The sticks were used by rubbing a brush on the stick or rubbing it in water until the proper color solution was achieved (Waters, 1940). Gum arabic is used in record ink as a protective colloid. Soluble inks have been used in the textile industry for the temporary marking of cloth for cutting or sewing operations. The advantage of this ink is that it can be removed by the hot-water wash used after finishing. Such ink is composed of a mixture of dilute acetic acid, albumin, basic dye, gum arabic, molasses and triethanolamine (Poschel, 1933). The question of whether ink jets will ever replace screens for textile printing is discussed in Dawson and Ellis (1994). The many problems involved in developing an ink-jet printer suitable for woven and knitted textiles are outlined. Efforts have been concentrated on adapting existing systems used for paper or carpet substrates. Work devised to produce a modular array of drop-on-demand valves to meet the required criteria is described, together with details of the selection of dyes and chemicals for a practical jet-printing process (Dawson and Ellis, 1994).

Watercolor inks can be easily prepared by maintaining a pigment in suspension with gum arabic. A typical ink base is composed of a 1:9:1 ratio of gum arabic to glycerol to water, followed by evaporation until a mixture with a specific gravity of 1.28 is achieved. Colloidal ground pigment is added to this mixture to yield the preparation. Gum arabic can even be used in quick-drying ink where the main solvent is ethanol and not water. Constituents of such formulae include water, a preservative (lysol), sodium nitrate, gum arabic and a water-soluble dye. Fabric- and laundry-marking inks contain gum arabic as a viscosity former and thickener. Pigmented inks such as white (produced with titanium dioxide) or gold (produced with bronze powder) ink contain gum arabic as a suspension aid in the formulation. Currently, the use of electrically conductive inks is on the rise because they can be used in the manufacture or repair of printed circuits or in the activation of electronic calculators. A typical formula contains hydrochloric acid, oxalic acid, lactic acid, tannic acid and gum arabic to obtain the proper viscosity (Ellis, 1940). In lithography, gum arabic can be used as a sensitizer for lithographic plates, on elements in light-sensitive compositions, as an ingredient of fountain solutions used to moisten plates during printing and as a protector during plate storage (Meer, 1982). In textiles, gum arabic can be used as a sizing and finishing agent in printing formulations for imparting designs or decorations to fabric. Gum arabic can be used as a coacervate with gelatin for ink microencapsulation (Green and Schleicher, 1957) and can be utilized for protective films for the storage of printing plates.

16.8.4 Tragacanth and polyacrylic acid

Gum tragacanth, a natural vegetable gum, is used in both the textile and paper industries as part of the printer's ink. The gum serves as an emulsification agent by increasing aqueous-phase viscosity and lowering the interfacial tension between oil and water, as a suspending agent via the repelling action of the galacturonic acid salt found in tragacanthic acid, and as a shear thinner via pseudoplastic flow. Polyacrylic acid and some of its derivatives serve as grinding media for the preparation of printing inks and paint pastes. A discussion on sulfopolyesters – new resins for water-based inks, overpoint lacquers and primers – can be found in Barton (1994). This author discusses the transition from solvent- to water-based inks in the various sectors of the printing industry. Special consideration is given to improvements that have been made since the breakthrough of water-based acrylic resins in 1970. Finally, the author deals with the improvements that are still needed, such as faster drying speeds, better foam control and lower odor retention, which, it is maintained, will be possible only via the introduction of a new range of resins (Barton, 1994).

16.8.5 Hydroxypropylcellulose

The primary application of HPC as a manufacturing aid in the ink industry is via its thickening, binding and suspension abilities, in addition to its solubility in alcohols and glycols.

16.8.6 Polyethylene oxide

PEO can be used for microencapsulated inks. Non-aqueous printing inks can be microencapsulated via an association complex between PEO and polyacrylic acid. The products of this process can then be used as dry, free-flowing powders to produce ‘carbonless’ carbon papers. If pressure is applied to the paper, the capsule wall is fractured and the ink is released. Various water-immiscible liquids can be encapsulated by taking advantage of the pH-controlled inhibition of the PEO–polycarboxylic acid association complex. Another application is reducing the volatility and misting of lithographic-press-dampening solutions by incorporating low-molecular-weight PEO resin in the solution.

References

Aginsky, V.N. (1994) Determination of the age of ballpoint pen ink by gas and densitometric thin-layer chromatography. *J. Chromatogr.*, **678**, 119–25.

Barton, K.R. (1994) Sulphopolyesters – new resins for water based inks, overprint lacquers and primers. *JOCCA – Surface Coatings Int.*, **77**(5), 180–2.

Belsito, G.C. (1994) Improved wet-printing techniques. *Am. Ceram. Soc. Bull.*, **73**(12), 58–60.

Benkreira, H. and Davie, C.M. (1994) The effect of aggregate structure on the dispersion behavior of highly aggregated alpha-beta copper phthalocyanine pigments for use in gravure printing inks. *Colloids Surf. Physicochem. Eng. Asp.*, **90**(1), 37–43.

Bourachinsky, B.V., Hugh, D. and Ely, J.K. (1982) Inks, in *Kirk Othmer Encyclopedia of Chemical Technology*, 3rd edn, vol. 13, Wiley Interscience, New York, pp. 374–98.

Braun, J.H. and Fields, D.P. (1994) Gloss of paint films. Effects of pigment size. *J. Coatings Technol.*, **66**(828), 93–8.

Dawson, T.L. and Ellis, H. (1994) Will ink jets ever replace screens for textile printing? *J. Soc. Dyes Colourists*, **110**(11), 331–7.

Ellis, C. (1940) *Printing Inks*, Reinhold Publishing, New York, pp. 230, 334, 346, 398–9, 417.

Erhan, S.Z. and Bagby, M.O. (1994) Polymerization of vegetable oils and their uses in printing inks. *J. Am. Oil Chem. Soc.*, **71**(11), 1223–6.

Gaylord, N.G. (ed.) (1963) *Polyethers in Polyalkylene Oxides and Other Polyethers*, vol. 13, part 1: *High Polymer Series*, Wiley Interscience, New York, pp. 169–89, 239–74.

Green, B.K. and Schleicher, L. (1957) US Patent No. 2,800,457.

Meer, W. (1982) Gum arabic, in *Handbook of Water-soluble Gums and Resins* (ed. R.L. Davidson) McGraw-Hill, New York, pp. 8.1–8.24.

Pachuta, S.J. and Staral, J.S. (1994) Nondestructive analysis of colorants on paper by time of flight secondary-ion mass-spectrometry. *Anal. Chem.*, **66**(2), 276–84.

Peng, X.L. (1994) The effect of ballpoint pen ink coating on the nucleation enhancement of low-pressure diamond. *J. Mat. Res.*, **9**(6), 1573–7.

Poschel, A. (1933) British Patent No. 393,132 (*Chem. Abstr.*, **27**, 5553).

Waters, C.E. (1940) *Circular No. C426*, National Bureau of Standards, pp. 3, 34–5, 45, 53.

17 Paper

17.1 Introduction

Paper originated in China in 105 AD. It was produced from flax and hemp, or the bark fibers of certain trees. Paper sheets are composed of small cellulosic fibers held together by secondary (hydrogen) bonds; the sheets are formed by passing a dilute suspension through a screen. The word 'paper' comes from the Egyptian word papyrus: a sheet made by pressing together strips of the reed *Cyperus papyrus*. Nevertheless, papyrus sheets are not considered to be paper because the fibers were not separated first before being reformed into sheets (Baum *et al*., 1982).

The knowledge of how to manufacture paper from bark and bamboo was passed from China to Japan (manufacture began there in *c*. 610). Paper-making spread through Central Asia, the Middle East, Europe (Spain 1150, France 1189, Germany 1320, England 1494), and finally, the USA in *c*. 1700 (Hunter, 1947; Smith, 1970).

The very large tonnages of paper produced annually have encouraged investigations into the possible uses of gums and resins in the various stages of paper production. Paper has some well-known defects: it is permeable to water, various vapors, oils and other fluids; it requires considerable processing before it can be used for printing. This processing takes the form of additions, which can be divided into those occurring during the paper's manufacture (beater addition), and impregnation and various types of coating. The additions during manufacture and impregnation are designed to improve the paper's strength (tensile, tear and bursting strength and folding endurance). Impregnation to ensure good printability makes use of synthetic resins instead of the formerly used natural products such as starch and casein. Sizing is a general term for the addition of any material that fills the paper pores. Natural sizes include casein and starch, and synthetics used for imparting wet strength such as urea-formaldehyde and melamine-formaldehyde resins. In some cases, surface treatment is referred to as sizing and this can involve coating procedures or a chemical reaction on the surface of the paper (Warson, 1972).

17.2 Raw materials

Paper is designed for different purposes, such as writing, printing, packaging, towels and tissues. The main components of paper and paperboard are mechanical/semi-mechanical wood pulp, unbleached kraft chemical wood pulp, white chemical wood pulp (including unbleached sulfite), waste fiber, non-wood fibers, fillers and pigments (FAO, 1977).

Pulp is produced mainly from hard and soft woods. Processing can involve mechanical methods to separate the fiber from the wood matrix and chemical processes to remove the lignin, which is the bonding material. Combinations of these two methods are also employed. Pulps produced by mechanical means contain ingredients similar to those of native woods, such as lignin and hemicellulose, in addition to the cellulose (since there is no removal of wood components). The process includes stone-grinding the wood, and separating and fracturing the fibers. The resultant pulps are used where opacity and good printability are needed. These pulps are bleached (by alkaline hydrogen peroxide or sodium hydrosulfite) to improve their brightness, light stability, permanence and strength (Baum *et al.*, 1982).

Newer mechanical methods (using disk refiners to produce pulp from wood chips) produce pulp with less debris and longer fibers. Thermomechanical processes include steaming at 120°C prior to fiberization in a pressurized disk refiner, resulting in less damaged fibers. Thermomechanical pulps have the potential to reduce the use of pulps obtained by chemical means. The addition of a chemical treatment (hydrogen peroxide and sodium sulfite) to the mechanical process is important when higher strength (a *c*. 50% increase) is desired. Pulp yields achieved by mechanical processing are higher than those of chemical pulps, since significant amounts of material are removed by chemical activity (Baum *et al.*, 1982).

Chemical methods dissolve lignin and hemicellulose, reducing their content within the pulp. Less mechanical energy is, therefore, needed to separate the fibers, which are undamaged and strong. Such treatments are necessary where paper strength and performance are important. Chemical pulps are produced by the kraft process, in which sodium sulfide and sodium hydroxide mixtures are used as pulping chemicals; yields are 46–56%. The higher-yield pulps (with 10% lignin) are used in bags or lineboard (where strength is important). The lower-yield pulps are bleached (by multi-stage bleaching with chlorine, hypochlorite and chlorine dioxide) to remove the lignin entirely and to produce high brightness (90% or more). Changes in the method result in different pulp and paper properties. Small amounts of pulp are still produced in the acid sulfite process. Small quantities of non-wood fibers are also used. They include cotton linters for filters and writing paper, esparto for filter paper and Manila hemp for tea bags, among other specialized purposes (Baum *et al.*, 1982).

17.3 Physical properties

The paper's physical properties depend on machine direction. Strength is greater in the direction of manufacture than in the cross-machine direction.

Moreover in the thickness direction, paper is quite weak. In other words, paper has three perpendicular symmetry planes with different properties. When producing sheets of wood pulp, the higher the quantity of fibers arranged in the direction of the moving wire, the higher the strength of the formed sheets (more fibers lined up in the same direction).

The basis weight of a paper is measured as the mass in grams per square meter. It is also expressed, in pounds, as the weight of a ream of 500 sheets of a given size and can differ from one paper to the next. Fine papers are 43.2×55.9 cm (with a basis weight of 60–150 $g\,m^{-2}$), newsprint 61.0×91.4 cm (49 $g\,m^{-2}$) and book papers average 63.5×96.5 cm. The thickness of a single sheet measured under specified conditions is expressed in micrometers. Several examples are facial tissue (65 μm), newsprint (85 μm), offset bond (100 μm) and book cover (770–7600 μm) (Baum *et al.*, 1982).

Tensile strength is the force per unit width parallel to the plane of the sheet that is required to produce failure in a specimen of specified width and length under specified conditions of loading. Stretch is the strain resulting from the application of a tensile load under specified conditions. Bursting strength is the hydrostatic pressure required to rupture a specimen being tested under specified conditions. Tearing strength is the average force required to tear a single sheet of paper. Stiffness is related to bending resistance. Folding endurance is the number of folds a paper can withstand before failure. It is important to mention that all tests are performed under standardized conditions. Other important tests include moisture content, water resistance and water-vapor permeability. Important optical properties include brightness, color, opacity, transparency and gloss (Baum *et al.*, 1982).

17.4 Chemical properties

Paper is made from cellulose fiber obtained from wood pulping. The fiber's chemical composition will affect color, opacity, strength, permanence, electrical properties and interfiber bonding. Since residual lignin in the fibers inhibits bonding, ground-wood pulp is used when a highly bonded structure is not needed (e.g. newsprint, absorbent papers). By comparison, hemicellulose in wood pulps contributes to bonding and is, therefore, used in wrapping papers and other grades that require bonding for strength, or transparency (e.g. glassine) (Baum *et al.*, 1982).

The chemical properties of the paper determine the type of additive needed to obtain a desired end-product. Dyes are added to modify optical brightness, resin to impart wet strength, starch to reduce penetration of aqueous liquids, pigment coatings to provide a smooth surface for printing, fillers to increase opacity, polymers to impart mechanical or barrier properties and cationic polyelectrolytes for dielectric recording. Special paper coatings have been developed for photography, thermal paper and carbonless copy papers. Some papers include compounds to promote flame retardance. Papers that do not include electrolytes can be used as insulators. Papers used for permanent documents must be low in acidity. Papers that receive metallic coatings or have antitarnish properties contain a reduced amount of sulfur compounds. If the paper is to be used in food applications, all chemicals related to it must be FDA approved (Baum *et al.*, 1982).

17.5 Manufacture and processing

For paper manufacture, a stock preparation consisting of an aqueous slurry is desirable, because the slurry can be conveyed and mechanically and chemically treated before delivery to a paper machine. In the case of adjacent papermaking, pulps are delivered to the paper mill directly after pulping. Purchased pulps and waste paper received as dry sheets or laps must be dispersed and separated by water (slushing), with a minimum of mechanical work (by Hydrapulper or Sydrapulper) so as not to alter the fiber properties. Mechanical operations such as beating are necessary to improve the physical properties of the paper (beating increases tensile strength, bursting strength, folding endurance and sheet density, and reduces tearing resistance). Beating (synonymous with refining) swells, cuts, macerates and fibrillates the cellulose fibers. The fiber becomes more flexible, thereby increasing interfiber contact during paper formation. Omission of beating results in a low-density, soft and weak paper. As a result of the refining, the wet specific surface and flexibility of pulp fibers increase, resulting in an increased ability of the fibers to bond when dried from a water suspension and thus increasing the strength of the sheet (Baum *et al.*, 1982).

As the industry developed, refining batch systems (such as the Hollander, developed in Holland in *c.* 1690) were replaced by continuous equipment (such as the Jordan, developed in *c.* 1860). Further information on these and other modified refiners (e.g. Hydrafiner, Claflin, the disk refiner) and their development can be found elsewhere (Claflin, 1907; Haskell, 1932).

17.6 Filling and loading

Slurry ingredients such as mineral pigments (e.g. titanium oxide and zinc pigments at 2–40% of the final sheet) are added to the pulp for furnishing papermaking. Fillers (e.g. kaolin, China clay) improve brightness, opacity, softness, smoothness and ink receptivity, and also serve as pigment coatings. With costly pigments, systems are designed to be closed in order to reduce losses. Retention aids are also used, especially with the costly titanium oxide.

17.7 Sizing, coloring and beater additives

Sizing is a process whereby materials are added to the paper in order to render the sheet more resistant to penetration by liquids. Facial tissue and blotting paper are generally unsized. Sizing agents include rosin (refined from pine trees or stumps), waxes, starches, glues, casein, asphalt, emulsions, synthetic resins and cellulose derivatives, among others. To control paper color, dyes and other colored chemicals (mostly water-soluble synthetic organic dyestuffs) are added to the stock during its preparation or during calendering. The degree of refining a pulp receives affects its optical properties and thus affects its capacity to hold dye. In general, refining does not change the amount of dye retained but rather deepens the shade of a given application of water-soluble dye. Refining also generally increases the retention of pigments and other water-insoluble dyes, but it may also change their depth and hue by decreasing pigment-particle size. Pigments are concentrated on the top side of a sheet. Other beater additives include adhesives, which are widely used to enhance fiber-to-fiber bonding. Starches are widely used, as are gums such as guar, LBG, modified cellulose and polymers such as urea-formaldehyde and melamine-formaldehyde. Other natural and synthetic materials are used to influence paper properties and the system's behavior during sheet formation and drying (Baum *et al.*, 1982).

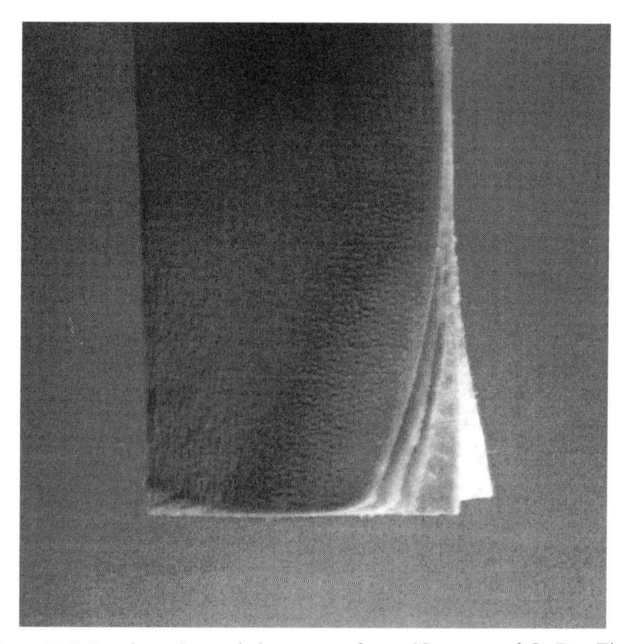

Figure 17.1 Laminated recycled paper surfaces. (Courtesy of O. Ben-Zion.)

The German paper industry consumes over 220 000 tons starch per year; this corresponds to a specific input of 18 kg ton^{-1} paper (Borchers *et al.*, 1993). Most of this (80%) is applied to the surfaces (Fig. 17.1), the remainder is in the pigment coating (8%), as a beater additive (7%) and applied as a spray (5%). Additionally, about 70 000 ton starch per year is used as adhesives in the corrugated-paper and paper-finishing industries. A considerable amount of the papers that have been treated with starch return to the raw material cycle in the form of production waste and waste paper. The starch used by the paper industry, therefore, presents considerable potential for microbiological growth in the water systems of paper machines. The size of the burden in a water system depends on the extent to which the starch has been modified. The application technology and the kind of starch used should, from this point of view, be selected not only on the basis of economy but also with a consideration of the general ecology (Borchers *et al.*, 1993; Myreen, 1994).

17.8 Sheet formation, pressing and drying

Continuous sheet formation and drying has been used for ~200 years. Two types of machine have been invented for this purpose: the cylinder and Fourdrinier machines (further information on these can be found elsewhere (Baum *et al.*, 1982)). Newer continuous paper machines are still being developed but their principles remain similar. Since there are many paper products, machine speeds vary. Heavy paperboards require a long drying time, with machine speeds of 50–250 m min^{-1}. With very dense papers (e.g. glassine, grease-proof paper, condenser tissue) speeds range from 20 to 300 m min^{-1}. Brown grades (paper bags and lineboard) are produced at 200–800 m min^{-1}, depending on their basis weight and the site of the paper machine. Newsprint machine speeds are 600 m min^{-1}. Most machines operate at 600–900 m min^{-1}. Drying-capacity restraints and difficulties in reeling the product limit modern tissue machine speeds to 1500–1800 m min^{-1}, with most operating at lower speeds. Novel designs for web handling, reeling and roll changing aim to permit tissue machine speeds of up to ~2000 m min^{-1} in continuous operation mode (Baum *et al.*, 1982).

At a water content of ~1.2–1.9 parts water per 1 part fiber, additional water removal is performed by evaporative drying, which is the bottleneck in papermaking. A dryer section is commonly composed of a series of steam-heated cylinders. Both sides of the wet paper are exposed to hot surfaces as the sheet passes from cylinder to cylinder. Water vapor is removed by way of elaborate air systems. The resultant moisture content of the dry sheet is usually 4–10% by weight.

The process in which the paper passes through further treatment after manufacture is called converting. These operations include embossing, impregnation, saturation, lamination, and the formation of special shapes and sizes.

17.9 Coatings

Pigment coatings are formulations of pigments, adhesives and additives that are applied to one or both sides of a paper sheet. They are applied for several reasons, such as changing the appearance of the base stock, improving opacity, imparting a smooth and receptive surface for printing, or providing special desired properties. Papers need to be bright to increase the contrast between the print and background, or dull to emphasize printing with glossy ink. The coating is porous and its opacity depends on the degree of porosity. Coating smooths the surface of the paper and allows full contact between it and the inked image. The coating absorbs the ink and several tests exist to determine ink absorbency (Baum *et al.*, 1982).

Pigments are applied as water suspensions. The total solids content varies from 35 to 70%. After application, water is removed and calendering is sometimes used for three purposes: smoothing the surface, controlling its texture and developing its glossy appearance. Pigment particles are small. Many pigments are less than 0.5 μm in size. The particles are smaller than the spaces between the fibers and they can, therefore, fill these spaces to manufacture a uniform surface mat. Pigments also control opacity, gloss and color. Adhesive is added to the pigment for binding purposes. If clay [$Al_2(OH)_4(Si_2O_5)_2$] is used for coating, then 10–15% casein or 15–25% starch is used as an adhesive to obtain the desired binding. Information on clay properties, commercial coatings, etc. can be found elsewhere (TAPPI Press, 1962, 1977).

Other coatings are based on titanium oxide (with an average particle size of 0.3 μm). This substrate is chemically inert, easily dispersed in water and has a unique brightness ($>95\%$). Its high cost precludes its use at more than one-quarter of the total pigment. Other coatings include anatase and rutile crystals. Precipitated forms of calcium carbonate are used with clays to improve brightness. Aluminum oxide dihydrate is also used as an ingredient in paper coating. Slaked lime and alum are used to produce satin white. Other pigments, such as barium sulfate, silica, calcium sulfate and zinc oxide, are used for different varieties of coatings (such as electrophotographic reproduction papers).

The primary function of the adhesive (added in quantities of 5 to 25%) is to bind the pigment particles together and to the raw stock. Thus strength is achieved, provided that the printing ink does not pick the coating up off the paper. Animal glue was the first adhesive used, replaced by casein in the late 1800s. Soy protein as a binder is similar to casein and is used as a substitute in many instances. Starches must be modified (e.g. by thermal conversion) to produce lower viscosities. Hypochlorite-oxidized starches are available in different viscosities and are used in sizing operations as well as in pigment coatings. Starches are quite hydrophilic. During their preparation they are heated at 93°C to cause breakdown then mixed with the pigment and used to coat the paper at high temperatures to reduce viscosity and permit easy coating. Starches are used especially in newspaper and magazine paper. They are not resistant to moisture but this quality can be changed by using aldehyde-donor cross-linking agents such as urea-formaldehyde and melamine-formaldehyde. PVA is a strong adhesive (fourfold stronger than starch and threefold stronger than casein). Therefore, only small amounts are required; its application is somewhat limited because of its high viscosity. Other adhesives include rubber latices and other emulsions that exhibit high gloss and a good response to calendering. Styrene–butadiene latex, usually supplied in a 40:60 mix, is used with starch, primarily in publication-grade papers (Baum *et al.*, 1982).

Additives for better coating control or to alter the properties of the final product include dispersing agents (e.g. polyphosphate, protein, casein) to transform pigments into a slurry, foam-control agents, adhesives, lubricants, plasticizers and flow modifiers. For packaging (Fig. 17.2), paper barriers against water, gases (water vapor, carbon dioxide, hydrogen sulfide), fatty materials (greases, fats, oils) and odors are necessary.

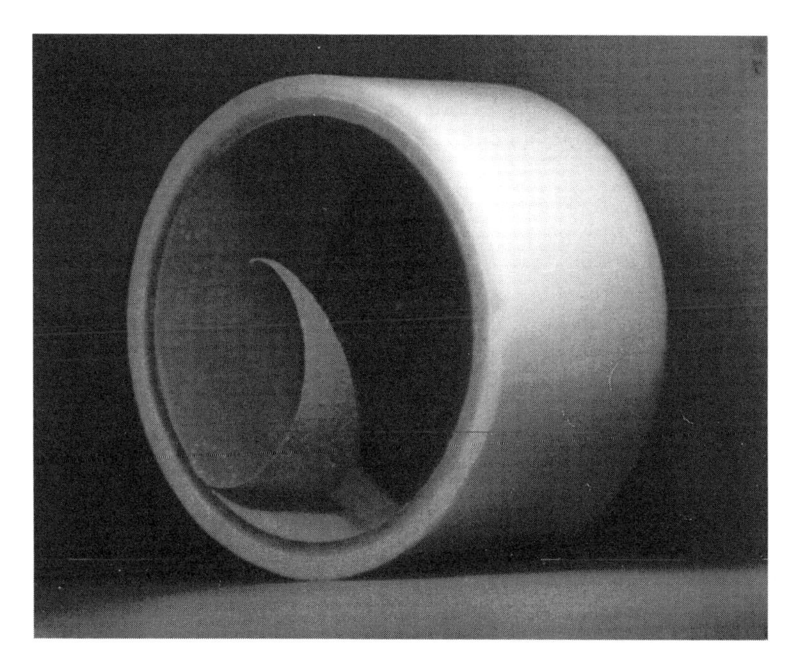

Figure 17.2 Laminated cylindrical cardboard (courtesy of O. Ben-Zion).

Many hydrocolloids are used to obtain effective paper coatings. Agar has been found suitable for use in photographic papers when esterified with succinic or phthalic anhydride and after enzymic hydrolysis. Agar can also be used as an adhesive in the gloss-finishing of paper products. HPC is used for coating because of its solvent solubility and thermoplasticity. HPC serves as an oil and fat barrier and is responsible for thermoplastic coating. PEO is used for paper coating and sizing. As a processing additive, it is used as a fiber formation aid.

The associative behavior of CMCs, HECs and hydrophobically modified cellulosic thickeners (HMCTs) in clay-based coatings was determined (Young and Fu, 1991) and their effects on coating rheology and coated-paper properties identified. The anionic CMCs are less adsorbed to kaolin clay than the non-ionic HECs and HMCTs. Adsorption onto styrene–butadiene latex was only observed with the hydrophobe-containing HMCT. The degree of adsorption to clay depends strongly on the polymer's molecular weight. The polymer size and the tendency to associate with clay influence the state of particle flocculation, viscoelasticity, high-shear rheology and the low-shear structure of the coating. Under high shear, low-molecular-weight cellulosic polymers behave as stabilizing colloids, whereas larger, clay-adsorbing polymers can lead to increased hydrodynamic volume of the dispersed particles. Coatings containing the less clay-adsorbing CMC are less resistant to flow under high shear than those containing the clay-adsorbing HECs, but they appear to have a weaker after-blade structure. Coatings thickened with HMCT have both low flow resistance under high shear and a strong after-blade structure (Young and Fu, 1991).

PVA (synthetic resin) is used in paper sizes and coatings. PVP is used in all types of paper manufacture, mostly as an economical fluidizer and antiblock agent in paper coating. The starches used in coating colors are enzyme-converted starches, thermochemically converted starches, oxidized starches, dextrins, hydroxyethyl starch ethers and starch acetates (Davidson, 1982). A full-scale treatment system has been reported to remove 96% of the influent resin acids (Zender *et al.*, 1994).

17.10 Papermaking additives

17.10.1 Formation and drainage aids

Lists of chemicals that have been cleared for use with food by the FDA, lists of the manufacturers of all types of product used by the paper industry, and surveys of chemical suppliers for the USA and Canadian pulp and paper industries are available (Buyers' Guide, 1980; Code of Federal Regulations, 1980; Kline, 1980).

17.10.2 Flocculants and defoamers

Repulsion of negatively charged fibers in water is not always sufficient to prevent all flocculation, which can result in uneven fiber density in the paper. Prevention of flocculation prior to immobilization of the wet fiber mat on the wire can be achieved by formation aids such as polyacrylamide, PEO, and natural gums (e.g. guar and LBG). The long fibers in Japanese art paper are dispersed with a polysaccharide from *Hibiscus manihot* root. Drainage aids alter the surface properties of the fibers to hold less water, resulting in limited fiber flocculation (Britt and Unbehend, 1980). Foam is a common problem in papermaking systems. It is caused primarily by surface-active agents in the pulp slurry or partially hydrophobic solid materials that function as foam stabilizers. Numerous defoamers are available, examples being paste defoamers (high solid content of emulsions of fatty acids, fatty alcohols and fatty esters), solid or brick defoamers (mixtures of fatty acids, fatty alcohols and an emulsifier) and liquid defoamers (solutions of fatty acids, fatty alcohols and fatty esters in oil).

17.10.3 Wet-web-strength additives, pitch-control agents, slimicides and creping aids

Wet web strength is an important paper property, when the wet sheet is transferred from the forming wire to the press section of the paper machine. The sheet must have enough wet strength at this stage to prevent breakage. Some water-soluble gums are added to improve wet web strength. They include LBG, guar gum and anionic polyacrylamides, which contribute to strength at solids contents $> 35\%$ (Putman *et al.*, 1982).

The wicking behavior of water into wet, formed paper strips consisting of cellulose fibers and varying proportions of powdered, superabsorbent CMC were investigated and contrasted with the performance of similar composites made with the superabsorbent CMC in fiber form. The degree of pore blocking caused by the swelling of the powdered superabsorbent CMC was found to be significantly greater than that produced by the fibrous form, at the same superabsorbent loading (Schuchardt and Berg, 1991; Wiryana and Berg, 1991).

LBG is added to the pulp suspension just before sheet formation on either a Fourdrinier or cylinder machine (Seaman, 1982). Galactomannans replace or supplement the natural hemicellulose in paper bonding. Because of their structure, galactomannans with primary and secondary hydroxyl groups are capable of bridging and bonding to adjacent fibers. In the paper industry, guar gum is believed to be the most efficient additive. Guar hull is almost white (LBG hulls are reddish-brown). Therefore, if guar hull is present in the paper sheet it is almost unnoticeable. The addition of galactomannan to pulp produces a regular distribution of pulp fibers within the sheet, and, therefore, fewer fiber bundles are formed. The result is an increase in the mechanical properties of the paper. An important aspect of galactomannan addition is the resultant increased binding of water; this decreases the amount of refining necessary, which in turn lowers power consumption. Machine speed and the retention of fines during processing are increased, without losing desirable end-product test properties (Seaman, 1982).

The main source of pitch (sticky pitch deposits may form at the wet end of the paper machine and can result in off-color spots or sheet breakage) in papermaking systems is natural wood resins, and other sticky materials such as asphalts, waxes and latices from waste paper, as well as resinous supplements to the papermaking system. Pitch dispersants maintain the pitch as a fine dispersion to eliminate agglomeration. Clay and other adsorbent fillers can also be added to the furnish to control the pitch problem. Slimicides are added to prevent slime produced by the bacteria and fungi that flourish on hemicellulose, starches and other organic additives. With the exception of toxic materials such as mercury compounds, the uses of which are limited, synthetic, organic slime-control agents are used (Putman *et al.*, 1982).

Disposable sanitary papers, used in napkins, towels and facial and toilet tissues, are designed to absorb water. This property can be improved by some resins, as well as by mechanical creping (which improves softness and absorbency by peeling the sheet from a steel roll with a sharpened doctor's blade maintained at an angle to the surface of the roll). Creping adhesives include animal glues, aminopolyamides and aminopolyamide–epichlorohydrin. Combinations of polyamide resin and release agents are produced to obtain the appropriate balance between poor and excessive adherence of the sheet to the roll, yielding suitable mechanical creping. Creping agents are applied by spraying an aqueous solution or emulsion in front of the dryer. Resins such as aminopolyamine–epichlorohydrin yield a coating on the dryer with the required degree of adhesion for optimum creping. Excessive adhesion to the dryer may be overcome with a release agent, such as a silicone-oil emulsion (Putman *et al.*, 1982).

17.10.4 Sizing agents

Sizing agents provide the paper with resistance to wetting by liquids. Sizing is usually aimed at producing water repellency. Proper sizing eliminates ink spreadability. Well-sized paper has an initial contact angle of 91–100°, permitting only limited wetting and spreading, and there is no tendency for water to penetrate the pores. At contact angles of $< 90°$, water wets, spreads over and penetrates the sheet. The high contact angle necessary for sizing is achieved with hydrophobic materials such as wax and polar–non-polar materials. Since the paper producer governs the radius and length of the pores within the paper (in other words governs the basis weight, bulk density and porosity of the sheet), only the chemical composition of the sizing agent can be changed to obtain the contact angle needed to produce the requested sizing.

Sizing agents include sodium or potassium soaps of unmodified rosin, followed by the addition of alum to accomplish the sizing by forming aluminum resinates. Rosin emulsion sizes based on fumaric-acid–fortified rosin are more efficient than the ordinary rosin size and have partially replaced the latter since the early 1970s (Putman *et al.*, 1982).

Cellulose-reactive sizing agents based on emulsions of alkylketene dimers of long-chain fatty acids can form covalent bonds with cellulose below 100°C, to produce very hydrophobic fiber surfaces. Other commercial cellulose reactive sizing agents exist. Sizing can be achieved at neutral to alkaline pH. Increased paper strength is achieved under slightly alkaline conditions, and such papers retain their strength with time much better than acid-sized papers. The action of MHPC on aqueous penetration into matrices containing MHPC of varying viscosity and concentration was studied (Wan *et al.*, 1991). The incorporation of MHPC into ibuprofen matrices improved wetting and enhanced water uptake into the matrices. The greater the amount of MHPC used, the larger the volume of water uptake. A higher-molecular-weight MHPC had greater intrinsic water-uptake properties than its lower-molecular-weight counterparts. Thus the action of MHPC on aqueous uptake depends on its molecular weight. MHPC can be divided into two groups according to this parameter. Depending on which group is used in the matrix, increasing the viscosity of MHPC within each group can either increase or decrease the water uptake into the matrices (Wan *et al.*, 1991).

Alkaline sizes are used to produce 'permanent' papers, to be used in books and documents that are destined to be preserved for tens to hundreds of years. Alkaline sizing permits the use of calcium carbonate as a filler to achieve higher brightness. A high degree of sizing can be achieved with wax emulsions, produced from paraffin, waxes or blends of these waxes with rosin. Wax emulsions are added to the pulp furnish after the addition of a rosin-based size and alum, in front of the paper-machine headbox. Wax emulsions are used in many applications and the choice of a wax or wax–rosin emulsion is based on the need to develop a high degree of sizing at minimum cost. Oil-resistant paper and paperboard can be obtained with the help of fluorochemical agents. The most commonly used fluorochemicals are those from the phosphate family, which have been approved by the FDA for direct-food-contact packaging. These agents provide excellent oil and grease resistance but do not provide sizing against aqueous penetrants (Putman *et al.*, 1982).

The ability of alginates to hold water, to gel, to emulsify and to stabilize has led to many industrial applications. The water-holding property of alginates is useful for paper sizings and to improve paper surface properties, smoothness and ink acceptance. Moreover, this same property is used in paper coatings to control their rheology as well as to prevent dilatancy at high shear (Cottrell and Kovacs, 1982).

An aqueous latex of a poly(beta-hydroxyalkanoate) (PHA) used as a paper imparted water imperviousness without changing mechanical properties. Blends of PHA latices with sodium CMC, polystyrene latex, carboxylated styrene/butadiene latex, natural rubber latex and starch powders form satisfactory films at room temperature (Lauzier *et al.*, 1993).

17.10.5 Dry-strength additives

Natural gums and starches are used to a large extent as dry-strength additives. Complementation is achieved with the corresponding anionic and cationic derivatives, sodium CMC and synthetic water-soluble polymers (e.g. anionic and cationic acrylamide polymers and PVA). The value of the dry-strength additive is assessed by measuring the responses of individual strength properties. Starches and modified starches (oxidized, enzyme-converted, hydroxyethylated and cationic) are used not only to improve surface strength, but also for surface sizing. Starches can be applied to the finished sheet by size press (Fig. 17.3), spraying or a foam application. In general, starch is not used alone, but in combination with other surface-sizing agents. These combinations result in increased surface strength and a better finish. They also improve the printing quality of the paper as a result of the increased surface strength and reduced linting (Putman *et al.*, 1982).

Cationic starch ethers have been established as additives in the paper industry for many years. In the past, starches were predominantly etherified

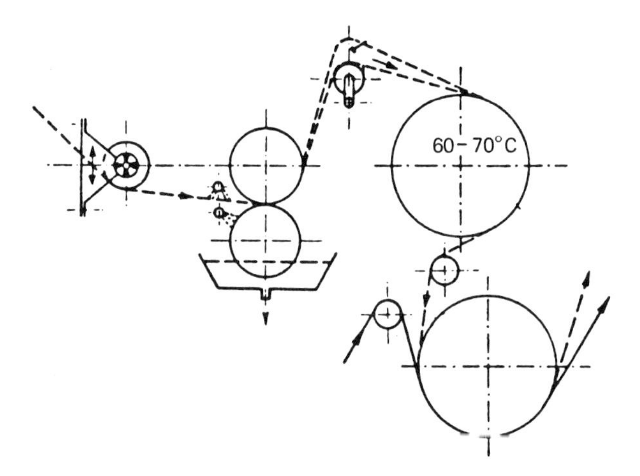

Figure 17.3 Basic design of a size press. (Adapted from *The Applications of Synthetic Resin Emulsions* by H. Warson, Ernest Benn Ltd, London, 1972.)

in aqueous suspensions; however, a trend towards 'dry' cationization has lately been observed (Hellwig *et al.*, 1992). In addition to the commercial aspects, there are also ecological advantages of dry-reaction control. The advantage of the dry process for producing cationic starch ethers was discussed and compared with the slurry process (Hellwig *et al.*, 1992). Production of starch derivatives by cooking extrusion is described elsewhere (Narkrugsa *et al.*, 1992).

Economic and legislative forces are behind the recent increase of secondary fiber use in paper products (Hipple, 1991). The effective use of deinked secondary fiber for printing and writing papers is currently limited because of runnability, strength and brightness problems. The level of wet-end cationic starch may need to be adjusted in order to obtain high strength equal to that of virgin paper and still maintain sizing (Hipple, 1991).

The first Chinese papermakers used mucilage extracted from roots and stems to strengthen their products. Natural gums such as LBG, guar gum and tamarind gum are presently used for the same purpose, with guar gum being used in the largest quantities. The gums are added as aqueous solutions and are adsorbed onto cellulose pulp fibers, thus augmenting the hemicellulose formed during refining to yield stronger paper (Putman *et al.*, 1982).

The Chinese papermakers also used starch as a strength additive. Corn, wheat and potato starches, as well as modified starches, are effective for this purpose. Starches have to be cooked in water before their addition as a solution to the papermaking system. Anionic starches are used with alum, creating a mechanism for starch retention as a result of the positively charged complex. Cationic starches are self-retaining. Starches are capable of increasing the paper's internal strength, as well as improving the paper's surface strength. CMC is used in alum-containing systems as a strength additive similar to anionic acrylamide polymers. CMC can also be used with an aminopolyamide–epichlorohydrin wet-strength resin. Advantageous substitutes for natural gums are synthetic acrylamide polymers, which can be easily modified to produce agents with the desired degree of substitution and molecular weight. These polymers provide improved retention of the fillers and rosin-sizing efficiency, thereby facilitating processing. The acrylamide polymers can be used as anionic or cationic agents and can, therefore, be involved in combinations with other anionic or cationic dry-strength additives to provide strengths that cannot be achieved using either polymer alone. Dry-strength additives improve the degree of bonding between fibers to achieve the desired effect. Dry strengths are increased by hydrogen-, ionic- and covalent-bonding polymers or their combinations. Wet strengths are improved with polymers that are covalently bonded to eliminate water disruption of ionic and hydrogen bonds. Urea-formaldehyde resins were the first wet-strength resins to require impregnation (cross-linking of cellulosic hydroxyl groups through methylene bridges) in the sheet. Melamine-formaldehyde resins can be used to obtain much higher wet and dry strengths than those achieved with urea-formaldehyde resins. Both resins require mineral acid or alum for catalysis of the thermosetting reaction. Aminopolyamide–epichlorohydrin resins are more expensive but pass curing at neutral to alkaline pH, which has the advantage of eliminating acid-catalysed degradation and paper embrittlement and achieves a softer, more absorbent product with less corrosion of the production machinery. Many other resin-agent families can be found commercially. Wet strength is presumably developed by the formation of a network, by cross-linking either with itself or the cellulose, which limits the swelling of cellulose and hemicellulose by water. The network may conserve part of the hydrogen bonds in the dry sheet or may be responsible for new covalent bonds formed by the wet-strength resin that remained unbroken by the water. With high-molecular-weight polyethylenimine, which lacks reactive functionality, wet strength may be achieved via the effects of entropy, i.e. simultaneous breakage of hydrogen and/or ionic bonds. However, no experimental support for this hypothesis exists (Putman *et al.*, 1982).

17.10.6 Fillers

High opacity of paper is important to prevent seeing images on the back side of the page. Opacity can be improved by incorporating fillers within the sheet to increase scattering of the light that passes through the sheet. Fillers are mineral pigments such as clay, calcium carbonate, silica, hydrated alumina and talc. Kaolin clay is the largest-volume pigment for this application. The clay provides a better-quality coating and is slightly more expensive than that used as filler clay. Coating clays are higher in brightness and have a smaller particle-size distribution than filler clay. Fillers are often added when the basis weight of the paper is lower than $65\,g\,m^{-2}$. Mineral fillers increase the surface area of the paper sheet and thus increase the scattering of light. Titanium oxide is an expensive filler that increases surface area and as a result of its high refractive index scatters light within the particle. It is used in cases in which high brightness and opacity are requested. Additional synthetic polymeric pigments are based on polystyrene latices and highly cross-linked urea-formaldehyde resins. The synthetic pigments are less dense than mineral fillers and can, therefore, be used to produce lightweight grades of paper (Putman *et al.*, 1982).

17.10.7 Binders and dispersants

Binders and dispersants can be natural or synthetic polymers. The most commercially successful binder is starch, in its derivatized form or unmodified. Studies of polymer–paper adhesion illustrated the thermodynamic nature of the bondability of polymers to plain, uncoated paper surfaces (Borch, 1991). The bond strength depends strongly on the chemical nature of the polymer surface and on that of the fibrous paper surface. Adhesion to paper may be characterized indirectly through thermodynamic analysis of the paper substrate, or directly through paper laminate or adhesion-tape peel testing (Borch, 1991). The effects of particle size, particle shape and surface chemistry on the cohesion of pigmented coatings based on clay, calcium carbonate and polystyrene latex, using CMC as a binder, were determined (Inoue and Lepoutre, 1992). The results indicated that coating cohesion, assessed at minimum binder content, increases with increasing pigment-particle size. Weak acid–base interactions were formed between pigment and the CMC binder in the case of polystyrene latex (Inoue and Lepoutre, 1992). Mineral particles intended for the coating of paper and board are surface-treated with a copolymer based on styrene and acrylic acid. The polymer is grafted to the mineral by first adsorbing the acrylic acid to the surface and then allowing this monomer to take part in the polymerization with the styrene. An improvement in the surface strength of papers coated with the modified particles was observed with this pigment (Skeppstedt *et al.*, 1993).

Folding boxboards, which count among the most important packaging materials, are manufactured from special board qualities that are printed and upgraded. After punching and creasing, the board material is glued at its longitudinal seams so that a three-dimensional hollow space may be formed from the flat blanks on special machines. To increase the viscosity of the adhesive pseudoplastic, liquids with optimized viscosity-relaxation characteristics can be utilized (Wilken and Baumgarten, 1992).

Other natural binders include proteins extracted from soy meal. Casein was often used as a binder until its high cost became prohibitive. Latices consisting of 48–52% (per weight) polymer dispersed in water are used as synthetic binders. The largest-volume binder is styrene–butadiene copolymer latex. Most later latices are carboxylated, i.e. they contain copolymerized acidic monomers. Other latex binders are based on polyvinyl acetate and on polymers of acrylate esters. Another synthetic binder is PVA. Dispersants are used to prevent particle association and the formation of viscous pastes.

Common dispersants are polyphosphates and sodium polyacrylate. Other water-soluble polymers are added to coating colors to act as water-retention agents and rheology modifiers. CMC, HEC, guar gum and its derivatives and sodium alginate are utilized to improve the rheological properties of coating colors and to help keep them on the surface of the paper (Putman *et al.*, 1982).

High-speed blade-coating (Fig. 17.4) of paper imposes severe constraints on the rheological properties of pigmented coating colors. Removal of the cobinder CMC was reported as a possible way to maximize the runnability of certain pigments, if the coating color is applied using a short-dwell coater and contains a pigment that imparts a high level of intrinsic water retention

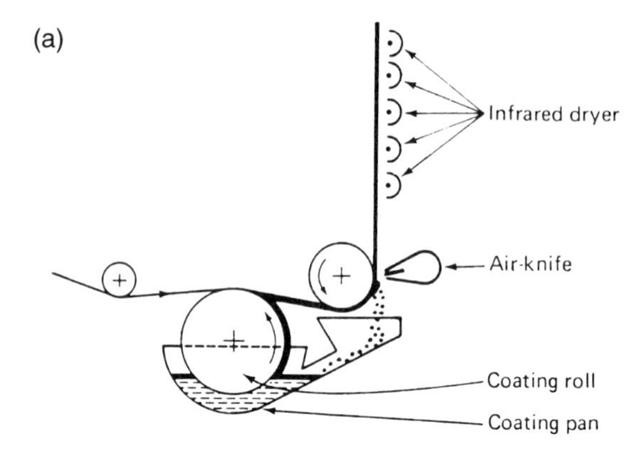

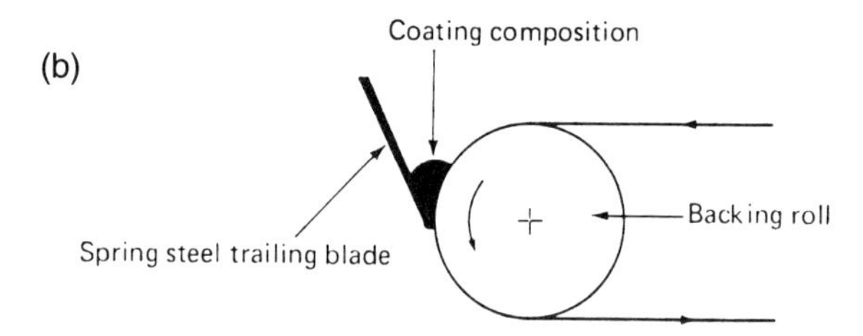

Figure 17.4 (a) Air-knife coater without drying tunnel; (b) trailing-blade coater. (Adapted from *The Applications of Synthetic Resin Emulsions* by H. Warson, Ernest Benn Ltd, London, 1972.)

(Gane *et al.*, 1992). There are no apparent advantages to using the more fluid pigments under optimized conditions if the overall paper properties and the requirement of a broad range of applied coat weights are considered.

References

Baum, G.A., Malcolm, E.W., Wahren, D. *et al.* (1982) Paper, in *Kirk Othmer Encyclopedia of Chemical Technology*, 3rd edn, Vol. 16, Wiley Interscience, New York, pp. 768–803.

Borch, J. (1991) Thermodynamics of polymer-paper adhesion – a review. *J. Adhes. Sci. Technol.*, **5**(7), 532–41.

Borchers, B., Forkel, H. and Ritter, K. (1993) The action of starch in the new paper–waste paper cycle. *Papier*, **47**(10), 40–6.

Britt, K.W. and Unbehend, J.E. (1980) *Proc. Tech. Assoc. Pulp Paper Ind., Papermaker Conf.*, p. 5.

Buyers' Guide – Chemicals (1980) *Lockwood's Directory of the Paper and Allied Trades,* Vance Publishing, New York, p. BG-80.

Claflin, G.D., Jr (1907) US Patent No. 864,359.

Code of Federal Regulations (1980) *Title 21,* Subchapt. B, Part 176, Sections 170, 180, April 1, US Government Printing Office, Washington, DC.

Cottrell, I.W. and Kovacs, P. (1982) Alginates, in *Handbook of Water-soluble Gums and Resins,* ch. 2 (ed. R.L. Davidson), McGraw-Hill, New York, p. 2.25.

Davidson, R.L. (ed.) (1982) *Handbook of Water-soluble Gums and Resins,* McGraw-Hill, New York.

FAO (1977) *World Pulp and Paper Consumptions Outlook – Phase II,* Industry Working Party for the Forestry Department of the Food and Agricultural Organization of the United Nations.

Gane, P.A.C., Mcgenity, P.M. and Watters, P. (1992) Factors influencing the runnability of coating colors at high speed. *TAPPI J.,* **75**(5), 61–73.

Haskell, J.D. (1932) US Patent No. 1,873,199.

Hellwig, G., Bischoff, D. and Rubo, A. (1992) Production of cationic starch ethers using an improved dry process. *Starch-Starke,* **44**(2), 69–74.

Hipple, B.J. (1991) Fine paper properties and the effects of wet-end starch when using deinked recycled fiber in an alkaline system. *TAPPI J.,* **74**(5), 79–84.

Hunter, D. (1947) *Papermaking*, 2nd edn, Alfred A. Knopf, New York.

Inoue, M. and Lepoutre, P. (1992) Influence of structure and surface chemistry on the cohesion of paper coatings. *J. Adhes. Sci. Technol,* **6**(7), 851–7.

Kline (1980) *The Kline Guide to the Paper Industry*, 4th edn, Kline, Fairfield, NJ, p. 68.

Lauzier, C.A., Monasterios, C.J., Saracovan, I. *et al.* (1993) Film formation and paper coating with poly(beta-hydroxyalkanoate), a biodegradable latex. *TAPPI J.,* **76**(5), 71–7.

Myreen, B. (1994) Pulp and paper manufacture in transition. *Water Sci. Technol.*, **29**(5–6), 1–9.

Narkrugsa, W., Berghofer, E. and Camargo, L.C.A. (1992) Production of starch derivatives by cooking extrusion. *Starch-Starke*, **44**(3), 81–90.

Putnam, S.T., Espy, H.H., Spence, G.G. *et al.* (1982) Papermaking additives, in *Kirk Othmer Encyclopedia of Chemical Technology*, 3rd edn, Vol. 10, Wiley Interscience, New York, pp. 803–825.

Schuchardt, D.R. and Berg, J.C. (1991) Liquid transport in composite cellulose-superabsorbent fiber networks. *Wood Fiber Sci.,* **23**(3), 342–57.

Seaman, J.K. (1982) Locoust bean gum, in *Handbook of Water-soluble Gums and Resins*, ch. 14 (ed. R.L. Davidson), McGraw-Hill, New York, p. 14.11.

Skeppstedt, A., Malhammar, G., Engstrom, G. *et al.* (1993) Use of polymer-treated mineral pigment in paper coatings. *J. Mat. Sci.*, **28**(21), 5819–25.

Smith, D.C. (1970) *History of Papermaking in the United States (1691–1969),* Lockwood, New York.

TAPPI Press (1962) *Pigmented Coating Processes for Paper and Board,* Technical Association of the Pulp and Paper Industry, Atlanta, GA.

TAPPI Press (1977) *Physical Chemistry of Pigments in Paper Coating*, Technical Association of the Pulp and Paper Industry, Book No. 38, Atlanta, GA.

Wan, L.S.C., Heng, P.W.S. and Wong, L.F. (1991) The effect of hydroxypropylmethyl cellulose on water penetration into a matrix system. *Int. J. Pharm.*, **73**(2), 111–16.

Warson, H. (1972) *The Applications of Synthetic Resin Emulsions,* Ernest Benn, London, pp. 739–823.

Wilken, R. and Baumgarten, H.L. (1992) The importance of the rheology of dispersion adhesives for folding boxboard manufacture. *Papier*, **46**(10), 81–7.

Wiryana, S. and Berg, J.C. (1991) The transport of water in wet-formed networks of cellulose fibers and powdered superabsorbent. *Wood Fiber Sci.*, **23**(3), 457–64.

Young, T.S. and Fu, E. (1991) Associative behavior of cellulosic thickeners and its implications on coating structure and rheology. *TAPPI J.,* **74**(4), 197–207.

Zender, J.A., Stuthridge, T.R., Langdon, A.G. *et al.* (1994) Removal and transformation of resin acids during secondary treatment at a New Zealand bleached Kraft pulp and paper-mill. *Water Sci. Technol.,* **29**(5–6), 105–21.

18 Spongy hydrocolloid matrices

18.1 Introduction

A cellular solid is an interconnected network of solid struts or plates that forms the edges and faces of cells (Gibson and Ashby, 1988). Cellular materials such as cork, first mentioned for use as bungs in wine bottles in Roman times, have been used for centuries. Recently, different artificial cellular solids have been developed. They include honeycomb-like materials and polymeric foams used for a variety of purposes, from disposable coffee cups to crash padding in aircraft cockpits. Foaming techniques are used to create foamed polymers, metals, ceramics and glass and these foams can be used for insulation, cushioning and absorbance of impact kinetic energy.

The structure of cellular solids ranges from the near-perfect order of the bee's honeycomb to the disordered, three-dimensional networks of sponges and foams (Gibson and Ashby, 1988; Jeronomidis, 1988). There is a clear distinction between open cell edges and closed-cell foams. In the first, the solid material has been drawn into the struts that form the cell edges. These struts join at vertices, usually giving an edge connectivity of four. In closed cells, solid membranes close off the cell faces, but the solid is rarely uniformly distributed between the edges and faces. When foaming takes place, surface tension may be the dominant force responsible for drawing the solid material into the cell edges, leaving a thin face framed by thicker edges. If surface tension shapes the structure, then the four edges meet at 108° at each vertex. Metal, ceramic and glass foams are good examples of this phenomenon (Gibson and Ashby, 1988).

18.2 Edible cellular solids

18.2.1 Breads

Many foods are foams (Fig. 18.1). Bread usually consists of closed cells, expanded by the fermentation of yeasts or by carbon dioxide from bicarbonate (Nussinovitch *et al.*, 1996). Breads display an enormous range of phase volumes, air-cell dimensions and anisotropy. Meringue consists of foamed egg white and sugar. The size of the gas cells, the maximal included phase and ease of processing are influenced by the content and character of the surface-active agents used.

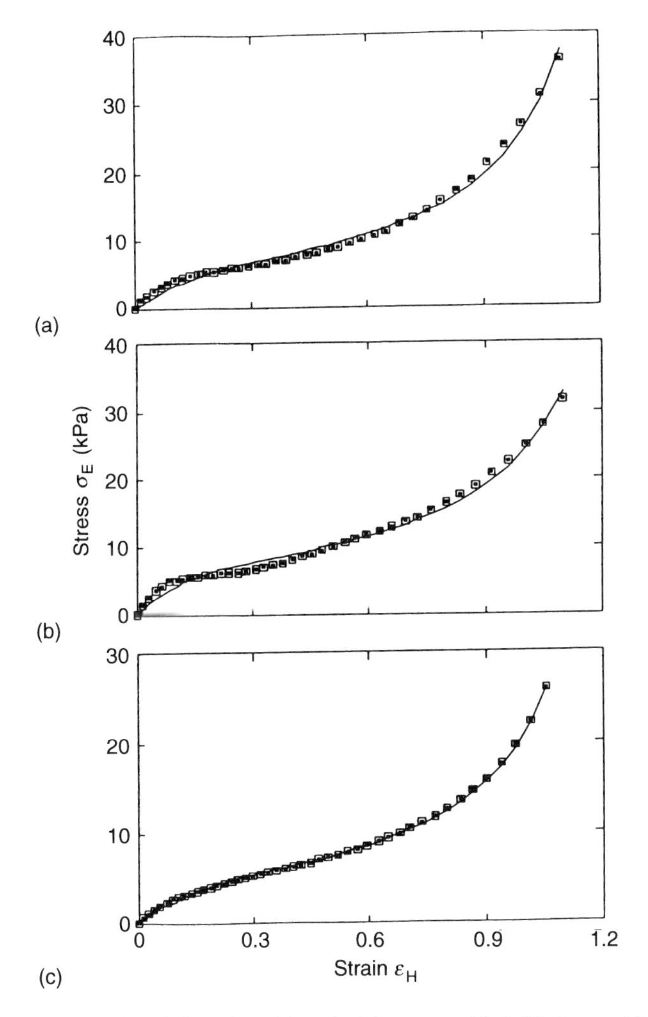

Figure 18.1 Stress–strain relationships of breads: (a) pumpernickel; (b) dense white bread; (c) country oat bread. (From Nussinovitch *et al.*, 1991.)

18.2.2 Foamed chocolates and sweets

Foamed chocolate represents a food that has been expanded to change its texture. Lillford (1989) described an X-ray projection of aerated chocolate. The air-phase volume included within the bar comprises interconnecting air cells of various sizes. This has a major influence on the bar's mechanical properties. In products such as sugar confections, the air-phase volume is high, and cell walls are thin and composed of very brittle sugar glass. The air is the primary cause of the textural difference between boiled sweets and this foamed structure, although both are made from sugar and water (Lillford, 1989). In fact, almost any material capable of solidification can be aerated.

Other hard brittle candies are often expanded to make them attractive to the consumers or, if they are sold by volume, to make them less expensive.

Other important foods are breakfast cereals and snack foods, which are steam-foamed to produce a different texture and crunchiness.

There are two major types of foam. The first, in terms of mechanical properties, is defined when the geometry of the structure, owing to the inclusion of a second phase, does not influence the basic deformation mechanism of the matrix material. Examples are high-density foams with small included phase volumes, or structures with lower densities built of very brittle material, where the deformation prior to fracture is very small (Lillford, 1989). Most food foams have larger phase volumes and make up the second foam type: those in which the geometry influences the deformation mode and, therefore, influences the modulus and strength of the structure.

18.2.3 Sponge cakes

The properties of sponge cakes have been studied by Attenburrow *et al.* (1989). They have an open cell structure. Water is included within the cell walls, which by their presence and content influence the mechanical properties. In other words, the mechanical properties can only be determined if the products being tested have been equilibrated in an atmosphere of controlled relative humidity. For sponge cakes, major changes in the stress–strain curve are observed if samples have been equilibrated at $a_w < 0.33$ and $a_w > 0.57$ (a_w is water activity) (Attenburrow *et al.*, 1989). At lower moisture contents, non-recoverable brittle failure occurs, whereas at higher moisture contents, the mode of collapse appears to be recoverable elastic buckling. Samples showed a square dependence of both modulus and fracture stress on bulk density. The plasticizing effect of the moisture can be concluded from the proportionality constants for both parameters. Among the parameters studied, the modulus of the samples showed the highest statistically significant correlation to sensorially perceived hardness.

18.2.4 Filled foams

In open cell structures, the contribution of entrained gas is effectively negligible. Under other conditions, the included phase significantly influences the mechanical properties of the foam. Closed cells filled with gas present an extra contribution to the modulus of the structure, the mathematical terms of which can be found elsewhere (Lillford, 1989). For open-cell foams containing a fluid, the viscous flow of the liquid strongly influences the strain-rate dependence of mechanical properties and hence maximal stresses and stress relaxation times. The contribution of the fluid to the strength of the open-cell foam is directly proportional to the liquid's viscosity, the strain rate and the sample area over which the load is applied, and it is inversely proportional to $1 - \varepsilon$ and the cell area (cross-sectional area of the interconnecting holes). Polymer foams contain interconnections of $\sim$0.1 mm, compared with $\sim 10\,\mu$m in tissues. In polymer foams not only liquid viscosities, but also gas viscosities become significant.

18.2.5 Sensory evaluation of foams

A sensory evaluation of foamed products depends on their initial deformation in the mouth upon fracture and comminution, moistening with saliva, enzyme action, partial warming to mouth temperature, followed by swallowing (Hutchings and Lillford, 1988). The evaluation is influenced by the entire process, including the transient structures formed during mastication (Lillford, 1989). Of course, brittle foams undergo crushing during mastication, and the size distribution of the resultant fragments is important to the evaluation. When the matrix of the product is built of polysaccharides, the saliva can plasticize the structure to induce a partial evaluation of 'melt in the mouth', which is also dependent upon geometry, wetting rate of the surface and solubility and swelling capacity of the matrix. The influence of the catabolic enzymes in the saliva on the matrix is dependent on the ingredients used to build the matrix. For instance, gelatinized starch is more influenced by the enzymes than any composition based on protein or fat. Temperature can influence the mechanical properties of solid foams composed of a matrix that melts between serving and swallowing temperatures, as in the case of gelatin, some polysaccharides and where triglycerides play a major structural role, as in whipped creams and aerated chocolate. If the product is a solid foam that includes ice (e.g. ice cream, sherbet), then melting is the major size-reduction mechanism.

18.3 Relative and solid densities

Relative density (ρ/ρ_s) is the most important structural characteristic of a cellular solid, where ρ is the density of the foam and ρ_s that of the solid from which it is made. Porosity is the fraction of the pore space in the foam and is defined by $1 - (\rho/\rho_s)$ (Peleg, 1982).

In addition, equations can be found relating cell dimensions and shapes to the relative density. The choice of equation depends on whether the structure is a honeycomb or a foam, and if it is a foam whether it has open or closed cells. Honeycombs are cellular materials that appear as a two-dimensional array of polygons which are packed to fill a plane area, like the hexagonal cells of the beehive, hence the name. True honeycombs are relatively rare. Examples include hexagonal aluminum, paper-resin structures, silicone rubber and ceramic honeycombs. (For example: alumina ceramic has a density of 1400 kg m^{-3}, an edge connectivity of 4, a mean edge $cell^{-1}$ of 4, the cell shape is square, the symmetry of structure is square, the cell-wall thickness is 0.48 mm with a relative density of 0.36; for comparison, rigid polyurethane, which is a typical foam, has a density of 32 kg m^{-3}, closed cells, an edge connectivity of 4, a face connectivity of 3, a cell-edge thickness of 30 μm, a cell-face thickness of 3 μm, the fraction of material in the cell edges is 0.70, standard deviation of cell size is 0.075 mm (Gibson and Ashby, 1988).)

18.4 Stress–strain behavior

Figure 18.2 shows a typical compressive stress–strain curve for a cellular material, composed of three regions (Nussinovitch *et al.*, 1989). The first is a linear-elastic region, followed by a plateau of roughly constant stress (second region) and leading to a final region of steeply rising stress. Each region is associated with a particular deformation mechanism. When a specimen is loaded the cell walls are bent, giving linear elasticity if the cell-wall material is linear-elastic. When a critical stress is reached, cells begin to collapse and eventually, at high strains, collapse is sufficient to allow opposing cell walls to touch (or their broken fragments to pack together), and further deformation compresses the cell-wall material itself. This gives the final, steeply rising portion of the stress–strain curve and is called densification.

18.5 Sponge creation and evaluation

Gels can also be used to create sponges. Four different gels – those with internally produced gas bubbles, those aerated by a fermentation process,

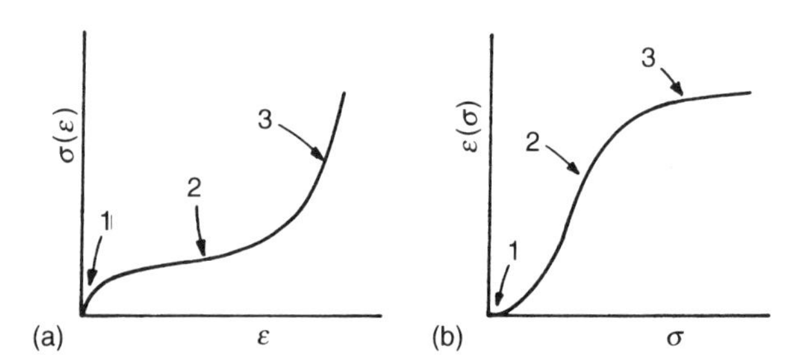

Figure 18.2 Schematic representation of the typical compressive stress–strain (a) and strain–stress (b) relationships of sponges. 1, deformation of the original matrix; 2, densification; 3, compaction of the collapsed cell wall material..

those made by including oil within the gum mixture, and those created by an enzymatic process – were dried to create solid cellular structures (Nussinovitch and Gershon, 1997a). They were compressed to 80% deformation between parallel lubricated plates at a constant deformation (displacement) rate of 10 mm min^{-1}, using an Instron universal testing machine. The Instron's continuous voltage versus time output was converted into stress versus engineering or Hencky's (natural) strain:

$$\sigma = \frac{F}{A_0} \tag{18.1}$$

and

$$\varepsilon_H = \ln[H_0/(H_0 - \Delta H)] \tag{18.2}$$

where σ and ε are stress and strain, respectively; F is the momentary force; ΔH is the momentary deformation ($H_0 - H(t)$); A_0 and H_0 are the original specimen's cross-sectional area and height, respectively; and $H(t)$ is the height at time t. Since the cross-sectional area of a compressed solid sponge specimen rarely expands to any significant extent (Gibson and Ashby, 1988), the engineering and 'true' stresses can be treated as equal for all practical purposes (Swyngedau *et al.*, 1991a,b).

The individual relationships were fitted to a compressibility model previously developed for the sigmoid stress–strain relationships of cellular solids (Nussinovitch *et al.*, 1989; Swyngedau *et al.*, 1991a,b)

$$\sigma = \frac{C_1\varepsilon}{[(1 + C_2\varepsilon)(C_3 - \varepsilon)]} \tag{18.3}$$

where C_1, C_2 and C_3 are constants. The constant C_1 is primarily a scale factor and has stress units. The dimensionless constant C_2 is a measure of the shoulder's prominence in the stress–strain curve; i.e. when $C_2 = 0$, the relationship has no shoulder and its slope increases monotonically. The constant C_3, also dimensionless, is a rough measure of the steepness of the stress–strain curve in the high-strain region. According to equation 18.3, when the value of ε goes to C_3, σ goes to infinity. This is determined to a large extent by the strain level at which collapse of the open structure has been completed and most of the resistance to deformation has been shifted to the compacted solid cell wall material. All the mechanical tests were performed in duplicate.

18.5.1 Sponges created by drying gels filled with internally produced carbon dioxide gas bubbles

It has been suggested that mechanically stable solid hydrocolloid sponges can be produced by freeze-drying agar and alginate gels (Nussinovitch *et al.*, 1993; Nussinovitch, 1995), where internally produced gas bubbles have been formed by immersing bicarbonate-containing gels in an acid bath. The resultant sponges (cellular solids) have characteristic compressive stress–strain curves and their properties are dependent on their composition as well as on the way they were prepared. In addition to this procedure, it is possible, at least in principle, to create hydrocolloid sponges, using three different non-related methods.

To produce these sponges, alginate (or other gelling agents) can be used for gel preparation. To fill the gel with internally produced carbon dioxide gas bubbles, its powder, calcium hydrogen orthophosphate ($CaHPO_4$) and calcium carbonate are added slowly to cold distilled water (10°C) and stirred until the ingredients are completely dissolved. A freshly prepared solution of glucono-δ-lactone (GDL) is then admixed with this solution with vigorous stirring. The alginate solution is poured into a container and allowed to set. After 48 h (Nussinovitch and Peleg, 1990), specimens can be taken from the slab using a cork borer. The gels are then immersed in a citric acid solution. The volume of the citric acid solution should be ~ 100 times the volume of a single gel specimen to guarantee excess acid. The acid diffusion into the gels can be monitored using phenolphthalein as an indicator.

A gel filled with carbon dioxide is shown in Fig. 18.3. The picture was taken after the process has been completed, that is after the acid had reached the gel center by diffusion. The acid's motion in the gel was diffusion controlled, as evidenced by the linearity of penetrated distance (X) versus $t^{1/2}$. As previously mentioned, the gels used in these experiments contained phenolphthalein (dark area in Fig. 18.4) so that the distance could be measured directly with a caliper after the specimen had been dissected. The slope of the X versus $t^{1/2}$ curve was about 0.7 mm min$^{-0.5}$.

Small gas bubbles were created inside the alginate gels (Fig. 18.4). After 2.5 h, about 900 bubbles cm^{-3} were counted. This number increased to about 2000–2700 after 24 or 36 h, depending on carbonate concentration. Bubble formation lowered the density of the gels, causing them to float. After some time bubbles began to leave the gel, causing some damage to its integrity, and the space they vacated gradually filled with liquid, at which point the gels began to sink again.

Figure 18.3 A gel filled with carbon dioxide. (From Nussinovitch, 1995; by permission of Oxford Press.).

Figure 18.4 Creation of gas bubbles in gel cylinders. The diffusion process can be backed using a suitable indicator (here phenolphthalein, the dark area).

Typical stress–strain relationships of ordinary and gas-filled alginate gels are shown in Fig. 18.5. Immersion in acid increases gel strength and deformability. This is because acid-induced cross-linking helps the gel retain its mechanical strength, even in the face of structural disruption caused by bubble formation. The presence of carbonate, however, has a disruptive

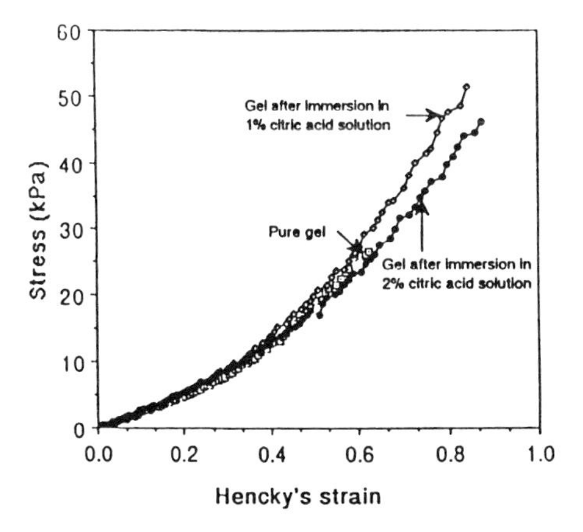

Figure 18.5 Stress–strain relationships for alginate gels before and after immersion in citric acid solution to produce gas bubbles within the gel. (From Nussinovitch, 1995; by permission of Oxford University Press.)

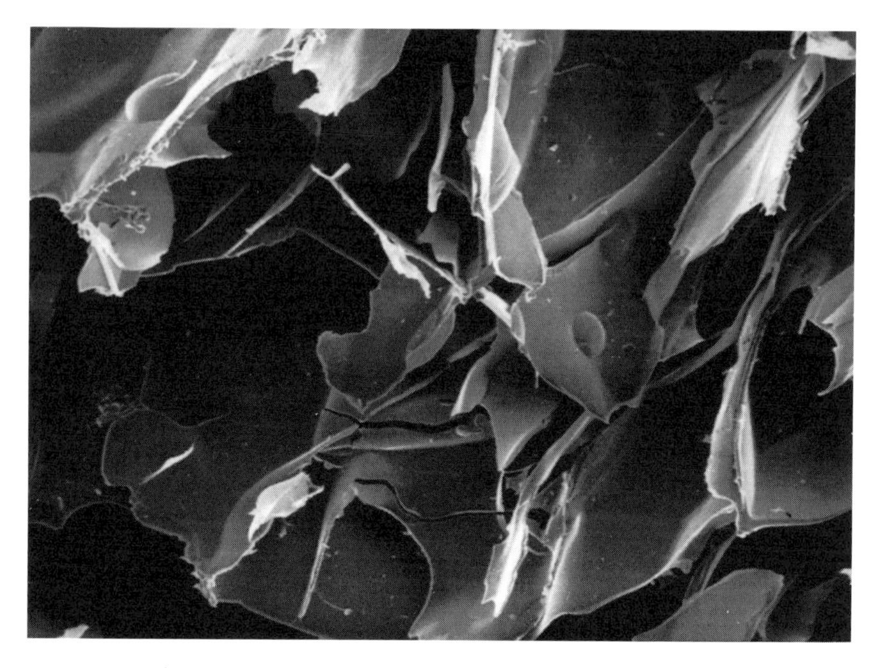

Figure 18.6 Cellular structure of a freeze-dried gel (mag. × 10 000).

effect, manifested primarily in lower stiffness. This is mainly the result of the pH increase beyond the level required for optimal cross-linking. The gel strength depends on both the acid and calcium carbonate concentrations.

Typical cellular structures of freeze-dried gel specimens are shown in Fig. 18.6 and typical stress–strain relationships of the dry sponges are shown in Fig. 18.7. They all exhibit the sigmoid-shape characteristic of cellular solids,

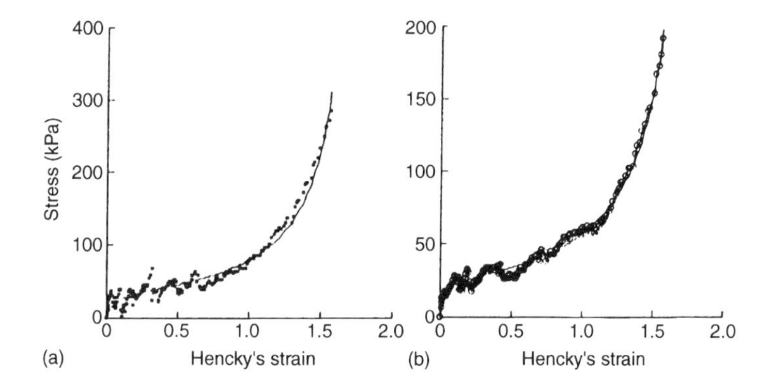

Figure 18.7 Typical stress–strain relationships for dried alginate gels (a) without treatment and (b) after 1.8 h immersion in citric acid solution. The solid lines are the fits of equation 18.3. (From Nussinovitch, 1995; by permission of Oxford University Press.)

Table 18.1 Compressibility parameters of freeze-dried gel sponges (1% alginate) with and without internally produced gas bubbles

Immersion time (h)	C_1 (kPa)	C_2	C_3	MSE
0	1180	18	1.8	104
0.8	3520	56	2.0	21
1.8	1690	36	1.8	16

MSE, mean square error. C_1, C_2 and C_3 are constants from equation 18.3.

reflecting three deformation mechanisms. The first part of the curve, i.e. the almost instant rise in stress, represents deformation of the intact sponge. Since all the tested specimens were brittle, failure occurred after a very small, hardly measurable strain. The region of the curve with moderate slope represents progressive rupture and collapse of the cell walls and densification of the compressed specimen. This is followed by compression of the compacted cell-wall solid material, reflected by the rapidly increasing slope of the stress–strain curve. The stress–strain curves of the alginate sponges, especially in the second region, were irregular, a phenomenon which is quite common in brittle solid foams (Attenburrow *et al.*, 1989; Rohde *et al.*, 1993). The fit of equation 18.3, which was originally developed for spongy baked goods and polymeric sponges and is represented by a solid line in Fig. 18.7, indicates that it too is a suitable compressibility model for this type of sponge. The regression parameters of equation 18.3 are summarized in Table 18.1. The higher magnitude of the mean square error of the alginate sponges is a reflection of the inherent ruggedness of the stress–strain relationships, rather than a reflection of the fit of equation 18.3 as a model. In the alginate sponges, the immersion of the gels in an acid bath did not result in a drastic loss of mechanical integrity. This is apparently because the disruptive effects of bubble formation were at least to some extent offset by the more extensive cross-linking. The constants C_1, C_2 and C_3, as previously mentioned, were determined by a non-linear regression procedure based on minimizing the mean of the squared deviations (Nussinovitch *et al.*, 1995).

18.5.2 Creating sponges by immobilization

Hydrocolloid sponges can be obtained using an immobilization process. *Saccharomyces cerevisiae* (yeast) is cultured in PDA broth for 2 days at 30°C. The cells are harvested by centrifugation and washed twice with sterile deionized water. The yeasts are diluted in sterile deionized water and stirred at 5°C for no longer than 15 min until immobilization is achieved. (Of course other microorganisms that can produce gas by fermentation are possible sources for immobilization.) Microorganisms can be

Table 18.2 Influence of time of sucrose fermentation with different concentrations of entrapped *Saccharomyces cerevisiae* on the mechanical properties of 2% agar gels

	Fermentation time at 30°C								
	0 days			3 days			7 days		
Yeast conc. ($CFU\,g^{-1}$)	Stress at failure (kPa)	Strain at failure	ED (kPa)	Stress at failure (kPa)	Strain at failure	ED (kPa)	Stress at failure (kPa)	Strain at failure	ED (kPa)
0	39.5 ± 1.1	0.26 ± 0.01	136.7 ± 4.1	39.7 ± 2.2	0.25 ± 0.03	142.2 ± 2.2	44.9 ± 2.0	0.25 ± 0.03	193.6 ± 3.0
10^8	28.9 ± 1.2	0.25 ± 0.02	134.9 ± 2.8	22.7 ± 0.8	0.23 ± 0.02	111.4 ± 1.8	20.4 ± 0.8	0.21 ± 0.01	101.8 ± 2.0
10^9	22.2 ± 1.4	0.21 ± 0.01	129.7 ± 2.2	11.5 ± 0.5	0.19 ± 0.01	59.9 ± 1.0	9.4 ± 0.6	0.17 ± 0.01	44.4 ± 0.6

Each result is the average of at least six determinations ±SD taken from two separate gel batches.

counted directly in a Neubauer chamber, by plating or by any other accepted means. Immobilization is performed by thoroughly mixing a hydrocolloid solution with the yeast dilution to obtain gels with a predetermined number of immobilized yeasts per unit volume. The microorganism suspension is added immediately after bringing the cell suspension to the desired temperature. The gel–cell suspension is mixed prior to pouring into stainless-steel molds. Reference gels with no microorganisms can be prepared in parallel. The gels are immersed in 5% sucrose solution to induce fermentation. The volume of the sucrose solution is ~ 100 times the volume of the immersed gels. Gels with entrapped carbon dioxide bubbles are taken after a predetermined time to study their mechanical properties in parallel to gels that have been freeze-dried to analyse their sponge properties and structure (Nussinovitch and Gershon, 1997b).

The influence of yeast immobilization on the mechanical properties of an agar gel at time zero is presented in Table 18.2. The higher the concentration of the microorganisms within the gel, the higher the disturbance to the gel's integrity. In other words, stress at failure and deformability modulus values decreased after the addition of the yeasts. Moreover, the brittleness of the

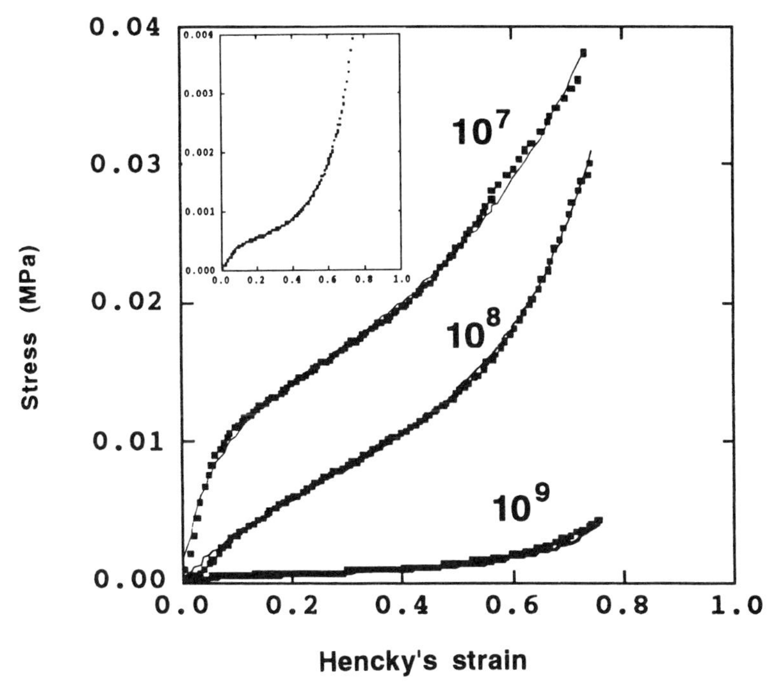

Figure 18.8 Stress–strain relationships of 'yeast' sponges formed by immobilizing three different initial concentrations of yeast. Inset: enlargement of the stress–strain curve formed by the highest yeast concentration (From Nussinovitch and Gershon, 1997a; by permission of Oxford University Press.)

gels increased, in agreement with previous results (Nussinovitch *et al.*, 1994). After immobilization, when the gels were immersed in the 5% sucrose solution, slow fermentation occurred, since no nitrogen source was added. Carbon dioxide bubbles and ethanol were produced. The gas bubbles travelled to the surface of the gel, causing some minuscule cracks therein and influencing the structure of the resultant sponge. The longer the fermentation period the less strong and stiff the gel. The shape of the stress–strain relationship of the 'yeast' sponge is showed in Fig. 18.8. The constants achieved by non-linear regression are presented in Table 18.3. The structure of these sponges is presented in Fig. 18.9. An interesting phenomenon can

Table 18.3 Model constants of non-linear regression in compression stress–strain relationships of yeast gels after 3 days of fermentation in freeze-dried sponges

Yeast conc. ($CFU\,g^{-1}$)	C_1 ($MPa \times 10^{-2}$)	C_2	C_3	MSE ($\times 10^{-7}$)
10^7	31.7	19.1	1.13	2.16
10^8	4.4	4.7	0.98	0.74
10^9	0.4	7.8	0.96	0.02

MSE, mean square error; CFU, colony-forming units. C_1, C_2 and C_3 are constants from equation 18.3.

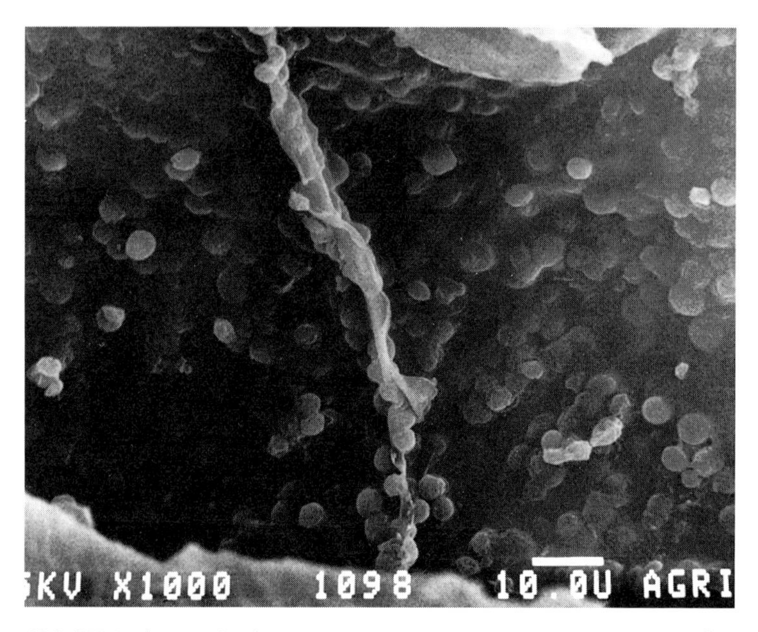

Figure 18.9 SEM micrograph of a 'yeast' sponge resulting from entrapment of $10^9\,CFU\,g^{-1}$ agar gel introduced into sucrose solution for 7 days before freeze-dehydration. (From Nussinovitch and Gershon, 1997a; by permission of Oxford University Press.)

be observed. The longer the fermentation period, the higher the biomass observed and in the compressed sponge there appears to be compaction of the yeast network replacing the hydrocolloid network (Nussinovitch and Gershon, 1997b).

18.5.3 Enzymatically produced sponges

It has recently been postulated that enzymes can be used to produce and change the structure of hydrocolloid sponges. The idea is quite simple: a substrate is included within a gel matrix and the gel is then embedded within an enzyme solution with a cut-off molecular weight that enables the enzyme's diffusion through the gel. Enzyme activity is dependent upon its and the substrate's concentration, temperature, pH, etc. An agar–starch sponge was obtained (Nussinovitch *et al.*, 1995) by letting the enzyme α-amylase (500 to 1500 ppm) diffuse into an agar–starch gel at 55°C for different times. The enzyme, which is a metalloenzyme, attacks starch molecules by randomly hydrolysing any of the α(1–4) linkages, except those near a branching point or the end of the starch molecule. Later the gel is freeze-dried to produce different textures of cellular solids. The remaining terminal sugar is in the α-configuration, which is why the enzyme is called α-amylase. The hydrolysis of the embedded starch by the enzyme can be divided into two phases. The first includes rapid dextrinization, during which the linear amylose is randomly degraded to smaller dextrins; this is followed by a slower saccharification resulting in mainly maltose and glucose, together with some short oligosaccharide polymers. After drying, the gels were analysed for their stress–strain relationships and textural properties. The resultant cellular solids produced typical sigmoidal stress–strain curves, and the higher the enzyme concentration, the higher the porosity of the resultant sponges and the lesser their stiffness.

18.5.4 Oil gels and sponges

Oil-alginate gels are produced as already described, except that up to 40% soybean oil is homogenized into the gum solution, to which a GDL solution is added later. Gels are kept at 5°C for 24 h. They are then equilibrated to room temperature before being heat treated for 5–15 min in warm water (60°C) to induce oil removal from the gels into the warm water in which they are immersed. The warm water is replaced several times: thus the oil–alginate gels are immersed in water for about 10 min before they are either compressed as gels to study their mechanical properties or freeze-dried to obtain alginate sponges. It should be emphasized that oil can be homogenized with various gelling-agent solutions to obtain later drying of the cellular solids. Since alginate creates a thermostable gel, additional heat treatments are possible, enabling the creation of other structures that differ from those resulting from gels which are not heat stable.

Table 18.4 Physical properties of soya vegetable oil[a]

Specific gravity ($kg\,m^{-3}$)	0.91
Iodine number	120–141
Refractive index	1.467–1.470
Saponification number	188–195
Unsaponified matter (%)	0.5–1.6
Free fatty acids (%)	0.02–0.1
Peroxide value (mequiv. kg^{-1})	1.0 max.
Moisture (%)	0.1 max.
Melting point (°C)	-21 ± 1
Smoking point (°C)	232–237
Flash point (°C)	335–340
Fire point (°C)	371–377

[a]Provided by the manufacturer.

In general, oil is included within alginate sponges to change their properties, such as structure, density, porosity, etc. The properties of soybean oil are presented in Table 18.4. The mechanical properties of alginate–soybean oil gels are shown in Table 18.5. The higher the content of the oil within the gel, the lower its stress at failure and stiffness, as reflected by the deformability modulus. The higher the content of the oil, the smaller the Hencky's strain at failure; or in other words, the gel is more brittle.

Two systems of alginate–oil gels and sponges are dealt with in this chapter: in the first, gels with or without oil are simply freeze-dried directly after preparation; in the second, the gels are heat-treated at 85°C for 15 min in water, three times in succession. Each time the water and the extracted oil is discarded. Oil content in the gels and sponges can be estimated by Soxhlet extraction or other method and is given in Table 18.6. After heat treatment, 40–50% of the oil has 'left' the gel. After freeze-drying, the percentage of oil within the sponge increases.

The stress–strain relationships of the oil sponges are presented in the untreated form in Fig. 18.10 and after heat-treatment and partial oil extraction in Fig. 18.11. The results of the non-linear regression for deter-

Table 18.5 Compressive mechanical properties of alginate–soybean oil gels

Oil (%, w/w)	Failure stress (kPa)	Hencky's strain	Deformability modulus (kPa)	R^2 (σ versus ε)
0	13.1 ± 0.2	0.65 ± 0.02	10.3 ± 0.3	0.993
10	11.9 ± 0.6	0.55 ± 0.08	8.5 ± 0.5	0.981
20	9.5 ± 0.5	0.54 ± 0.04	6.0 ± 0.6	0.988
30	3.9 ± 0.1	0.50 ± 0.01	3.4 ± 0.6	0.996
40	3.6 ± 0.4	0.47 ± 0.02	2.1 ± 0.1	0.991

Each result represents the average of six determinations ± SD.

Table 18.6 Oil content in oil–alginate gels and sponges

Oil content within alginate gels (%, w/w)	Oil content in gels after heat treatment (%, w/w)	Oil in sponges (%, w/w)
0	0	0
10	6	40
20	12	57
30	18	66
40	28	75

mining the constants of equation 18.3 for the curves in Fig. 18.10 are presented in Table 18.7. In the Fig. 18.10 inset, the ruggedness of the curves up to 60% deformation is shown. Two facts can be observed. The higher the oil content within the sponge, the smoother the curve. In addition, the C_3 constant of the non-linear regression decreases. The higher the bulk density of the sponge (from 0.074 g cm^{-3} for those without oil inclusion to

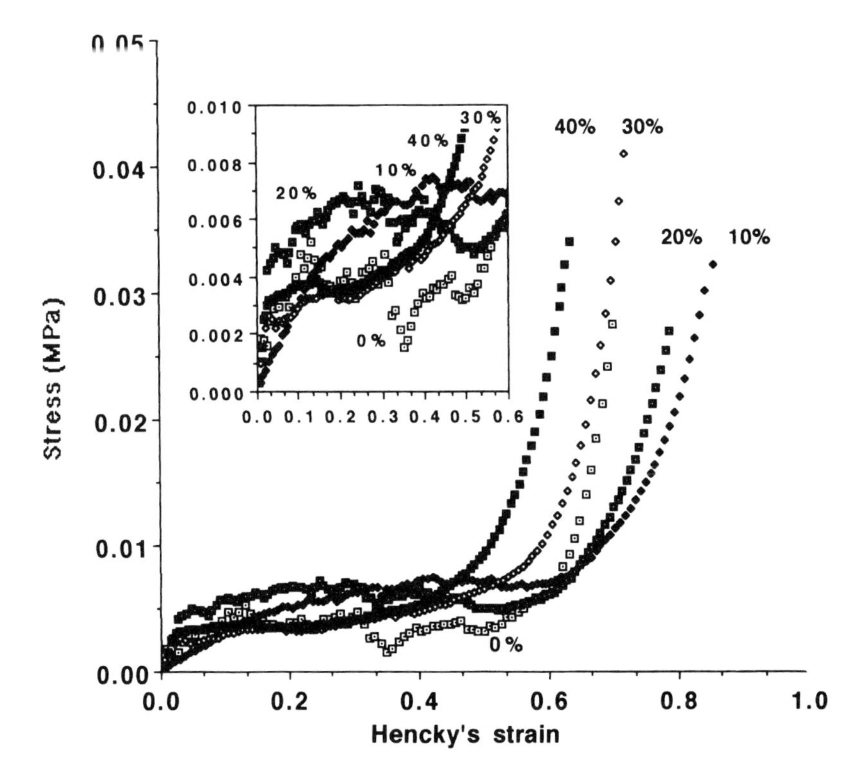

Figure 18.10 Stress–strain relationships of oil sponges from untreated gels. Initial oil concentrations (0–40%) are as mentioned on the figure. (From Nussinovitch and Gershon, 1997b; by permission of Oxford University Press.)

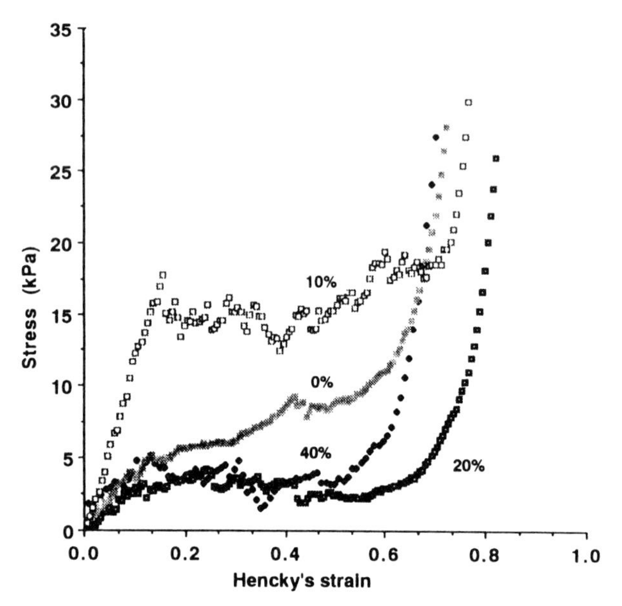

Figure 18.11 Stress–strain relationships of sponges (containing varying initial amounts of oil) from heat-treated gels. (From Nussinovitch and Gershon, 1997b; by permission of Oxford University Press.)

0.29 g cm^{-3} for 40% oil inclusion before freeze-drying), the more the stress tends to steepness at smaller deformations. The heat treatment used is one of several possible oil-extraction methods. After extraction (and see Fig. 18.2 and the previous discussion), the stress–strain curves become more rugged. The heat treatment may be disrupting the gel structure, physically damaging the specimen surface.

The mechanical properties of the oil sponges were determined to show that even with a high proportion of well-dispersed and embedded oil within the hydrocolloid matrix, the integrity of the sponges is maintained. In other

Table 18.7 Model constants of non-linear regression in compression stress–strain relationships of freeze-dried oil sponges

Oil in gels (%, w/w)	C_1 (MPa)	C_2	C_3	MSE[a] ($\times 10^{-6}$)
0	0.32	444	0.94	1.39
10	0.19	65	0.93	1.35
20	4.10	1719	0.88	3.60
30	0.01	5	0.76	0.78
40	0.52	304	0.68	0.72

[a]MSE, mean square error. C_1, C_2 and C_3 are constants from equation 18.3.

words, if these sponges are found to be useful in the future, inclusion of a high percentage of oil is not an obstacle to the sponge's production.

The porosity of the sponges changed dramatically after including the oil. Porosities changed from ~95% (no oil included) to ~80% (40% oil included in the gel before freeze-drying). The porosity of the other heat-treated system was not studied. Electron microscopy of the sponges (Figs 18.12–18.14) revealed that the higher the oil content in the sponge, the more closed cells within its structure. In addition, the structure of the cells changed from having big openings to having rounder, smaller ones. Since oil sponges can be used as carriers of oil-soluble vitamins and colors and are suitable for human consumption, the color of such sponges is important. If no other natural or artificial colors are added then the higher the oil content within the sponge, the lower its L value (lightness decreases) and the higher the a value. A small decrease in b value was also found. These results indicate that as the oil content increases there is a tendency for the color to become more yellowish. Various sources of vegetable oils (olive, soybean, corn, sunflower, cotton, canola and others) can be included in the sponges, giving different intensities and chromas. For example, olive oil's hue varies between yellowish and greenish. Therefore, oil entrapment within the sponge (especially in high proportions) influences the appearance of the product and plays a major role in its sensory evaluation, a potentially important consideration in its future marketing (Nussinovitch, 1995).

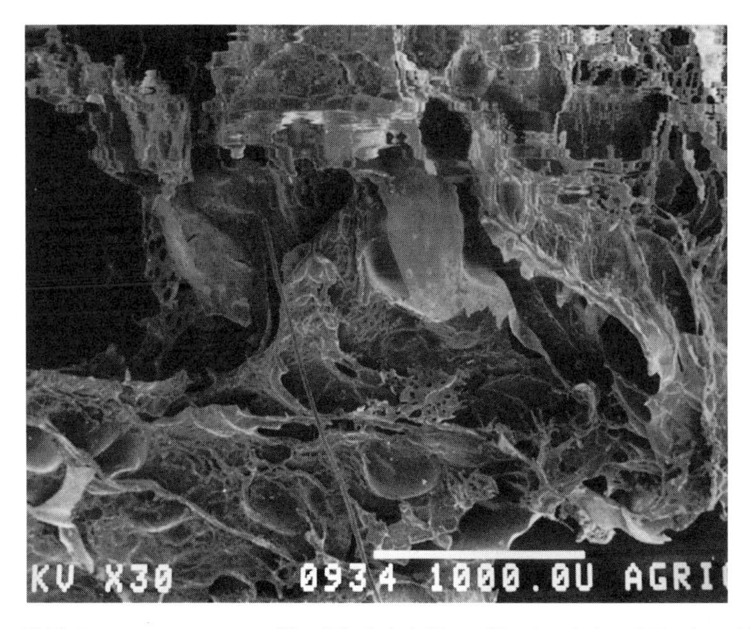

Figure 18.12 Agar sponge structure. No oil included. (From Nussinovitch and Gershon, 1997b; by permission of Oxford University Press.)

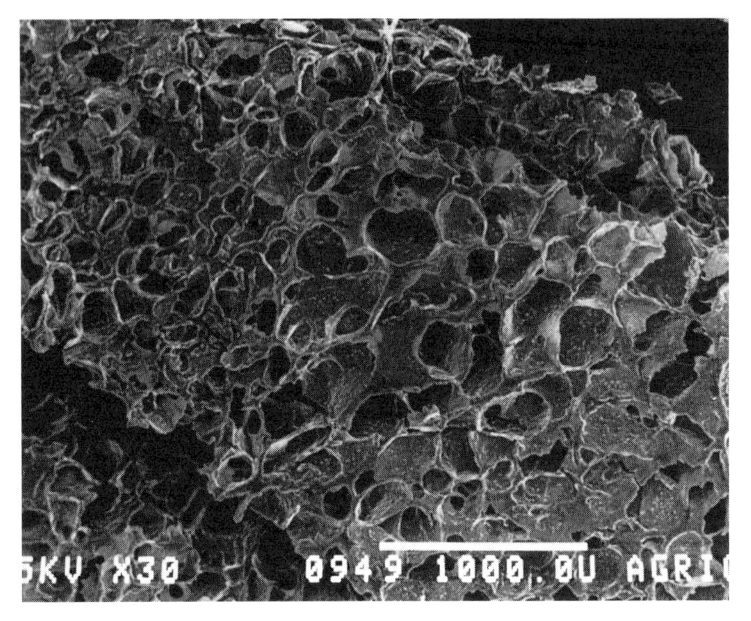

Figure 18.13 Agar sponge with 30% oil (compare with Fig. 18.12 to note how the structure changes and the cells begin to close up). (From Nussinovitch and Gershon, 1997b; by permission of Oxford University Press.)

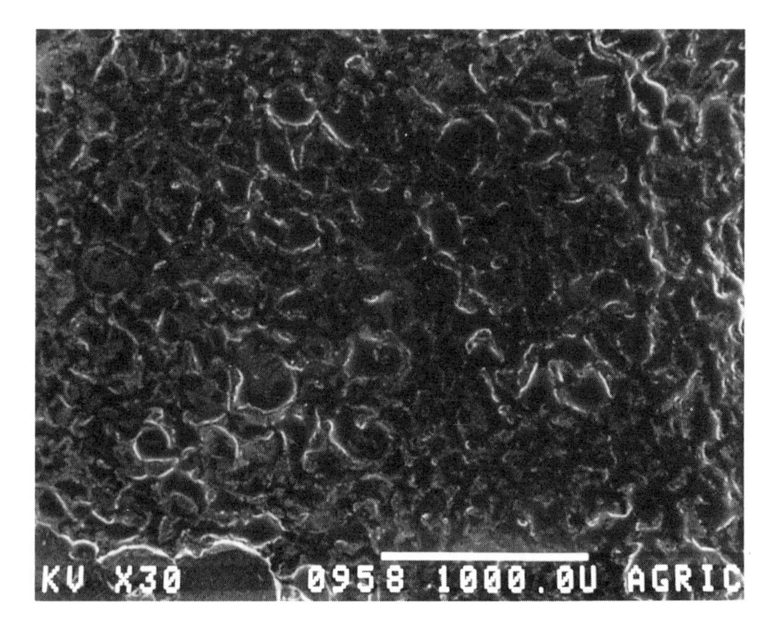

Figure 18.14 Agar sponge with 40% oil (compare with Fig. 18.12 to note how the total structure of the sponge has changed, becoming smoother). (From Nussinovitch and Gershon, 1997b; by permission of Oxford University Press.)

The concept of oil-containing hydrocolloid sponges is new. As reported previously (Nussinovitch, 1995), the oil is embedded in the sponge cell walls and coated by the dried hydrocolloid network, resulting in a possible decrease in oxygenation. In other words, the sponge is another way of encapsulating oil. Moreover, after its incorporation into the sponge, oil at concentrations below 20% does not migrate at all and is, therefore, not prone to rapid oxidation. We believe these types of sponge have further usefulness as carriers for vitamins, minerals and/or colors and could serve as high-energy rations for military purposes, among other things.

References

Attenburrow, G.E., Goodband, R.M., Taylor, L.J. *et al.* (1989) Structure, mechanics and texture of a food sponge. *J. Cereal Sci.*, **9**, 61–70.

Gibson, I.J. and Ashby, M.F. (1988) *Cellular Solids: Structure and Properties*, Pergamon Press, Oxford, UK.

Hutchings, J.B. and Lillford, P.J. (1988) The perception of food texture – the philosophy of the breakdown path. *J. Text. Studies*, **19**, 103–7.

Jeronomidis, G. (1988) Structure and properties of liquid and solid foams, in *Food Structure* (eds J.M.V. Blanshard and J.R. Mitchell), Butterworth, London.

Lillford, P.J. (1989) Structure and properties of solid food foams, in *Foams, Physics, Chemistry and Structure* (ed. A.J. Wiloson), Springer Verlag, London, pp. 149–66.

Nussinovitch, A. (1995) Compressive characteristics of hydrocolloid sponges, in *8th Int. Conf. Ind. Exhibit. Gums and Stabilisers for the Food Industry*, July 1–14, The North East Wales Institute, Cartrefle College, Wrexham, UK.

Nussinovitch, A. and Gershon, Z. (1997a) Alginate oil sponges. *Food Hydrocolloids* (in press).

Nussinovitch, A. and Gershon, Z. (1997b) Physical characteristics of agar-yeast sponges. *Food Hydrocolloids*, **11**(2), 231–7.

Nussinovitch, A., Nussinovitch, M., Shapira, R. and Gershon, Z. (1994) Influence of immobilization of bacteria, yeasts and fungi spores on the mechanical properties of agar and alginate gels. *Food Hydrocolloids*, **8**(3–4), 361–72.

Nussinovitch, A. and Peleg, M. (1990) Strength–time relationship of agar and alginate gels. *J. Text. Studies*, **21**, 51–60.

Nussinovitch, A., Peleg, L. and Gershon, Z. (1995) Properties of agar–starch sponges, in *8th Int. Conf. Ind. Exhib. on Gums and Stabilisers for the Food Industry*, July 1–14, The North East Wales Institute, Cartrefle College, Wrexham, UK.

Nussinovitch, A., Peleg, M. and Normand, M.D. (1989) A modified Maxwell and a non-exponential model for characterization of the stress relaxation of agar and alginate gels. *J. Food Sci.*, **54**, 1013–16.

Nussinovitch, A., Steffans, M.S. and Chinachoti, P. (1991) Exponential model for the characterization of the compressive stress–strain relationships of bread. *Lebbensm. Wiss. Technol.*, **24**, 266–9.

Nussinovitch, A., Velez-Silvestre, R. and Peleg, M. (1993) Compressive characteristics of freeze-dried agar and alginate gel sponges. *Biotechnol. Progr.*, **9**, 101–4.

Peleg, M. (1982) Physical characteristics of food powders, in *Physical Properties of Foods*, ch. 10 (eds Peleg, M. and Bagley, E.B.), Avi, Westport, CT, pp. 293–321.

Rohde, F., Normand, M.D. and Peleg, M. (1993) Effect of equilibrium relative humidity on the mechanical signatures of brittle food materials. *Biotechnol. Prog.*, **9**, 497–503.

Swyngedau, S., Nussinovitch, A. and Peleg, M. (1991a) Models for the compressibility of layered polymeric sponges. *Polym. Eng. Sci.*, **31**, 140–4.

Swyngedau, S., Nussinovitch, A., Roy, I. *et al.* (1991b) Comparison of four models for the compressibility of breads and plastic foams. *J. Food Sci.*, **56**, 756–9.

19 Textiles

19.1 Introduction

The many uses of textile fabrics include clothing, protection and decoration. These fabrics can be found as coverings for furniture and floors, and as curtains, in kitchens, bathrooms, bedrooms and living rooms, as well as in public buildings. They are used for many different purposes. Fibers (the fundamental unit in the fabrication of textile yarn) are used as insulators and in filter cloths, pulley belts and conveyor belts.

We come into contact with textiles many times daily. In fact, textile fibers and their use predate recorded history. Early fibers comprised some source of natural origin, wool, flax, cotton and silk being some examples. Most of the early fabrics were composed of a simple, plain weave interlacing groups of fibers and yarns (yarn is a generic term for a continuous strand of textile fibers or filaments suitable for knitting or weaving). With time, more complex fabrics were developed and, after the industrial revolution, their production in factories began to replace home manufacture. Early in the 20th century, the first artificial fiber, rayon, came into being, followed by the introduction of cellulose acetate in the 1920s. Since the late 1930s, many new fibers have been developed (Mark *et al.*, 1967–9; Goswami *et al.*, 1977). The aggregate demand for textile fibers and the factors affecting the market shares of individual fibers in four major end-use markets in the USA – apparel, home furnishings, carpets and industrial – were analysed by Kirby and Dardis (1993), based on annual data from 1956 to 1986.

The word textile comes from the Latin word *textilis* and verb *texere*, which means 'to weave'. Today the word 'textile' is defined as 'any product made from fibers', and when the word 'textiles' is used with the term 'fiber' it refers to 'any product capable of being woven or otherwise made into fabrics'. The textile industry is large and of the highest importance in terms of economic value.

19.2 Textile fibers

All fabrics are composed of textile fibers. These fibers may be short or long, fine or coarse, soft or stiff, smooth or rough. Fibers are the building blocks in fabric manufacture. They are generally formed into yarns by knitting, weaving, knotting and braiding. After manufacture, finishing procedures (e.g. 'wet processing') are applied to achieve the desired product. Color is sometimes included in this processing step, or added independently. Thus textile production includes fiber content, yarn structure, fabric structure, finish and color application (Moncrieff, 1963).

A textile fiber must possess several primary properties such as a suitable length-to-breadth (width) ratio, adequate strength, flexibility or pliability, cohesiveness or spinning quality, and uniformity. Secondary properties that may heighten demand include physical shape, specific gravity, luster, moisture regain, elastic recovery, elongation, resilience, thermal behavior and resistance to organisms, chemicals and environmental conditions (Kaswell, 1953).

Typical length-to-width ratios for natural fibers are 1400×10^6, 3000×10^6, 170×10^6, 3000×10^6 and 33×10^6 for cotton, wool, flax, ramie and silk, respectively. Tenacity is a primary property that usually refers to the strength of individual fibers. It is defined as tensile stress when expressed as force per unit linear density of the non-strained specimen. Tenacity may be given in grams per Tex or grams per denier. Tex is the weight in grams of 1000 m of yarn. Denier is a yarn-number unit and is equal to the number of grams of fiber required to reach 9000 m. Typical tenacities (g denier^{-1}) of selected fibers are raw cotton 3.0–4.9, silk 2.4–5.1, wool 1.0–1.7, acrilan acrylic 2.0–2.7, regular nylon 4.5–5.9, dacron polyester 4.0–4.3 and lycra spandex 0.6–0.8. Fiber strength is also measured in pounds per square inch (tensile strength) and is obtained by determining the force required to break a fiber cross-sectional mass equivalent to one square inch. For most fabrics, a minimum fiber tenacity of ~ 2.5 g denier^{-1} is desirable. High-strength fibers are important in producing cloths for industrial (work) purposes (Joseph, 1966).

Textile fiber flexibility is an important quality in clothes because it enables freedom of movement through fibers that are bendable, pliable or flexible. Cohesiveness is the fourth requisite property of a fiber, and is a measure of the fibers' ability to stick together during the manufacture of yarn. A different essential property is fiber uniformity.

Secondary properties include physical shape, specific gravity (e.g. cotton 1.54–1.56, silk 1.34, wool 1.30–1.32, nylon 1.14, saran 1.72, lycra spandex 1.00, viscose rayon 1.52), luster, moisture regain (examples of regain at 70°F (21°C) and 65% relative humidity are raw cotton 8.5, silk 11, wool 16, nylon 4–5, viscose rayon 11.5–16) and moisture absorption, elastic recovery and elongation, resiliency, flammability and other thermal reactions (Gordon, 1964).

Natural fibers include cellulosic fibers (including seed hairs, bast fibers, leaf fibers and nut-husk fibers) and protein fibers (including animal-hair fibers, animal secretions, mineral fibers and natural rubber). Manufactured (often called 'manmade') fibers include cellulosic fibers, modified cellulosic fibers, protein fibers, synthesized fibers, mineral fibers and others (alginate, tetrafluoroethylene).

Yarn is a continuous strand of textile fibers, filaments or materials in a form suitable for knitting, weaving or otherwise intertwining to form a textile fabric (ASTM, 1965). Details on the processing of yarns using cotton or wool systems can be found elsewhere (Gurley, 1959).

19.3 General finishes

Before the consumer buys the fabric it receives one or more finishing treatments. Color (except for white fabrics) is also applied. The converting industry is responsible for the application of finishes to fabrics, and these can be classified into various groups: mechanical or chemical, permanent or non-permanent, general or functional, and those that can change the fabric's appearance or not. Mechanical finishes are applied to fabrics by mechanical means. Chemical finishes make use of alkalis, acids, detergents, bleaches, resins or other chemical substances. The finish is designed for durability at least in terms of average use and care. Finishing operations include the use of hydrocolloids, and, therefore, some details pertaining to routine finishes are included here (Joseph, 1966).

Beetling is a mechanical finish applied to cotton and linen fabrics. The fedded fabric is rolled and rotated in a machine, where large hammers pound its face, providing it with improved luster. The pounding also flattens the yarns and closes the weave, ensuring that the finish will withstand wear and maintenance if laundered carefully and ironed to restore its appearance. This treatment is used for table coverings, or in any situation in which flat surfaces are to be covered with attractive, smooth, flat fabrics (Joseph, 1966).

Bleaching is applied to produce white fabrics or as a preparatory step for dyeing or printing. Bleaching is a chemical finish that should be repeated by the consumer if color retention is desired. The bleaching solution contains sodium hypochlorite and hydrogen peroxide. The fabric is immersed in the bleach at the desired temperature for the prescribed time, then the fabric is thoroughly rinsed and dried. Optical brighteners alter the reflectance characteristics of the fabric surface and are, therefore, used to change the visual effect of whiteners (Joseph, 1966).

Mechanical finishes such as **brushing** involve the removal of short, loose fibers from the surface of the fabric. **Calendering** is applied to different fabrics (calendering is a finishing process for fabrics that produces a flat, shiny, smooth surface produced by passing the cloth through hollow cylinder rolls or a friction and glazing calender). **Pressing** is the term used for wool fabrics.

These mechanical processes need to be repeated after each laundering or cleaning. For such a finish, high pressures are applied to smooth the fabric's surface. There are more complicated calendering processes such as schreinerizing and embossing (embossing is the production of raised designs in relief on the surface of fabrics to give a three-dimensional effect). **Carbonizing** is applied to wool fabrics. The process includes immersion of the wool fabric in a solution of sulfuric acid at high temperatures for a short time to convert vegetable matter within the wool yarns to carbon, which is then removed by scouring and brushing. **Crabbing** is a mechanical finish applied to wool fabrics. It includes immersion in hot water then cold water, and then passing between rollers. Thus a proper crabbing permanently sets the weave. Another mechanical finish is **decating**, which is applied to wool, silk, rayon and blends of these fibers to obtain luster, and in wool to develop permanent luster. **Fulling** is an additional mechanical finish applied to wool in order to produce a compact fabric by shrinking. Moistening, heating and pressure cause yarns to shrink, the weaves to close up and the fabric to develop body and closure. **Heat setting** is heating a fabric at selected temperatures for a specified length of time. The temperatures used should be less than those used for finishing, and the duration should be predetermined and short. Heat setting causes molecular rearrangements, relieves internal stresses, brings about molecular crystallinity and the individual fibers tend to permanently assume the structural arrangement in which they were held during the finishing procedure. Heat setting of thermoplastic fiber fabrics yields dimensional stability, resilience, elastic recovery and permanent design (Joseph, 1966).

Inspection of every fabric occurs before it leaves the manufacturing plant. Visual inspection (**perching**) is performed by passing the fabric over a perch illuminated by lights behind and above the frame. **Burling** is usually applied to wool and consists of removing knots or other imperfections in the yarn without creating inferior fabrics. **Mending** is the repair of imperfections; this application results in the fabric's classification as second quality (Joseph, 1966).

Scouring is carried out many times together with bleaching. Fabrics are scoured to remove foreign materials (natural waxes, dirt, processing oils and sizing compounds). Scouring makes use of soaps or synthetic detergents with alkaline builders. No molecular changes in the fiber appear to occur.

Shearing (a mechanical process) involves cutting or shearing off undesirable surface fibers on fabrics, leaving a clear view of the weave and giving the fabric a uniform appearance.

Singeing consists of burning the fuzz (fiber ends) to obtain a smooth surface. Before this procedure, the cloth is brushed to remove loose fibers, lint and dust. Two methods of singeing are used. One includes passing the fabric over heated plates when one plate is red hot and singes the surface. The second method consists of passing the fabric (at 100–300 yard min^{-1} (90–270 m min^{-1})) directly over gas flames. After singeing, the fabric is immersed in water to extinguish any sparks or afterglow on the fabric.

Sizing is the application of various materials to the fabric to produce stiffness or firmness. Cellulose fabrics are sized with starch or resins. Starch is applied to cellulosics, particularly cotton, to add luster and improve the body of the fabric. Starch adds weight and improves the look of the fabric, particularly inferior ones. Several techniques exist to apply the starch and to remove the excess. Gelatin and dextrins also may be used.

Tentering is a mechanical (by a machine in which the fabric is held with clips) straightening and drying of fabrics (Joseph, 1966).

19.4 Special finishes

Finishes that alter the appearance of fabrics include special calendering: schreinerizing, moire and embossing, among many others. Some fabrics receive special calendering that imparts a design to the surface. To achieve a permanent effect, thermoplastic fibers are softened and durability of appearance is imported. Resins are added to yield a durable design. If pressure alone is used on thermoplastic fibers, the design will usually be lost during the first cleaning. In order to detect and differentiate among a variety of durable press finishes, differential scanning calorimetric and thermogravimetric techniques were employed under dynamic nitrogen conditions. (Durable press is a textile-finishing chemical treatment that imparts wrinkle resistance to fabrics by forming cross-links between cellulose molecules (Horie and Biermann, 1994; Andrews and Collier, 1992).) Changes in residue, rate of weight loss, peak intensity and peak temperature were observed and varied with reactant, catalyst and washing procedure. The ability to distinguish among polycarboxylic acids, catalysts and/or other formaldehyde-based reactants is of value to the textile chemist (Traskmorrell *et al.*, 1993).

Schreinerizing is a finish produced on a schreiner calender. It yields a soft luster, especially with cellulosic fibers such as cotton, linen, tricot-knit lingerie, and fabrics made of nylon and polyester fibers. Flattening of the fabric produces a more opaque appearance.

Mercerization is a chemical finish applied to cotton or other cellulosic fibers. It increases strength and improves luster and dyeing characteristics. The yarns are immersed in 18–27% sodium hydroxide and held under tension during the finishing procedures. The fibers swell and the resultant round cross-section reflects light with a better sheen. The process also produces fibers with increased strength and affinity for coloring. Although chemicals are used, the change in the fiber is physical. Crystallinity decreases and rearrangement of the molecules increases fiber tenacity. After immersion in caustic soda (sodium hydroxide) the fabric is rinsed several times and the remains of the base are neutralized by acid, followed by an additional rinse to ensure complete removal of the acid from the fabric (Joseph, 1966).

Moire finishes impart a soft luster and a design is obtained from differences in light reflections. Once it was applied only to silk, but today it is used with a variety of fibers, including acetate and nylon. The ribbed fabric is doubled and fed through rollers that exert pressure and add heat. The ribs on one thickness impress images on the other thickness via flattening. If a pattern has been 'scratched' onto the metal roll it is reproduced on the fabric. If no definite design is found on the metal roll, an irregular or broken-bar effect is created. Two heated rolls are used, a large one coated with cloth and a smaller one (which holds the design). To achieve durability with non thermoplastic fibers (e.g. rayon, cotton), resins are added by impregnation before moire calendering (Joseph, 1966).

Finishes and laundering will affect the surface characteristics of cotton and polyester fibers (Rhee *et al.*, 1993). Changes in the surface of fibers are monitored through contact-angle determinations before and after treatment of the fabrics with durable-press, fluorocarbon stain repellent and antistatic finishes. The different effects of the surfactant treatments were attributed to different deposition patterns of the surfactant molecules on the fiber surfaces. The effect of laundering on the finished fibers was dependent on the retention of the finishes on the fabrics during the laundering process (Rhee *et al.*, 1993).

Embossed fabrics are characterized by their three-dimensional designs. Resins are added first for durability. A variety of calenders are used and different types of design are adapted to embossing.

Permanent polish can be regarded as a special finish, and modified calenders and the addition of special chemicals produce a degree of permanent polish. Glazed surfaces are achieved by friction calenders. Permanent glazing is achieved by impregnation with resins before calendering. Non-permanent glazes are achieved by using wax, glue, starch and/or shellac.

Cire ('wet look') is a high polish (accomplished by impregnating the fabric with wax or a thermoplastic substance and passing it through a

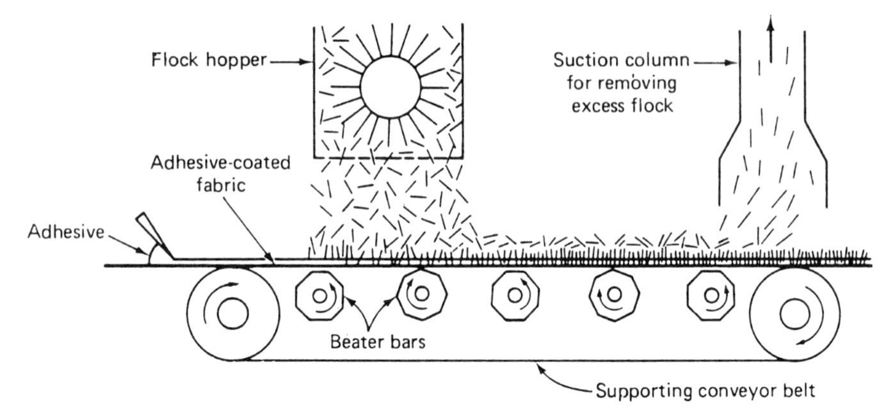

Figure 19.1 Beater bar flocking. (Adapted from The Applications of Synthetic Resin Emulsions by H. Warson, Ernest Benn Ltd, London, 1972.)

friction calender) applied to silk or silk blends. (Impregnated fabric is a fabric in which the interstices between the yarns are completely filled with an impregnating substance throughout the thickness of the material, as distinguished from sized or coated materials, where these interstices are not completely filled.) The polish can be durable although it is not considered permanent. Raised surfaces are obtained by mechanical **gigging** and **napping**. The nap hides the yarns and weave and produces a soft, hairy appearance. **Flocking** produces a raised surface by adding short fibers to the surface of the fabric (Figs 19.1 and 19.2). Gigging is used on wool, rayon and other fibers. Napping is used on cotton, rayon, and wool, among others. Flocking is somewhat similar to raised-surface finishes. It consists of attaching very short fibers to the surface of the fabric by means of an adhesive. It is

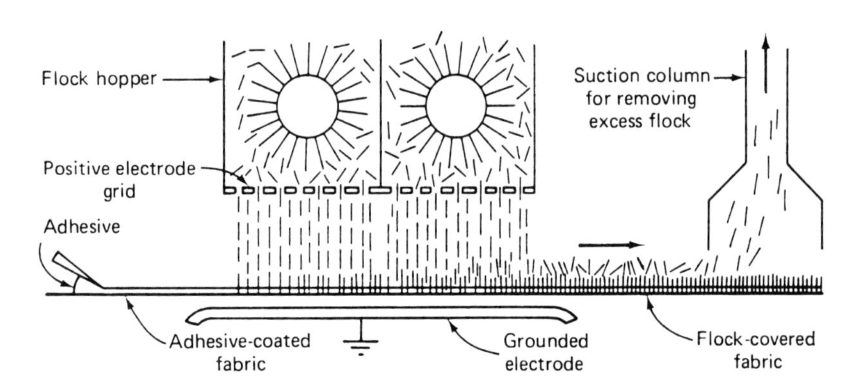

Figure 19.2 Electrostatic flocking.

composed of rayon fibers (low in cost) or fibers of cotton, wool or other short fibers. The best fibers are square on the ends. The adhesive is printed onto the fabric and the flock is applied by mechanical or electrostatic methods. It is comparatively permanent if laundering is not performed at high temperatures. Dry cleaning will soften or dissolve the adhesive (Warson, 1972).

Stiffness can be achieved in a number of ways. Acid (sulfuric) finishes produce transparent cottons with permanent stiffness. Immersion in acid under controlled conditions for short intervals followed by neutralization produces an acid finish. Alkalis (bases) produce certain basic finishes. Base in a paste form is spread onto the fabric in the desired design. The coated areas shrink and the untreated areas pucker as a result. After the base's removal, the crinkled effect is comparatively durable. Sizing contributes stiffness to fabrics, albeit only temporarily. Thermosetting resins and plastic compounds have also been used successfully in producing stiffness.

Other finishes, such as oil, fat, wax emulsions, soaps and synthetic detergents, substituted ammonium compounds and silicone compounds contribute to softening, adding body to fabric and facilitating application of other finishes. Optical finishes include those that contribute to luster and those that reduce fiber luster and sheen via the addition of a white pigment such as titanium dioxide. Other sheen reducers include heat treatment and a finish called chavacete, which softens tricot knits and produces a silk-like appearance in combination with other reducers. Functional finishes confer abrasion-resistance, absorbence, anti-slip or antistatic qualities, bacteriostats, fire (flame) inhibition, fume-fading resistance, heat setting, mothproofing, stabilization, water repellence, waterproofing, stain and soil resistance, minimum-care and permanent press finishes. Supplementary reading on the topic can be found elsewhere (Lynn and Press, 1961; Marsh, 1966; Valko, 1968; Petersen, 1971; Williamson, 1980).

19.5 Hydrocolloids and resins in the textile industry

The previous information was designed to provide an introduction, some definitions and partial details to those readers interested in textiles. Gums and resins are used to a large extent in the textile industry, in many products such as adhesives, antistats, backings (back-sizings are fillers added as part of the resin or gum composition to increase the weight or modify the handling of a fabric), as well as in carding, coating, for durable press, dyeing, in finishes, tinting, knitting, inks, laminating, laundering, lubricants, printing, sizing, etc. The following paragraphs give a brief overview of these applications.

19.5.1 Adhesives and printing

There are a number of major applications for water-based polymers in the textile industry, in which the adhesive properties of the polymers are essential to the final product. Water-based printing inks for textiles can also be conventionally included under this general heading.

Alkyls and hydroxyalkylalkylcelluloses are used in textiles for adhesion purposes, carpet back-sizing, dye thickening, flocking adhesives, latex coatings, printing pastes and warp sizes.

Gum karaya, which has only been produced since 1920, was found to be a superior adhesive when partially wetted with water (this gum absorbs water and swells to 60–100 times its original volume). The gum is modified for the textile industry to have increased solubility, enabling its use for printing operations. Better dissolution is achieved by pressure-cooking the gum karaya suspensions, with the rate of dissolution varying with pressure. Textile gum solutions contain 15–18% solids. Gum karaya can also be solubilized by treatment with sodium peroxide, persulfate or persilicate. The hydrocolloid solution is blended with the dye and later used for color printing on cotton fabrics (Knecht and Fothergill, 1924).

HEC serves as a thickening, flow-control and water-binding additive in textile adhesives, sizes and coatings. Acrylic latices utilized in the textile industry can be modified with HEC (which serves as a binder and adhesive) for viscosity regulation, and water-loss and penetration control. These products, which are formulated to perform under specific conditions, require different levels and different application methods of HEC; details are available from the manufacturers. Another application of HEC is in textile printing. A typical formulation contains ~0.74% HEC and 2.2% all-acrylic latex (60% solids by weight) in the premix in addition to 2.8% water and 0.09% polypropylene glycol. The master mix includes 84.26% acrylic latex, 0.09% polypropylene glycol, 5.3% oxalic acid, 0.96% tri(2-ethylhexyl) phosphate, 2.51% ammonium hydroxide and 1.05% water-soluble melamine resin. This formulation is designed for application in a rotary-screen printer. Because of its high water-holding capacity, HEC produces better flock acceptance and density and reduces the tendency during curing, resulting in excellent print definition (Bikales and Segal, 1971).

PVA forms strong bonds with porous, water-absorbing surfaces such as textiles, where it is used as an effective pigment binder. PVA adhesives are normally applied by wet-bonding. Hydrolysed grades of PVA are used to formulate adhesives that are water resistant and quick setting. PVA can create films with strength, flexibility and abrasion resistance, and it can therefore be used as warp sizes for filaments and spun yarns (Pritchard, 1970; Finch, 1973).

Polyvinyl acetate and its copolymers in emulsion form tend to be used as finishes rather than adhesives. Polyvinyl acetate emulsions can be used to coat interlinings so as to impart heat-seal properties. Because of their wide range of properties, various acrylic emulsions are of major interest as textile adhesives. Of particular interest are the type that self-cross-link upon drying in the presence of mild acid catalysts. Acrylic emulsions may contain a high percentage of solids, and the cured films are flexible and fairly soft. Such an emulsion is easy to formulate and handle. The resultant products are resistant to both dry cleaning and wet laundering and do not spoil draping quality of the finished goods. Thermosetting acrylics may also be used in preparing a thickened adhesive paste for combining fabrics with polyurethane foam. When polyester, polyacrylonitrile or polyamide fiber is coated with a polyvinyl resin, bonds are normally very weak. Moreover a blend of resorcinol–formaldehyde resin, containing free resorcinol and less than one-third of its weight (solids content) of a polyvinyl chloride latex, a vinyl chloride–vinyl acetate copolymer or a butadiene–vinyl pyridine copolymer, will give good adhesion between fabric made from one of these synthetic fibers and a polyvinyl chloride organosol (Warson, 1972).

Vinyl chloride copolymer latices are used (more satisfactorily than solutions) to bond vinyl films to diverse fabrics. The addition of synthetic rubber latex yields maximum flexibility.

Laminates, which began with the bonding of polyurethane foam to fabric to improve its warmth, resilience and sales appeal, were described as a growing development in textiles in the early 1970s. An acrylic formulation for lamination can be composed of an acrylic of the self-cross-linking type, requiring an acid catalyst, and an MC thickener. The MC is first dispersed in xylol to prevent lumping upon its addition to the emulsion. In flocking, short fibers are glued to a base fabric or other substrate, usually perpendicular to the base, to produce a pile fabric effect. The success of a fabric of this type depends to a large extent on the efficiency of the adhesives, which in ~ 80–85% of the cases are aqueous acrylic based. Since very high viscosities (50 000–100 000 cps) are required for this process, thickening agents such as MC can be used after dispersing as a slurry in xylene. A viscous solution of cellulose can also be added. Alternatively alkali-thickenable or soluble acrylic emulsions can be used to increase viscosity. Flocks can be applied by the early 'beater bar' method or by the more modern electrodeposition (Figs 19.1 and 19.2).

Pigments have been bound to textiles by means of polymeric adhesives, initially natural gums and resins, in the historic processes of the textile industry. In contrast to a pigment, a dye must have a definite affinity for the fibers themselves. Generally the pigment needs to be in an oil (resin) phase or dispersed in water with standard dispersants. Thickeners such as MC are recommended in such cases.

The major problem with pigment-dyed fabrics is crocking (i.e. the rubbing-off of color owing to unsatisfactory fixation or penetration). Crocking can be reduced by rubber-type latices and acrylic resins, which have the advantage of not being affected by light. These resins vary considerably in their resistance to wet and dry crocking, and self-cross-linking acrylic types are best for this purpose. In such cases, much higher binder to pigment ratios are required than with other applications. The tendency of most pigments to flush into the oil phase is decreased by adding $\sim$10–15% of a conventional wetting agent or colloid dispersant.

Printing-paste formulations are used on a wide variety of substrates including woven fabrics, knitted fabrics and non-wovens of almost all natural and synthetic fibers including asbestos, glass fibers and even leather. In such formulations, the emulsion thickener includes five parts (by weight) of a 6% gum tragacanth solution. In screen-printing-paste formulations, a solution of 5% CMC is used, and in flock-printing paste, a 10% MC solution is used.

19.5.2 Gums as antistats

Considerable charges of static electricity often tend to build up in synthetic textile fibers and fabrics because of their hydrophobic and non-conducting nature. A number of synthetic resin products can be used as durable antistatic agents. These are emulsion-derived, ionic products such as ammoniacal solutions of acrylate copolymers with acrylic or other polymerizable acid (Warson, 1972).

PEGs comprise low- to medium-molecular-weight, synthetic water-soluble polymers that play an important role in textile processing. They can be used as additives in viscose-spinning baths, thereby improving fiber properties and formation. In weaving and knitting, they function as lubricants. They can also serve as softeners, antistatic agents and fabric conditioners. Modifed PEGs are preferred for use as emulsifiers, lubricants and softeners in textile processing.

PEO resins can be used as dyeing aids, fugitive textile weft and textile antistat. Only small quantities are added to polyolefin, polyamide and polyester compositions, which are melt spun into textile fibers to impart an antistatic nature or improve dyeability. As a fugitive textile welt, water-soluble PEG monofilaments are used. They serve as periodic weft threads during weaving. Prior to severing, the weft thread is dissolved in an aqueous solution (such as a dyeing bath), thereby simplifying cutting of the fabric at predetermined locations.

19.5.3 Backings

Synthetic resins serve as fiber adhesives in the manufacture of upholstery and carpeting. Binders should be selected before use in the bonding of natural fibers for upholstery, spring-securing pads for sprung chairs, filters and packaging materials, as well as the backing of carpets and long-piled fibers. Polychloroprene latex binders impart stiffness to natural animal hair. The addition of zinc oxide sometimes gives the fibers better acid resistance. A vinyl toluene–butadiene copolymer latex has been recommended for upholstery and rug backing. With Axminster carpets, the pile is mechanically knotted into position and a binder is required to improve the body of the carpet and to provide additional pile anchorage. Starch and gelatin were originally used for this purpose, but synthetic resin latices are now used, sometimes as an additive to starches but often alone (Warson, 1972).

Latex systems need to be thickened for use as rug backings in the preparation of resilient, rubberized, non-slip floor coverings. Back-sizings are mineral materials (fillers), e.g. china clay, added as part of a resin or gum formulation in order to increase the weight or modify the handling of a fabric. These same characteristics are desired in the backing of fabric. Thickened latex is required for operations in which fabric is dipped into the rubber latex and the covered form is used to prepare products such as gloves.

Most of the standard synthetic polymer emulsions and rubber latices, whether natural or synthetic, can be applied to carpets, often with a heavy filling to reduce cost. The fillers require standard dispersants. Thickeners such as starches, starch derivatives or cellulose derivatives may be required to control the degree of penetration; a pH adjuster may also be required, especially with carboxylated latex. In tufted carpets, a continuous yarn is loosely tufted by needle action into an inexpensive hessian base. For example, a rayon tufted carpet was prepared by anchoring the tufts onto hessian with a binding formulation composed of whiting, sodium hexametaphosphate, sodium CMC, water and styrene–butadiene copolymer latex. Latices of copolymers of methylmethacrylate and butadiene can be used for primary as well as secondary backing of woven and tufted carpets. Emulsions of styrene with methacrylic esters as a textile coating have also been reported (Catalan *et al.*, 1994). Formulae for woven carpets may include latex, thickener (polyacrylate), starch solution, ammonia and water and are composed of 23.5% solids. For tufted carpets, a lick roller is used, whereby a roller rotates in a trough of latex compound (Fig. 19.3) and the carpet travels across the top of the roller, which usually applies an excess of material. This is removed and returned to the trough by means of a scraper knife. The coating weight should be sufficient to cover the jute yarn in the backing cloth and that part of the pile that is on the back of the carpet, but it should not be detectable on the face side. High viscosity is required so that the compound will not strike through the carpet. A secondary backing requires a high viscosity compound that will not penetrate. The drying stage includes passage at temperatures of up to 150°C, where the carpet does not get hotter than 100°C (Warson, 1972).

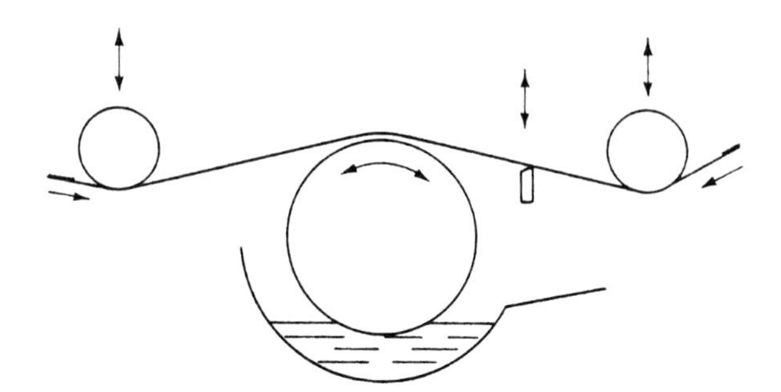

Figure 19.3 Latex application by lick roller. (Adapted from Nuessle *et al.* (1958), *J. Am. Dye. Rep.*, **47**, 765.)

Carpet foam backings make use of latex into which air has been injected. It is pumped into a foaming head and mixed to give the foam a fine structure. The curing and gelling dispersions are introduced into the latex foam through rubber grommets in the wall of the blender, ammonium acetate being suggested as the gelling agent. The sensitized latex is passed through a pipe onto the cloth, where it is spread to the desired foam thickness. Where foam is used to back tufted carpets, it is usual to give a primary backing to provide a stiffer handle and better binding of the tufts. Primary backing can be by means of a rubber latex (natural or synthetic), polyvinyl chloride, starch or a starch–polyvinyl acetate compound. Upholstery backings are similar to those described for carpets, except that they must be able to withstand scouring and dyeing without disintegration (Warson, 1972).

19.5.4 Binders

Karaya gum is less expensive than tragacanth and has been found to be superior to other gums for various purposes, e.g. as a binder, in the paper and textile industries. Acrylic latices are modified (thickened and penetration control added) with HEC to be used as non-woven fabric binders and fabric-back coatings, and for fabric- and carpet-dyeing purposes. HEC has been found to be advantageous over other gums and resins in achieving convenient thickening at low concentrations, for its easy dissolution at different temperatures, its tolerance for other dissolved materials and salts in the system, and its properties of good water-loss and penetration control. PVA can be utilized as a binder for textile fibers particularly in non-woven fabrics. Tamarind gum, obtained from the seeds of the tamarind tree, can also be used for sizing cotton and jute yarns.

19.5.5 Special uses of CMC in textiles

Sodium CMC is a water-soluble anionic linear polymer. Its use is estimated to be more widespread than that of any other water-soluble polymer. The textile industry's use of CMC is only exceeded by its use of detergents; the textile industry comes third after the drilling/mining and paper industries in the use of CMCs. Part of its uniqueness lies in its higher tolerance to alkali metal salts relative to other hydrophilic polymers. The purity range of CMC for textiles is classified as semi-refined: it contains 90–95% sodium CMC. CMC is shipped in bulk for certain end-use applications, including textile warp sizing. In plants that require large amounts of it, CMC is stored in silos and can then be conveyed via screw-feeding mechanisms or air-conveyance systems. If small amounts are needed, conveyance is not a problem. Generally, CMC is utilized in cloth manufacture and in fiber finishing. It has been used as low add-on warp sizing since the 1950s, and the demand for it has increased because of the lower amounts needed and its lower biological oxygen demands compared with starch (Stelzan and Klug, 1980).

19.5.6 Carding

PEGs and their fatty acid esters have many uses in the textile industry. They serve as sizes and lubricants for carding, spinning, weaving and knitting fibers and yarns. Later on in processing they can be easily removed. PEG, 1–3%, is regularly added to viscose spinning solutions along with ethoxylated fatty amines to modify fiber orientation, permit faster spinning and obtain higher-tenacity fibers or cellulose films. It is used in the manufacture of rayon tires and as a finishing treatment for fabrics; it contributes to softness and a pleasant feel.

19.5.7 Additional uses of hydrocolloids in the carpet industry

Alkyl cellulose and hydroxyalkylalkylcellulose are used as thickeners for carpet back-sizing formulations. They provide faster drying speeds, a broader effective-coating-viscosity range, more uniform coating and better holdout. Guar gum can help in dyeing and printing in the carpet industry. If yarns are package-dyed, 0.1–0.2% guar gum is used as a migration control agent to achieve good color distribution within the package. If the yarn is dyed after carpet formation, via the Beck methodology (exhausting dyes onto the fiber), no gum is actually needed, but with the new continuous-dyeing method, gum is added as a viscosity modifier. When fixation (in the steaming chamber) occurs, the gum within the dye solution serves as a controller of dye-molecule migration, and thus a level shade is obtained from surface to backing and side to side of the carpet. Guar gum is used at concentrations of 0.20–0.25%. The low concentration is an advantage because it is easy to wash the excess off. Carpet printing requires full control over dye viscosity, spreadability and wetting. This can be achieved by incorporating guar gum, which at the level added to the dye is easily dispersible, soluble and compatible with other ingredients. When ozone is used to decolorize textile dye solutions, the rates of reaction are not sensitive to pH and are only mildly affected by temperature. Guar gum used in the textile industry as an aid in dyeing carpets increased the consumption of ozone by 20–60% under the conditions studied and had a small effect on the reaction rate (Carriere *et al.*, 1993).

Another hydrocolloid used for carpet dyeing and printing is HEC, which promotes color acceptance and uniformity. Other galactomannans such as LBG are widely used in printing and dyeing. Similar to the contribution of other gums, they change the viscosity of the dye material and control its thickening effect.

19.6 Cellulose derivatives: archeological aspects

A recently proposed laboratory model for studying environmentally dependent chemical modifications in textile cellulose (Kouznetsov *et al.*, 1996) is based on a capillary zone electrophoretic/mass spectrometric estimation of modified glucose residues in cellulose enzymatic hydrolysates isolated from linen samples. The filtration of atmospheric air through the linen samples leads to statistically significant methylation and acetylation of textile cellulose, depending on air volume, the amount of aerobic bacteria cells, ecological conditions and the presence of bacteriostatic and protein-denaturing agents. The design of alkylation in the cellulose samples correlates with the measured ^{14}C and ^{13}C values and protein contamination levels. Applications for forensic and archeological chemistry can be envisioned (Kouznetsov *et al.*, 1995, 1996). In a previous manuscript, Kouznetsov *et al.* (1994) detected alkylated cellulose derivatives in several archeological linen samples by capillary electrophoresis/mass spectrometry. Fiber mineralization was also investigated using Fourier transform infrared (FTIR) microscopy (Gillard and Hardman, 1996). The mineralization of cellulose and protein fibers can be simulated in the laboratory using oxygenated aqueous solutions. The mechanism particular to a given solution has been shown to depend on the initial metal-ligand bonds formed and on kinetics factors relating to mineral-product formation. These experiments rationalize both the occurrence and the comparative rarity of mineralized organic fibers in archeological deposits. FTIR microscopy has revealed that traces of organic components can survive long-term burial and, under appropriate circumstances, permits their identification, even in highly mineralized samples (Gillard and Hardman, 1996). Other factors that affect the apparent radiocarbon age of textiles are discussed by Jull *et al.* (1996).

References

Andrews, B.A.K. and Collier, B.J. (1992) Finishing additives in treatments of cotton fabrics for durable press with polycarboxylic acids. *Ind. Eng. Chem. Res.*, **31**(8), 1981–4.

ASTM (1965) *Standards on Textile Materials*, part 24, American Society for Testing and Materials, Washington, DC, p. 39.

Bikales, N.M. and Segal, L. (eds) (1971) High polymers, in *Cellulose and Cellulose Derivatives*, vol. V, Wiley-Interscience, New York.
Carriere, J., Jones, J.P. and Broadbent, A.D. (1993) Decolorization of textile dye solutions. *Ozone Sci. Eng.*, **15**(3), 189–200.
Catalan, R., Farias, S. and Melo, R. (1994) Emulsion of styrene with methacrylic esters as coating of papers. *Bol. Soc. Chil. Quim.*, **39**(3), 227–35.
Finch, C.A. (ed.) (1973) *Polyvinyl Alcohol Properties and Applications*, Wiley, New York.
Gillard, R.D. and Hardman, S.M. (1996) Investigation of fiber mineralization using Fourier transform infrared microscopy. *ACS Symp. Ser.*, **625**, 173–86.
Gordon, J.C. (1964) *Handbook of Textile Fibers*, Morrow, London.
Goswami, B.C., Martindale, J.G. and Scardino, F.L. (eds) (1977) *Textile Yarns: Technology, Structure and Application*, Wiley, New York.
Gurley, M.H. (1959) *Man-made Textile Encyclopedia*, Textile Book Publishers (division of Wiley-Interscience), New York, p. 229.
Horie, D. and Biermann, C.J. (1994) Application of durable press treatment to bleached software Kraft handsheets. *TAPPI J.*, **77**(8), 135–40.
Joseph, M.L. (1966) Introductory Textile Science, Holt, Rinehart and Winston, Inc., New York.
Jull, A.J.T., Donahue, D.J. and Damon, P.E. (1996) Factors that affect the apparent radiocarbon age of textiles. *ACS Symp. Ser.*, **625**, 248–53.
Kaswell, E.R. (1953) *Textile Fibers, Yarns and Fabrics*, Reinhold, New York.
Kirby, G. and Dardis, R. (1993) Interfiber competition in the United States. *J. Text. Inst.*, **84**(1), 120–9.
Knecht, E. and Fothergill, J.B. (1924) *The Principles and Practice of Textile Printing*, 2nd edn, Griffin, London, pp. 123–4.
Kouznetsov, D.A., Ivanov, A.A. and Veletsky, P.R. (1994) Detection of alkylated cellulose derivatives in several archaeological mass-spectrometry. *Anal. Chem.*, **66**, 4359.
Kouznetsov, D.A., Ivanov, A.A., Veletsky, P.R. *et al.* (1995) A laboratory model for studies on the environmental dependent chemical modifications in textile cellulose. *New J. Chem.*, **19**(12), 1285–9.
Kouznetsov, D.A., Ivanov, A.A., Veletsky, P.R. *et al.* (1996) A laboratory model for studying environmental dependent chemical modification in textile cellulose. *Text. Res. J.*, **66**(2), 111–14.
Lynn, J.E. and Press, J.J. (1961) *Advances in Textiles Processing*, vol. I, Textile Book Publishers (division of Wiley-Interscience), New York.
Mark, H.F., Atlas, S.M. and Cernia, E. (eds) (1967, 1968 and 1969) *Man-made Fibers, Science and Technology*, vols I–III, Wiley Interscience, New York.
Marsh, J.T. (1966) *An Introduction to Textile Finishing*, 2nd edn, Chapman & Hall, London.
Moncrieff, R.W. (1963) *Man-made Fibers*, 3rd edn, Wiley, New York, p. 18.
Petersen, H. (1971) in *Chemical Aftertreatment of Textiles* (eds H.F. Mark, N. Woodling and S.M. Atlas), Wiley-Interscience, New York, p. 135.
Pritchard, J.G. (1970) *Poly (Vinyl Alcohol) Basic Properties and Uses*, Gordon and Breach, New York.
Rhee, H., Young, R.A. and Sarmadi, A.M. (1993) The effect of functional finishes and laundering on textile materials. Surface characteristics. *J. Text. Inst.*, **84**(3), 394–405.
Stelzer, G.I. and Klug, E.D. (1980) CMC, in *Handbook of Water-soluble Gums and Resins* (ed. R.L. Davidson), McGraw-Hill, New York, pp. 4.1–4.24.
Traskmorrell, B.J., Andrews, B.A.K. and Catalano, E.A. (1993) Thermoanalytical characteristics of durable press-treated cotton fabrics. *J. Appl. Polym. Sci.*, **48**(8), 1475–84.
Valko, E.I. (1968) in *Man-made Fibers*, vol. III (eds H.F. Mark, S.M. Atlas and E. Cernia), Wiley-Interscience, New York, p. 499.
Warson, H. (1972) *The Applications of Synthetic Resin Emulsions*, ch. 10, Ernest Benn, London, pp. 635–738.
Williamson, R. (1980) Fluorescent brightening agents, in *Textile Science and Technology*, vol. IV, Elsevier Scientific, New York.

20 Texturized products

20.1 Introduction

Many processed foods contain hydrocolloids (gums), which govern the product's functionality and organoleptic acceptability. Hydrocolloids are natural, modified natural, synthetic or biosynthetic polymers that dissolve in water and have the ability to thicken or gel aqueous systems. Examples include algin, starch, modified starches, agar, carrageenan, gelatin, xanthan and gum arabic, as well as some proprietary formulations. Hydrocolloids are frequently used in preparing gum-based foods, significantly affecting their appearance, physical properties and shelf life. Hydrocolloids can be selected to suit the manufacturing conditions, and specific manufacturing processes have been invented to take advantage of the functional properties of the various gums (Nussinovitch, 1993).

A gel is a rigid three-dimensional polymeric network capable of holding large quantities of water. Simple gels are basic structures that can be modified to achieve desired textures. Hence, most foods having a cellular structure are in essence modified gels (Silberberg, 1989). Moreover, a large proportion of food products are solids containing 50 to 90% water, and as such they can be regarded as multi-component gels (Tolstoguzov and Braudo, 1983). The microstructural appearance of certain foods leads to their proposed classification as composites. Examples of common foods viewed as composite gels are commercial frankfurters, commercial surimi, fish, bread, yogurt and processed Colby cheese (Aguilera, 1992). Raw meats are composed of fibers oriented in a protein–gel matrix and processed meats are composed of fat globules dispersed within a protein–gel matrix. Dairy products such as cheese and yoghurt are composed of a protein gel–fat globule composite. Cereals are composed of starch granules dispersed in a collapsed gel matrix.

20.2 Restructured foods

Many fabricated foods contain hydrocolloid gelling agents. When an alginate solution is dropped into a soluble calcium salt solution, an insoluble calcium alginate 'skin' forms almost immediately. This simple reaction provides the basis for the manufacture of many restructured foods.

The first patent based on this reaction was granted in 1946 for a process designed to create artificial cherries (US Patent No. 2,403,547), which contained no fruit.

Novel structured fruit products made of pulp, a wide range of hydrocolloid gels and other additives have been the subject of a number of patents and commercial applications (Glicksman, 1976; Lodge, 1981; Szczesniak, 1968; Tolstoguzov, 1971). These products are, in most cases, composite materials in which particulates are embedded in a polymeric gel matrix (Ring and Stainsby, 1982). Many patents discuss the possibility of using a combination of alginates with other hydrocolloids such as agar and carrageenan, together with fruit pulp and other traditional food additives to create simulated fruit products (Szczesniak, 1968; Tolstoguzov, 1971; Unilever Ltd, 1974). The available technological information about such products concentrates mainly on the methods of producing different gel systems containing pulp, sugar and acid.

The mechanical properties of composite fruit products based on agar and cold-set alginate have been previously studied (Mitchell and Blanshard, 1976; Nussinovitch *et al.*, 1989). The addition of orange pulp causes a reduction in gel yield stress to a minimum at a pulp concentration of about 20–30% in agar- and carrageenan-based composite fruit products (Fig. 20.1), and at ~10–15% pulp in cold-set alginate gels. Beyond this minimum, the trend reverses and some gel strengthening is observed (Nussinovitch *et al.*, 1991a). Pulp particle size and properties play an important role

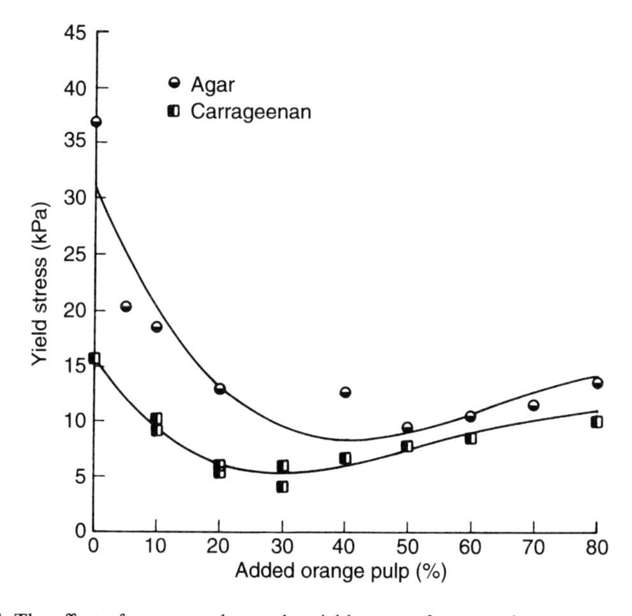

Figure 20.1 The effect of orange pulp on the yield stress of agar and carrageenan gels. (From Nussinovitch *et al.*, 1991a; by permission of Academic Press Ltd, London.)

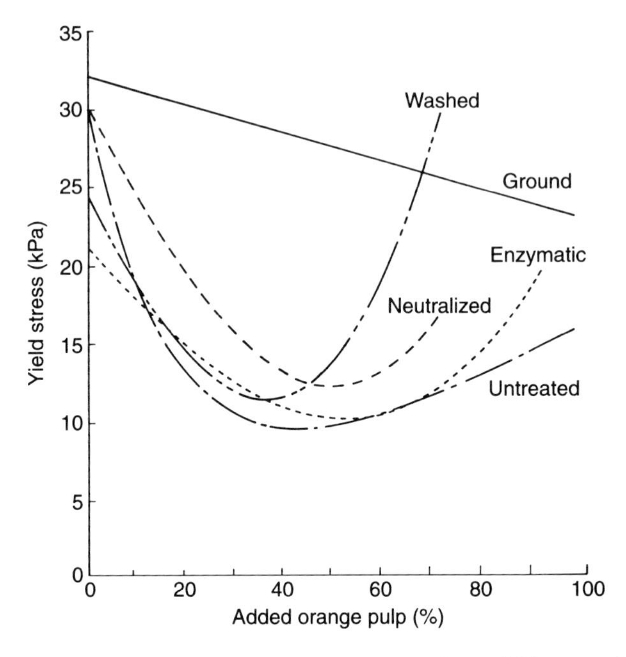

Figure 20.2 The effect of different treatments of orange pulp on yield stress of 2% agar gel. (From Nussinovitch *et al.*, 1991a; by permission of Academic Press Ltd, London.)

in determining gel strength (stress at failure). Added ground pulp affects agar gels the least, whereas washed pulp is capable of strengthening the gel to close to its original yield stress (Fig. 20.2). The effect of adding pulp on the gel's mechanical properties is also dependent on the pulp type. Orange pulp alone seems to have a much less dramatic effect on gel strength than banana pulp (Nussinovitch *et al.*, 1991a). The latter appears to cause a regaining of initial gel strength when a relatively high quantity of pulp is added. It should be noted that banana pulp contains sugar, and part of its effect may be attributable to this factor (Fig. 20.3). However, a similar strengthening phenomenon is observed with sugar-free, washed orange pulp, indicating that the cell-wall material plays an important role.

The first phase in the pulp's effect on gel strength is characterized by a weakening of the gel's structure via interference with matrix formation. Very fine grinding of the pulp minimizes this interference, indicating the major role played by steric effects. At a high enough pulp concentration, if a pulp structure is formed, the weakening trend is reversed. In this case, the mechanical properties of the pulp particles and their ability to form interacting structural elements determine the strength of the combined gel–pulp system. Pulp-structure formation appears to depend on the shape and size of the particulates. Ground orange pulp, lacking structure-formation ability, causes only a decrease in gel strength with no minimum at all,

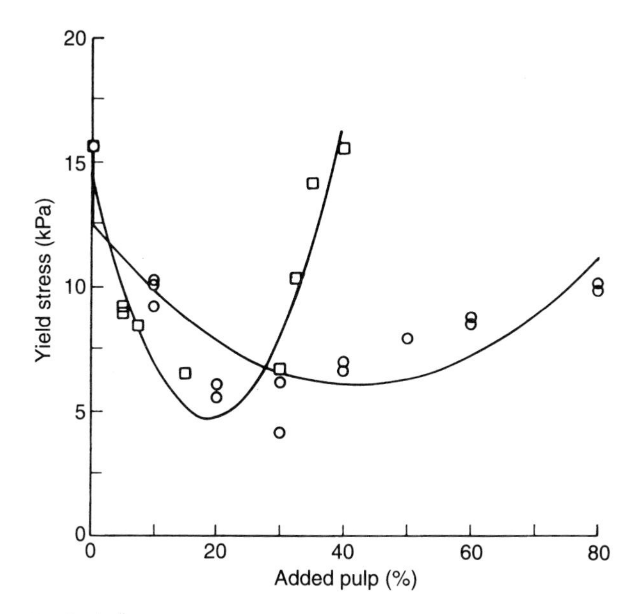

Figure 20.3 The effect of orange (○) and banana (□) pulps on the yield stress of carrageenan gels. (From Nussinovitch *et al.*, 1991a; by permission of Academic Press Ltd, London.)

but the gels obtained are stronger than those containing unground pulp samples (Nussinovitch *et al.*, 1991b).

Since the 1970s, many alternative processes for preparing fabricated fruits have been described in patents and technical publications (Wood, 1975; Luh *et al.*, 1976; Tolstoguzov and Braudo, 1983). Calcium alginate gels (1% algin), with or without agar (1%) as a presetting agent, were used to form a matrix for texturized (2–8%) raspberry pulp (Fig. 20.4) (products with over 10% highly acidic pulp are so weak that they collapse under their own weight) (Nussinovitch and Peleg, 1990). The agar did not have an appreciable effect on the product's strength, but it made it comparatively more brittle and stiff. Although full equilibrium of such texturized products took about 48 h or more, all of their principal mechanical features could be determined in tests performed after only 24 h (Nussinovitch and Peleg, 1990).

The combined effect of fruit pulp, sugar and gum on some mechanical parameters of agar and alginate gels was studied (Nussinovitch and Peleg, 1990). From a quantitative study of agar–sugar–pulp texturized systems and GDL–sugar–pulp systems, maxima and minima in the strength and deformability modulus of agar texturized products were observed at ~25% sugar and 20% pulp, respectively. In the GDL system, a minimum in these mechanical parameters was observed at a pulp concentration of about 22% (Fig. 20.5).

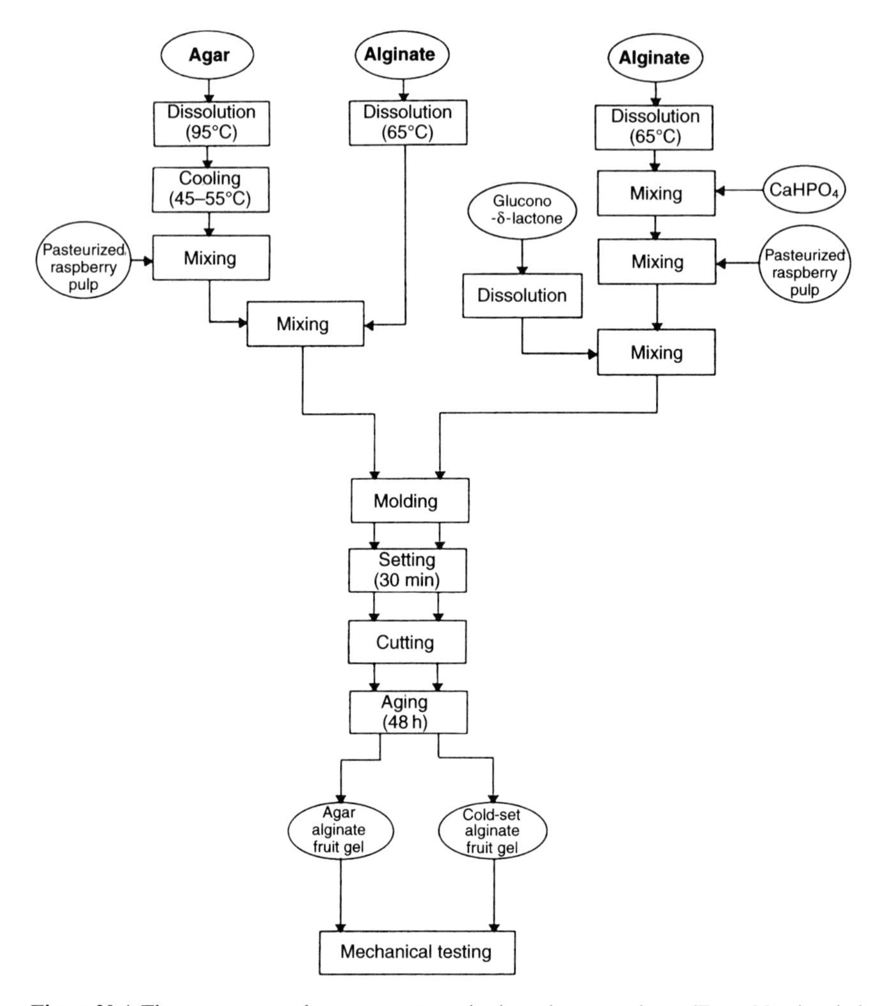

Figure 20.4 The processes used to prepare texturized raspberry products. (From Nussinovitch and Peleg, 1990.)

Fruit pulp is frequently used in creating fruit analogs. However, highly acidic pulps, with pH values of 3.0–3.5, tend to weaken gels. A procedure was, therefore, developed to prepare products containing up to 90% high-acidity pulp. The pulp is neutralized, enabling its addition at high concentrations. Later GDL is added, and an acid/calcium bath is used to acidify the system while strengthening the final product (Kaletunc *et al.*, 1990).

Although the technology involved in including a high proportion of fruit pulp or juice has improved (Kaletunc *et al.*, 1990), texturized fruits are still subject to some limitations. Their texture often does not resemble that of

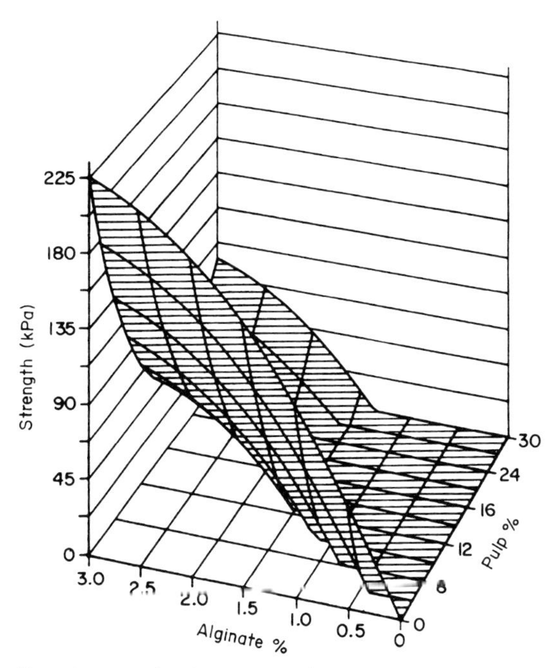

Figure 20.5 The effect of gum and pulp concentration on the strength of alginate–sugar–pulp gels. (From Nussinovitch *et al.*, 1991b; by permission of Oxford University Press.)

real fruit. One of the reasons for this is that during mastication, the fruit juice or liquid is only liberated to a small extent in texturized fruits, compared with that released by fresh citrus fruit. Recently, succulent agar and alginate-based texturized products were prepared by entrapping pasteurized grapefruit vesicles (juice sacs) within the gel structure (Weiner and Nussinovitch, 1994). In both matrices, strength decreased as the percentage of juice sacs increased. It was impossible to build an agar matrix with more than $350\,g\,kg^{-1}$ cells. With alginate, $50\,g\,kg^{-1}$ more cells (up to $400\,g\,kg^{-1}$) could be added before a very weak, texturally unacceptable product was produced. Nevertheless, use of this method resulted in a succulent product. With the percentage weight loss of liquid during compression increasing with juice-vesicle content, entrapment of juice cells within a matrix overcomes one of the limitations imposed on traditional pulp-based texturized fruit products, namely their inherent 'dryness'. The liberation of sweetened liquid upon biting into the product results in a texture that more closely resembles that of real fruit (Weiner and Nussinovitch, 1994).

Multi-layered hydrocolloid-based foods are important in the framework of future foods. The pertinent question here is whether the stiffness of multi-layered texturized fruits can be estimated from the deformability moduli of its individual layers. Recently, a mathematical model to predict the deformability modulus of a four-layered gel array of texturized fruit has been developed (Ben-Zion and Nussinovitch, 1996). Each layer was composed of 1.5% agar, 0.5% LBG and one of four pasteurized fruit pulps (50%): banana, apple, kiwi or strawberry. The model was based on the assumption that the uniaxial stress in the layers is the same and that their deformations are additive. No significant differences were found between experimental and calculated deformability moduli predicted by the model. The model provides a tool for estimating multi-layered gel stiffness and may be applicable to other food systems that behave similarly (Ben-Zion and Nussinovitch, 1996).

20.3 Types of fabricated food

Many different types of fabricated food can be created by using different main ingredients (such as fish, meat, fruit or vegetables), and simply changing the concentration of the matrix former (e.g. alginate) and selecting an appropriate source of soluble calcium. The variety of final products is huge and reasons for their production include the utilization of food stocks to increase food supplies, the creation of new textures, cost reduction and the design of new products in industrialized societies (Aguilera and Stanley, 1986).

Fabricated onion rings were introduced in the 1970s. A slurry, mainly of sodium alginate and diced onions, is extruded into a calcium chloride bath. After the skin forms, the rings are dusted with breading batter, fried in deep fat, blast-frozen and packed. Cellulose ethers, which form lipophobic protective films, are used to prevent oil absorption during frying.

Other hydrocolloids – carrageenan, gelatin (and recently gellan) and combinations of gums such as carrageenan and LBG – have been used to fabricate food products. Examples include reconstituted pimento strips (based on alginate and gum arabic), imitation caviar and restructured fish and shellfish (Nussinovitch, 1993).

Extrusion processes for the manufacture of snacks use a flour or starch matrix in combination with flavors, colors and spices. The starch component is transformed by heat and pressure to yield the desired functional properties in the finished product. Hydrocolloids such as cellulose ethers are used to improve thermostability and to provide lubrication during extrusion. Starches are also widely used in conventional batch-canning operations to maintain the consistency of the product after heat treatment. Such starches need to be stable over a wide range of storage conditions and must retain their physical and sensory qualities. This requirement is especially important for creamed vegetable soups, gravy-based products and pie fillings. Modified starches that do not thin out under continuous heating and do not gel when stored at low temperatures are used for this purpose (Nussinovitch, 1993).

20.4 Molded products

A common method of manufacturing gum-based texturized products is to pour a hot or cold solution of the gum, mixed with other ingredients, into a mold and to let it set. Agar forms a firm gel at low concentrations and is used in the preparation of several Japanese dessert products. These traditional desserts are molded into ingot-shaped bars. Generally, they are packed and sealed in a laminated-film container. These jelly-like confections are based on agar, sugar and mashed Azuki beans. Alternatives to the beans include agars, powders and fruit pulp.

A modern version of the dessert 'sweet agar jelly' is a highly sweetened, mildly flavored candy of low acidity. Processing includes dissolving the agar in a boiling solution of sucrose and invert sugar, followed by brief cooling and the addition of juice, acidulant, flavoring and coloring. Moisture content is reduced by drying in an oven. Since this product is sticky, it is coated with an edible agar and starch film (for more details see Chapter 10). The film is produced by brushing a hot solution of the agar–starch mixture onto a polished metal surface.

The process of slab hardening can also be used to produce canned multi-layered dessert gels. These can be made from commercially produced hydrocolloids. A so-called simultaneous method for preparing layered desserts from a xanthan gum–LBG blend and carrageenan has been recently reported. Solutions of the hydrocolloids (the blend and carrageenan) are poured into a mold at 158°F (70°C) and cooled, whereupon they separate into two distinct layers. These products can contain juice, coffee extract or dietary fiber as well as different colorings, flavorings and texturizing materials (Nussinovitch, 1993).

20.5 Drying

The goal in food drying is to remove water at the lowest possible temperature, thereby optimizing product quality. With fruits and vegetables, this sometimes results in a tough, leathery surface on the dried particle, which can adversely affect rehydration and eating qualities. The problem can be reduced by wetting the fruit or vegetable particles with cold solutions of water-soluble gums such as guar or CMC prior to drying. Several methods are used, for example soaking under vacuum to improve penetration of the hydrocolloid into the particles.

Drying techniques, such as spray-drying, are used to encapsulate food components such as flavorings and protect them from environmental influences. The mixtures or emulsions of the flavoring and a gum in water must be dried rapidly. During spray-drying, water is removed and the flavored particle becomes coated with a gum film. Gum arabic is best suited to this purpose, but ghatti gum, another plant exudate, is also used. Gum arabic reduces surface tension, enabling the emulsion to form more easily. It also coats dispersed oil droplets so that they are surrounded by an electrically charged surface. Since all the droplets then have the same charge, they repel each other, inhibiting collision and coalescence. Gum arabic is used widely for microencapsulation. Modified starches have been used for the same purpose when gum arabic is unavailable, but with less success.

20.6 Stabilization

Hydrocolloids are used as emulsion stabilizers in salad dressings and ice cream. Salad dressings are oil-in-water emulsions prepared by mechanical homogenization. The dispersed phase is stabilized by the increased viscosity and yield strength provided by the hydrocolloid. Tragacanth, PGA and xanthan can be used.

In ice creams and similar frozen desserts, stabilizers promote smoothness, reduce drip, improve body and increase resistance to heat shock. The hydrocolloids limit incorporation rates into crystals, thereby minimizing the growth of ice and sugar crystals, which adversely affect texture.

Freezing often causes undesirable changes in foods, and hydrocolloids are used to improve their quality. To produce a high-quality ice cream, a blend of guar gum or CMC with a smaller amount of carrageenan may be used. If xanthan and guar gum are used instead, viscosity is lower and faster processing is obtained. Karaya gum has been used in the past as a stabilizer for frozen desserts but has been replaced almost completely by other gums. Carrageenan, guar gum and CMC have also been used as stabilizers in other frozen products. In foods containing starch as the main ingredient, there is a tendency for water to exude from the gel. Therefore, starch-based products curdle and undergo syneresis (loss of water) after freezing and thawing. Modified starches have been developed to deal with the problem (Nussinovitch, 1993).

Freeze-texturization has been used for the production of porous starch-based sponges (Torres, 1977). Sponges prepared from high-amylose cornstarch had a 'fibrillar crab-meat-like' character, whereas those prepared from pure amylose are 'flaky and meat-like'. Sponges consist largely of gelatinized starch and, therefore, possess high water-holding capacity and low solubility.

Freezing is also used to modify the texture of sponge-like products produced from soybean curd. In freeze texturization, two phases (ice and concentrated solution/suspension) are initially formed by cooling below the freezing point of the solution. Thus a fibrillar structure is induced by the controlled growth and orientation of the ice phase favored by unidirectional freezing; this is followed by fixation via protein–protein interactions (Lugay and Kim, 1981). Lillford (1985) discussed the physiochemical principles involved in the freeze-texturization process. The rate at which freezing is induced determines the character of the formed structure (spongy or fibrous). Flow-freezing to achieve a crystal growth rate that exceeds that of nucleation is necessary to achieve a homogeneous texture. Mechanical means are also used to break ice crystals and for fiber formation and induction of alignment.

Many milk and juice products are processed aseptically, and specialized carrageenans are used in such products. These carrageenans are soluble at ultra-high temperatures, stabilize the milk protein, suspend fat globules and preserve the product's taste; they also permit hot or cold packaging of the finished product. Carrageenan is also widely used in other milk products, such as processed beverages, pasteurized milk, dry-mix beverage powders, milk puddings and cheese products (Nussinovitch, 1993).

The formation of stable foams is essential in preparing many food products, ranging from bread to whipped cream and mousses. Air is incorporated to produce the desired texture in the final product, a porous, thin-walled structure. Starches, and sometimes proteins, can be induced to foam by two main mechanisms: sudden vaporization of a solvent, or via the generation of a gaseous phase in proofing and dough-baking. Techniques such as vapor-induced puffing and explosion-puffing are widespread. The texture of solid starch-based foams depends on the amylose: amylopectin ratio. High-amylose products are hard and dense, whereas high-amylopectin products are fragile but can be modified by amylose supplementation to yield a structure that is more resistant to breakage (Matz, 1984).

Gums are used to stabilize these foams. Gum mixtures, such as carrageenan, sodium alginate, guar and CMC, are frequently used. Cellulose ethers are used to stabilize whipped toppings. Although protein-based foams can be stabilized by physical means (for example, heating or cooling), gums such as carrageenan, alginate and LBG are often added to react with the protein and form a stable foam (Nussinovitch, 1993).

References

Aguilera, J.M. (1992) Generation of engineered structures in gels, in *Physical Chemistry of Foods* (eds H.G. Schwartzberg and R.W. Hartel), Marcel Dekker, New York, pp. 387–421.

Aguilera, J.M. and Stanley, D.W. (1986) in *Food Engineering and Process Applications*, vol. 2 (eds M. Le Maguer and P. Jelen), Elsevier Applied Science, London, p. 131.

Ben-Zion, O. and Nussinovitch, A. (1996) Predicting the deformability modulus of multi-layered texturized fruit and gels. *Lebensm. Wiss. Technol.*, **29**, 129–34.

Glicksman, M. (1976) Fabricated food. *Cereal Foods World*, **21**, 17–26.

Kaletunc, G., Nussinovitch, A. and Peleg, M. (1990) Alginate texturization of highly acid fruit pulp and juices. *J. Food Sci.*, **55**(6), 1759–61.

Lillford, P.J. (1985) in *Properties of Water in Foods* (eds D. Simatos and J.L. Multon), Nijhoff, Dordrecht, p. 543.

Lodge, N. (1981) Kiwi fruit: two novel processed products. *Food Technol. New Zealand*, **16**(7), 35, 37–8, 41.

Lugay, J.C. and Kim, M.K. (1981) in *Utilization of Protein Resources* (eds D.W. Stanley, E.D. Murray and D.H. Lees), Food and Nutrition Press, Westport, CT, p. 177.
Luh, N., Karel, M. and Flink, J.M. (1976) A simulated fruit gel suitable for freeze dehydration. *J. Food Sci.*, **41**, 89–93.
Matz, S.A. (1984) *Snack Food Technology*, AVI Publishing, Westport, CT.
Mitchell, J.R. and Blanshard, J.M.V. (1976) Rheological properties of alginate gels. *J. Text. Studies*, **7**, 219–34.
Nussinovitch, A. (1993) Gum-based texturized products, in *Yearbook of Science and Technology*, McGraw-Hill, New York, pp. 138–40.
Nussinovitch, A., Kopelman, I.J. and Mizrahi, S. (1991a) Mechanical properties of composite fruit products based on hydrocolloid gel, fruit pulp and sugar. *Lebensm. Wiss. Technol.*, **24**, 214–17.
Nussinovitch, A., Kopelman, I.J. and Mizrahi, S. (1991b) Modeling of the combined effect of fruit pulp, sugar and gum on some mechanical parameters of agar and alginate gels. *Lebensm. Wiss. Technol.*, **24**, 513–17.
Nussinovitch, A. and Peleg, M. (1990) Mechanical properties of a raspberry product texturized with alginate. *J. Food Process. Preserv.*, **14**, 267–78.
Nussinovitch, A., Peleg, M. and Normand, M.D. (1989) A modified Maxwell and a nonexponential model for characterization of the stress relaxation of agar and alginate. *J. Food Sci.*, **54**, 1013–16.
Ring, S.G. and Stainsby, G. (1982) in *Progress in Food and Nutrition Science*, vol. 6, *Gums and Stabilisers for the Food Industry: Interactions of Hydrocolloids* (eds G.O. Phillips, D.J. Wedlock and P.A. Williams), Pergamon Press, Oxford, pp. 323–9.
Silberberg, A. (1989) in *Polymers in Aqueous Media Performance Through Association* (ed. J.E. Glass), *Advances in Chemistry Series No. 223*, American Chemical Society, Washington, DC, p. 3.
Szczesniak, A. (1968) *Simulated Fruits and Vegetables.* US Patent No. 3,362,831.
Tolstoguzov, V.B. (1971) *Method of Preparing Artificial Foodstuffs.* USSR Patent No. 296,554.
Tolstoguzov, V.B. and Braudo, E.E. (1983) Fabricated foodstuffs as multicomponent gels. *J. Text. Studies*, **14**, 183–212.
Torres, A. (1977) Textural properties of amylose and starch sponges. MSc thesis, Department of Food and Agricultural Engineering, University of Massachusetts, Amherst, MA.
Unilever Ltd (1974) *Simulated Fruit.* British Patent No. 1,369,198.
Weiner, G. and Nussinovitch, A. (1994) Succulent, hydrocolloid-based, texturized grapefruit products. *Lebensm. Wiss. Technol.*, **27**, 394–9.
Wood, F.W. (1975) *Artificial Fruit and Process Therefore.* US Patent No. 3,892,870.

Abbreviations

ADI acceptable daily intake
AFM atomic force microscopy
AG arabino–galactan
AGP arabino–galactan–protein complex
ANFO ammonium nitrate/fuel oil
BOD biological oxygen demand
CA controlled atmosphere
CIR cosmetic ingredient review
CMC carboxymethylcellulose
CTFA cosmetics, toiletry and fragrance association
DA degree of amidation
DE degree of esterification
DMSO dimethyl sulfoxide
DP degree of polymerization
DS degree of substitution
DSC differential scanning calorimetry
EC European community
ESR electron spin resonance
FDA food and drug administration
FIRA FIRA jelly tester
GDL glucono-δ-lactone
GI glycoprotein
GMP good manufacturing practice
GRAS generally recognised as safe
HEC hydroxyethylcellulose
HLB hydrophilic-lipophilic balance
HMP high methoxy pectin
HPC hydroxypropylcellulose
HPMC hydroxy propylmethylcellulose
IR infra red
LBG locust bean gum
LFRA Stevens LFRA texture analyzer
LMP low-methoxy pectin
LP letterpress
M mannuronic acid

MC methyl cellulose
MCC microcrystalline cellulose
MHPC methyl hydroxypropylcellulose
MG mannuronic-guluronic
NMR-nuclear magnetic resonance
PABA p-aminobenzoic acid
PCR polymerase chain reaction
PE pectinesterase
PEG polyethyleneglycol
PEL poly(e-lysine)
PEO polyethylene oxide
PETN pentaerythritol tetranitrate
PG polygalacturonase
PGA propylene glycol alginate
PVA polyvinyl alcohol
PVP polyvinylpyrrolidone
SHMP sodium hexametaposphate
SiC silicon carbide
TNT trinitrotoluene
TPA texture-profile analysis
TSPP-tetrasodium pyrophosphate
UTM universal testing machine
UV ultra violet

Index